AF553673

Integrated Pest Management in the Tropics

Integrated Pest Management in the Tropics

Part-II

Edited by
Dharam P. Abrol
Division of Entomology
Sher-e-Kashmir University of Agricultural Sciences & Technology
Jammu—180009, Jammu & Kashmir, India

NEW INDIA PUBLISHING AGENCY
New Delhi – 110 034

NEW INDIA PUBLISHING AGENCY

101, Vikas Surya Plaza, CU Block, LSC Market
Pitam Pura, New Delhi 110 034, India
Phone: + 91 (11)27 34 17 17 Fax: + 91(11) 27 34 16 16
Email: info@nipabooks.com
Web: www.nipabooks.com
Feedback at feedbacks@nipabooks.com

ISBN : 978-93-85516-11-5

Composed, Designed and Printed in India.

Foreword

Since the end of World War II with the consequent devastation of agricultural production areas in much of Europe and Asia, and the prospect of accelerated human population growth, sustainable food production has been a high priority in most countries. With pest (insects, plant pathogen, and weeds) damage accounting for over 30% of losses in total agricultural production, pre- and post-harvest, plant protection specialists and producers were eager to adopt any new technology that had the potential to reduce those losses and thus contribute to the fight against hunger. A technology arouse after the war based on the newly discovered organo-synthetic pesticides. It is estimated that since 1947, almost three billion kg of chlorinated hydrocarbon insecticides (CHIs) were used. Initial results were spectacular leading to predictions that "some pests will become extinct". Early on, there was little awareness or concern about the possible side effects of that technology. Soon, however, CHIs caused one of the most pervasive intrusions of man-made chemical molecules into the landscape with severe health and environmental consequences. Early success overshadowed obvious signs of problems and abuses. The need to keep food production apace with population growth overwhelmed any environmental impact consideration. Success of the green revolution varieties was, to a large extent, dependent on the availability of those organo-synthetic insecticides, herbicides, and fungicides, as many of the varieties were very susceptible to insect pests and pathogens. The health and environmental risks of the heavy reliance on pesticides were, however, too real to ignore. Rachel Carlson's 1962 book was a wake-up call that led concerned scientists to more aggressively pursue a path that led, 10 years later, to formulate the concept of Integrated Pest Management (IPM). Despite recent advances in pest control technologies and IPM program expansion worldwide, world crops in the early years of the 21st century continue to sustain up to 30 percent losses to the aggregate impact of pre- and post-harvest pests, a level similar to those suffered at the beginning of the 20th century. These losses persist even while pesticide use continues to rise worldwide and novel technologies involving genetically modified organisms (GMOs) have added new weapons to the arsenal in the fight against pests.The need to advance the IPM approach to pest control has never been greater and only through the education of young generations of agricultural researcher, extension specialists, and producers, can this goal be achieved. This book, written with that audience in mind, offers a significant contribution to the advancement of IPM in the tropics.

The tropics encompass most of Africa, South Asia, and Central and South America where much of the population explosion is occurring. The tropics account for about 33 percent of the world's crop land, and 40 percent of the human population, however, tropical agriculture produces no more than 25 percent of the world's cereals, 25 percent of fruits and vegetables, and 20 percent of meat (from a discussion paper prepared for the G20 Conference on Agricultural Research for Development (12-13 September 2011, Montpellier, France). It is also in this vast area where pest problems are most severe. Whereas temperate zone countries may have the benefit of harsh winters and relatively short growing seasons which moderate or disrupt the growth cycle of pest populations, the tropics often grow crops all year round, but the same weather conditions that allow successive crops also favour outbreaks of arthropod pests and plant diseases, and the rapid growth of weeds.

Without the thoughtful development of IPM systems, based on sound ecological principles, and the concerted effort to extend to growers the most effective systems, growers in the tropics resort to pesticides. Although the rational use of pesticides is not contrary to IPM, without a robust IPM strategy, growers tend to misuse those chemicals with negative health, environmental, and economic consequences. Much of the early development of IPM happened mostly in temperate zone countries of Europe and North America, and many of the key tropical crops, such as oil palm, coconut, coffee and cocoa, the long term plantation crops, were not included in that early development. Fortunately, since the latter part of the 20th century, national research programs in India, Indonesia, the Philippines, Brazil, among several others, combined with research streaming from the network of International Research Centers, have yielded an impressive body of information capable of supporting progress in the development and implementation of promising IPM systems specifically targeted to the tropics.

The present book 'Integrated Pest Management in the Tropics' edited by Professor Dharam P. Abrol, brings together an important body of information in support of efforts to develop IPM strategies for most of the key crops in the tropics: cereals, legumes, cotton, coffee, cocoa, vegetables, sugarcane, root and tuber and plantation crops. That information is covered in 26 chapters in 2 parts that include an introductory section with overviews of factors that are critical for IPM development, such as impact of emerging pests, cropping systems, and climate change, and reviews of biological and chemical controls, as well as other novel control tactics. The hapters are devoted to specific crops. Contributing authors bring expertize from years of field research from diverse states of India, as well as countries in Africa, Europe, North and South America.The book provides a thoughtful approach to advance IPM research and implementation under tropical conditions. The editor and his collaborators are to be congratulated for a needed addition to the IPM Literature.

December 25, 2015

—Marcos Kogan
Director Emeritus
Integrated Plant Protection Center
Oregon State University, USA

Preface

The growing human population at an alarming rate is a major challenge facing agriculture in the 21st century. It is expected to rise to 9 billion or more during the mid of this century requiring food productivity to be doubled as compared to present, posing a great challenge on the part of the agricultural scientists to develop high yielding production technology and intensification in crop produce practices. Intensification would have serious implications of increased use of pesticides to control incidence of emerging pests for quick results. The increased use of pesticide would pose more problems in the form of development of insecticide resistant populations, resurgence of pest populations and pollution of the environment. These challenges have resulted in reorientation of pest management strategies in the form of ecofriendly approach of integrated pest management (IPM). The IPM has been widely practised and accepted in countries with temperate climates, however, such a progress has been inconspicuous by its absence in areas with tropical climates which constitute around two-third of earth's surface.

The tropical areas have great diversity of land utilization and farming practices which have the potential of more ecological imbalances compared to temperate areas. Long activity periods in the absence of cold winter allows many insects to produce more number of overlapping generations. Furthermore, hot and wet conditions encountered in tropics affect the performance of pesticides and biology of the pests.

Evidently, tremendous opportunities exist that might lead to modification of some of the existing theories, or reinforce them using long-term observational and experimental data from tropical insects. This is because the warm temperatures of the tropics permit year-round growth and multiplication of insects. While many temperate insects pass through only one generation per year, many tropical insects pests pass through 12-14 generations in the same period. Wider dissemination of knowledge about the tropical insects and the research opportunities they offer will promote collaborative work among scientists from developed and developing countries, for the benefit of both and the science of entomology in general. This thought has been one of my main motivations for embarking on this work. The present book "Integrated Pest Management in the Tropics" contains 26 chapters on diverse aspects of IPM in tropics and is an attempt to cover integrated pest management from multidisciplinary, multi-country and multi-faceted components providing a unique overview of the efforts made to develop IPM strategies on a crop-by-crop basis for each continent. Chapter 1 gives an overview of the pest problems in tropics and losses caused by them. This is followed by an overview of new and emerging insects pests in the tropics (Chapter 2). Chapters 3 to 7 discuss general topics such as Biological and Chemical control, Impact of cropping systems and climate change and emerging technologies for tropical pest management. Chapters 8-18 deal with specific problems of tropical crops such as Integrated pest management in tropical vegetables, cereals, food legumes, root and tuber crops, bananas, citrus sugarcane, tea, coffee and cocoa, palms, cotton and other plantation crops. Chapter 19 provides detailed information on role of sex pheromones and IPM. Chapters 20 and 21 deal with role of plant chemicals and chemical ecology on pest management. Chapter 22 emphasizes the role of ecology in pest management. Chapters 23 deal with the risks of exotic pests on biodiversity, biosecurity issues and cultural and physical methods, respectively. Chapter 24 deals with the most challenging and pressing need of taxonomy for successful pest control. Chapter 25 provides information on statistical techniques in IPM. Chapter 26 Critically explores the future of integrated pest management in the 21st Century. This book provides multidisciplinary perspective about the different facets of IPM and shall serve as reference book for students, teachers, researchers, extension functionaries and policy planners. This book is the outcome of my personal experiences and the contributions of several workers which have been incorporated. All the contributors deserve special appreciation for writing chapters in their respective fields in great depth with dedication. I am particularly thankful to Professor Dr. Raghavendra Gadagkar, Centre for Ecological Sciences, Indian Institute of Sciences Bangalore who has always been a source of inspiration, needed help, guidance and encouragement. I thank my university authorities for the excellent working atmosphere and needed encouragement for compiling such a voluminous book. Dr. Uma Shankar and Dr. Debjyoti Chaterjee needs special mention for their help in various ways. My sincere thanks are due to my wife Professor Dr. Asha Abrol, daughter Er. Vitasta and son Er. Rajat for their endurance and help while writing this book. Last but not least I am extremely thankful to the great scientist and most influential thinkers in the field of IPM, Dr. Marcos Kogan Professor and Director Emeritus Integrated Plant Protection Center Oregon State University to write foreword for this book.

Place: Jammu **Dharam P. Abrol**

Date 31-01-2015 Editor

Contents

Part-II

Part-I

Integrated Pest Management in the Tropics; pp. 439-497

New India Publishing Agency, New Delhi (India)

CHAPTER - 16

Integrated Pest Management (IPM) of Palm Pests

Faleiro, J. R.[1], Jaques, J.A.[2], Carrillo, D.[3],R. Giblin-Davis[4], C. M. Mannion[3], E. Peña-Rojas[5] and J. E., Peña[3]

[1]Food and Agriculture Organization of the United Nations, Date Palm Research Centre, P. O. Box 43, Ministry of Agriculture, Al-Hassa-31982 Saudi Arabia

[2]Universitat Jaume I (UJI),Unitat Associada d'Entomologia Agrícola UJI-Institut Valencià d'Investigacions Agràries (IVIA),Departament de Ciències Agràries i del Medi Natural, Campus del Riu Sec, Av. de Vicent Sos Baynat, s/n.E-12071 Castelló de la Plana, Spain

[3]University of Florida, Tropical Research and Education Center Homestead, FL 33031, USA

[4]University of Florida, Fort Lauderdale Research and Education Center Fort Lauderdale, FL 33314 USA

[5]Corpoica, Palmira, Colombia

16.1 Introduction

According to Howard (2001) the major world crop palms are coconut palm (*Cocos nucifera* L), African oil palm (*Elaeis guineensis* Jacq.) and date palms (*Phoenix dactylifera* L). Many other palm species provide products for international commerce and also grown as ornamentals. Some palm species are of local or regional importance and have great potential for expanded development and distribution. In general, commercial production of palms regularly starts on or near natural habitats with little agricultural development. The constituent arthropod fauna in unexploited areas are likely invaders to palm- monoculture (Hassan, 1972; Forster *et al*., 2011). For instance, the degree of damage caused by coconut pests can be related to conditions in the environment, particularly to the composition of the vegetation associated with the palm as most of the original host plants of these pests are found adjacent to the cultivation (Lever, 1969). It is also generally agreed that most of the caterpillar and beetle pests of palms are native to the area where the palm production is located and that their populations are often regulated naturally (Tuck, 1998; Howard *et al*., 2001).

In plantation crops such as palms, pests and diseases have affected potential expansion to different areas of the world. For instance, according to Genty *et al.*(1982) the commercial exploitation of oil palms in South America began in earnest 40 years ago, but damage from diseases and arthropods rapidly discouraged further expansion of many plantations. Dhileepan (1991) suggested that oil, coconut and arecanut in India share a complex of arthropods. Human intervention and movement of pest-infested palms has caused a displacement of palm pests from their areas of origin to other areas of the world (*i.e.*, *Rhynchophorus ferrugineus* (Olivier), *Raoiella indica* (Hirst) and *Aceria guerreronis* (Keifer) resulting in a strong need for integrated pest management (IPM) of palm pests in palm growing regions.

The basis for IPM incorporates the pest's biology and ecology and pest sampling, monitoring, economic thresholds, and the application of the management tactics such as chemical, biological, autocidal, plant resistance, and microbial control. There are several texts that discuss IPM of palm pests (Menon and Pandalai, 1958; Dhileepan, 1991; Moura, 1994; Howard *et al*., 2001). Howard *et al.* (2001) classified pests of palms as defoliators, sap feeders, and stem borers. The defoliators included lepidopterans such as *Opisina arenosella* Walker (Oecophoridae), *Artona catoxantha* Hamps (Zyganeidae), and *Brassolis sophorae* L (Nymphalidae); leaf beetles (Coleoptera: Chrysomelidae), and longhorn grasshoppers (Orthoptera: Tettigoniidae). The sap feeders included several Hemipterans and Acarina and the stem borers included *Orcytes* spp. (Coleoptera: Scarabeidae), *Rhynchophorus* spp. (Coleoptera: Curculionidae) and *Coptotermes curvignathus* Holmgren (Isoptera: Rhinotermitidae). Information about pests of coconut palms can be found in Hartley (1967), Lever (1969), Menon and Pandalai (1958), Hassan (1972) and Howard *et al.* (2002). Description of pests of oil palms is addressed in Mariau (1982), Genty (1978), Genty *et al.* (1982); Tiong *et al.* (1982), Dhileepan (1991); Hoong and Hoh (1992). Pests of date palms are summarized in Gassouma (2004), Blumberg (2008), Al-Jboory (2007) among others.

The history of palm pest management began with sole applications of insecticides, but later evolved to include other IPM tactics. For example, Wood *et al.* (1974) suggested that injections of Monocrotophos (organophosphate insecticide) into oil palm trunks was effective for controlling the bagworm, *Metisa plana* (Walker) (Lepidoptera: Psychidae), but the residual effect was shortlived. Wood *et al.* (1973) also recommended that large field aerial application of insecticides in Malaysia should be mainly restricted to leaf-eating pests, particularly against bagworms (*Crematopsyche pendula* Joannis and *M. plana*). Genty (1978) reported on pests of oil palms in Latin America and emphasized the use of economic thresholds of pests and the condition of the plantation before initiating chemical control. Genty *et al.* (1982) reported that defoliators such as *Sibine fusca* Stoll occur sporadically and could be controlled by aerial application of a densonucleosis virus which was considered efficient for containing pest outbreaks. Sudtharto *et al.* (1990) reported that while the caterpillars *Calliteara* sp., *Dasychira* sp. (Lymantriidae), *Mathusia* sp. (Amathusiidae), *Arasada* sp. (Noctuidae), and *Ambadra* sp. (Notodontidae) cause damage to young oil palms in Sumatra, they are highly parasitized, predated upon and affected by viral diseases. Huger (2005) emphasized the effectiveness of using a baculovirus against *Orcytes rhinoceros* (L) (Coleoptera: Scrabaeidae) as part of an IPM program for control of *O. rhinoceros*.

Blumberg (2008) summarized the current knowledge on the distribution, natural enemies, economic importance and management of 16 major species of date palms in Israel. An IPM approach was the result of outbreaks of scale insects in the 1950s. Blumberg (2008) reported that the IPM approach included physical and cultural control by covering palm fruit bunches with plastic nets and early harvest of dates, respectively, to achieve efficient control of fruit moths (*i.e.*, *Cadra figulilella* Gregson, *Spectrobactes ceratoniae* (Zeller), *Arenipses sabella* Hampson). In Israel, microbial control primarily using *Bacillus thuringiensis* against date moths and the use of pheromone traps for controlling the red palm weevil *R. ferrugineus*, reduced insecticide applications and conserved natural enemies. Al-Jboory (2007) emphasized the use of biological control for pests attacking date palm in Iraq. Talhouk (1991) reported that pest problems of date palms in the Arabian Peninsula were mostly caused by faulty management and suggested the use of antifeedants against *Orycte* ssp. adults and sex pheromones and anti-steril ants against date fruit moths.

The following pests will be addressed as examples of palm IPM in this chapter: weevils (Curculionidae), mites (Tenuipalpidae and Eriophyidae), aphids, scales, mealybugs, whiteflies, and fulgorids (Hemipteran:Sternorrhyncha and Fulgoromorpha).

16.2 Weevils (Curculionidae)

Geographical distribution and host range

Weevils by large constitute the most serious group of Coleopteran insect pests attacking a diverse range of palm species worldwide. Among the

Rhynchophorus species the red palm weevil (RPW) *R. ferrugineus* is the only species that has significantly expanded its geographical range from its original center of origin in South and Southeast Asia to almost the entire globe as compared with other *Rhynchophorus* species previously described by Wattanpongsiri (1966). Ecological niche modeling predicts that this pest can expand its range still further (Fiaboe *et al.*, 2012). *Rhynchophorus vulneratus* and *R. schach* are synonyms of *R. ferrugineus* with *R. ferrugineus* comprising the red palm and/or the Asian palm weevils (Boheman, 1845; Hallett *et al.*, 2004).The following is the geographical range of nine *Rhynchophorus* species as reported by Wattanapongsiri (1966) and updated by Giblin-Davis *et al.* (2013) (Fig. 16.1).

The host range of *R. ferrugineus* has increased ten fold since the mid 1950s when Nirula (1956) reported the pest on only four palm species as compared to its present host range of 40 palm species worldwide including *Areca catechu* L., *Arenga saccharifera* Labill, *A. engleri* Becc., *A. pinnata* (Wurmb), *Bismarckia nobilis* Hildebrand and Wend, *Borassus flabellifer* L., *B.* sp., *Brahea armata* S. Watson, *B. edulis*, *Butia capitata* (Mart.) Becc., *Calamus merrillii* Becc., *Caryota cumingii* Lodd., *C. maxima* Blume, *Cocos nucifera*, *Corypha utan* Lamk., (= *C. gebanga*, *C. elata*), *C. umbraculifera* L., *Chamærops humilis*, *Elaeis guineensis*, *Livistona australis* (R.Br.) Mart., *L. decipiens* Becc., *L. chinensis*Jacq.R. Br., *L. saribus* (= *L. cochinchinensis*) (Lour.) Merr., *Metroxylon sagu* Rottb., *Oncosperma horrida* (Scheff.), *O. tigillarium* (Ridl.), *Phoenix canariensis* (Chabaud), *P. dactylifera*, *P. roebelinii* O'Brien, *P. sylvestris* Roxb, *P. theophrastii* Greuter, *Pritchardia pacifica* Seemann and Wendland, *P. hillebrandii* (Kuntze) Becc., *Ravenea rivularis* Jumelle and Perrier, *Roystonea regia* (Kunth.), *Sabal umbraculifera* (Jacq.) Martius, *Trachycarpus fortunei* (Hook), *Washingtonia filifera* (L. Lindl), *W. robusta* H. Wendl., and *Syagrus romanzoffiana* (Cham.) (Anonymous, 2013).

Although there are nine *Rhynchophorus* species as mentioned above, this chapter describes the biology, damage, seasonality and management of two major species, *R. ferrugineus* and *R. palmarum*, which have their homes in the Asian and American continents, respectively.

16.2.1 Rhynchophorus ferrugineus

Biology, damage, economic thresholds and seasonality

The female of *R. ferrugineus* oviposits in holes made with its rostrum (Nirula, 1956). Eggs are laid in clutches which hatch in 1-6 days. Larvae are known to live for 25-105 days (Wattanapongsiri, 1966; Avand Faghih, 1996). Eggs of *R. ferrugineus* are 2.5 mm long and light yellow (Wattanapongsiri, 1966). Oviposition in *R. ferrugineus* is strongly affected by temperature.

Dembilio *et al.*(2012) reported the lower temperature thresholds for oviposition and egg hatching to be 15.45°C and 13.95°C, respectively, and under these circumstances, no new infestations would be expected during most of the winter in the Mediterranean basin (Dembilio and Jacas, 2012).

The larvae feed on palm tissue while moving toward the interior of the palm leaving behind chewed-up plant fibers which have a distinctly fermented odor. Infested palms often exhibit protruding chewed-up fiber where larval tunnels can reach over a meter in length. In date palm infestations,the galleries are mostly confined within a meter of the collar region at the base of the trunk (Abraham *et al.*, 1998; Sallam *et al.*, 2012), versus in *P. canariensis* infestations which occur in the crown of the palm (Dembilio *et al.*, 2012). The larval stage of *R. ferrugineus* is reported to have 3-17 instars under laboratory conditions (Nirula, 1956; Martin-Molina, 2004) and is completed in about two months with larvae increasing in size and becoming 2-3 cm long and 0.5 to 1 cm in diameter (Wattanapongsiri, 1966). Mature grubs migrate to the periphery of the palm trunk and prepare a cocoon made of palm fibers. After covering themselves with the cocoon, larvae enter a prepupal stage followed by a pupal stage (Murphy and Briscoe, 1999). Studies in live *P.canariensis* from Spain showed the existence of 13 larval instars in *R. ferrugineus*. Depending on mean temperatures, larval development can be completed in about 40 days in summer and up to 160 days in winter-spring in the Mediterranean region (Dembilio and Jacas, 2011). Detecting palms in the early stage of attack with early instar larvae is vital for the success of an IPM program for *R. ferrugineus* (Faleiro, 2006). The availability of efficient and cost effective detection devices that are easy to handle and sensitive enough to detect *R. ferrugineus* will significantly enhance the efficiency of management against this cryptic pest (Faleiro, 2006).

The pupal period is reported to range from 11-45 days. Salama *et al.* (2002) estimated the lower and upper temperature thresholds that the pupae of *R. ferrugineus* cannot survive are 2.3^{0}C and 44 to 45^{0}C, respectively. Similar results have been reported from Spain (Cabello, 2006). A new generation of adults that emerge may remain within the same host and reproduce until the palm meristem is destroyed resulting in the palm death. Subsequently, adults will fly away and look for new hosts (Dembilio and Jacas, 2012). In the Mediterranean region, less than one generation per year can be expected in areas with mean annual temperatures (MAT) below 15°C and more than two generations in areas with MAT above 19°C (Dembilio and Jacas, 2012). The length of the life cycle (egg to adult) is known to vary among different rearing hosts and ranges from 48-82 days on palm lumps (Ghosh, 1912) to 105-210 days on sago palm pith (Kalshoven, 1950). Adult weevils are known to live for about three months. Under laboratory conditions *R. ferrugineus* requires 40.4 degree days (DD) for egg hatching, 666.5 DD for complete larval development in living

P. canariensis and another 282.5 DD to reach adulthood (Dembilio *et al.*, 2012). Grove humidity due to enhanced soil moisture and flood irrigation provides temporary harborage to adults and increases the possibility of *R. ferrugineus* infestation in date palm (Aldryhim and Khalil, 2003; Aldryhim and Bukiri, 2003).

*Rhynchophorus ferrugineus*can infest palms of high economic value where, if not detected early, can become lethal to the host. Hence, an assumed low action threshold of 1% infested palms has been recommended to implement areawide IPM programs for *R. ferrugineus* (Faleiro, 2006). The population of *R. ferrugineus* is also known to be highly aggregated (Faleiro *et al.*, 2002) with infestations occuring in clusters in and around heavily infested plantations. Sampling plans incorporating the action threshold, aggregation index, and error of making the wrong decision have used to develop and validate areawide *R. ferrugineus* IPM programs (Faleiro and Kumar, 2008; Faleiro *et al.*, 2010).

Weevil captures in traps baited with *R. ferrugineus* pheromone (ferrugineol) have been used to assess seasonal variations in flight activity. In the Middle East, peak weevil activity is recorded during the period of March through May while a second smaller peak in trap captures occurs in October and November in Egypt (El-Garhy, 1996; Sebay, 2003; Kaakeh *et al.*, 2001; Vidyasagar *et al.*, 2000; Soroker *et al.*, 2005). In India, highest trap captures occurred in October and November, while the lowest captures occurred in June and July (Faleiro, 2005). In the more temperate climate of southern Japan, weevil activity was observed in March through mid December (Abe *et al.*, 2009). Dembilio and Jacas (2011) studied the oviposition pattern throughout the year in the Mediterranean basin and found it to extend in the north from early April to mid October and early November and the egg hatching period from mid March to late October. However, these periods were much shorter in the south and although oviposition would stop during the coldest winter months, egg hatching would continue during the entire year in the southwestern part of the Basin (i.e. Egypt). Based on these findings, Dembilio and Jacas (2011) recommended that any cultural management practice producing wounds on the palm should be best performed in winter, when oviposition is notably reduced and immature mortality is highest. Pillai (1987) suggested that fresh wounds on the palm tissue should be protected from invading female weevils, whereas Abraham (1971) proposed cutting coconut fronds one meter away from the frond base to prevent easy entry of the pest into the trunk which occurs when fronds are pruned close to the trunk.

Trapping as a control method : Current IPM strategies for *R. ferrugineus* rely on mass trapping of adults using pheromone lures. Traps baited with insecticide-laced food attractants were used to trap adult palm weevils in Indonesia, the West Indies and India prior to the discovery of palm weevil pheromones (Kalshoven, 1950; Hagley, 1965; Kurian *et al.*, 1979; Abraham *et al.*, 1989). With the discovery of male-produced aggregation pheromones that attract both male and female palm weevils, trapping of adult weevils with food-baited pheromone traps has been widely used to monitor and control both *R. palmarum* and *R. ferrugineus.*

In the case of *R. ferrugineus*, the aggregation pheromone, 4-methyl-5-nonanol (Hallett *et al.*, 1993) along with 10% of a related ketone (4-methyl-5-nonanone) increased weevil captures by 65% (Abozuhairah *et al.*, 1996; Oehlschlager, 2006). The optimal release rate for the pheromone was found to be three milligrams per day (Hallett *et al.*, 1993). However, release rates as low as 0.48 milligrams per day have been shown to be effective in Indian coconut plantations (Faleiro, 2005).

Plant volatiles along with insect-produced chemicals are known to generate synergy particularly in the order Coleoptera (Borden, 1985). Bucket traps (5-20 l capacity) containing the pheromone lure along with fermenting food bait laced with a non-repellant insecticide are currently used in several countries to monitor and mass trap *R. ferrugineus* (Faleiro, 2006). Emission of the pheromone alone from a trap does not attract as many weevils as compared with a fermenting food source plus pheromone which attracts many more weevils into the trap (Hallett *et al.*, 1993; Al-Soaud, 2011). Dates and sugarcane are among the best food baits recommended for use in *R. ferrugineus* pheromone traps. Trapping efficiency is reduced if the food baits are not changed every 7-15 days (Faleiro, 2006).Dark coloured traps (red or black) with a rough exterior are known to increase weevil captures (Hallett *et al.*, 1999; Abuagla and Al-Deeb 2012). Also, ethyl acetate significantly increases weevil trap captures when incorporated into *R. ferrugineus* pheromone plus food traps (Oehlschlager, 1998; Sebay, 2003; Abdallah *et al.*, 2008). Contingent upon the intensity of the pest in the field, trap density for mass trapping could range from 1trap per ha at relatively low weevil activity to 10 traps per ha in heavily infested plantations where weevil activity is high (Faleiro *et al.*, 2011). It is recommended to set the trap under shade to sustain field longevity of the lure. The need to replace the food bait periodically in food-baited pheromone traps becomes cumbersome and costly, especially at a higher trap density and in an areawide operation where traps have to be set deep inside the plantation without motorable roads. In this context, a bait-free method to 'attract and kill' *R. ferrugineus* adults was tested and found to sustain the trapping efficiency at levels comparable to the traditional food-baitedpheromone traps, during a test period of three months in date plantations of Saudi Arabia (El-Shafie *et al.*, 2011).

Low bait-lure synergy due to inappropriate trapping protocols could diminish the trapping eficiency of *R. ferrugineus* pheromone traps and lead to increased infestations around such traps. In Spain, mass trapping is only allowed under direct supervision of the local Department of Agriculture to limit the possibility of attracting weevils and also to help prevent the risk of new infestations in an uninfested area (Dembilio and Jacas, 2012). In several Middle Eastern countries of the Gulf region, the IPM program for *R. ferrugineus*, which includes pheromone trapping, is implemented by officials of the Ministry of Agriculture. This ensures proper maintenance of the traps thereby making the traps highly attractive to the adult weevils. Weevil captures in *R. ferrugineus* pheromone traps are female skewed (usually two female weevils for every male captured) which is desirable because these female weevils have been shown to be young, gravid and fertile (Abraham *et al.*, 2001; Faleiro *et al.*, 2003).These females could generate new infestations if not trapped and killed, as recorded in several countries prior to commencement of trapping adult weevils with pheromone traps. Repellents against *R. ferrugineus* offer vital protection to palm sites that are most prone to attack and could be used to enhance captures in *R.ferrugineus*pheromone traps through a push-pull strategy. A recent report from Italy indicates that α-pinene, singly or in combination with methyl salicylate or menthone, could serve as a potential repellent against *R. ferrugineus* (Guarino *et al.*, 2013).

Chemical control : Preventive and curative chemical treatments, mostly in the form of sprays (showers/drenches) and stem injection, respectively, form an important component of the IPM strategy against *R. ferrugineus*. Application of organophosphates and carbamates to control *R. ferrugineus* since the 1970's starting in India indicates that these chemicals were widely used against *R. ferrugineus* (Murphy and Briscoe, 1999; Faleiro, 2006; Dembilio and Jacas, 2012). In the recent past new generation insecticides like imidacloprid and fipronil have been used in both preventive and curative treatments (Barranco *et al.*, 1998; Kaakeh, 2006; Llácer *et. al.*, 2012). Semi-field trials with imidacloprid applied by soil injection to *P. canariensis* showed 100% and 94% efficacies, respectively (Llácer*et al.*, 2012). In a field assay, two applications of imidacloprid per year successfully reduced mortality of *P. canariensis* palms to less than 27% compared with more than 84% for untreated control palms (Dembilio *et al.*, 2010a). Insecticide application on fresh wounds of palm tissue to prevent oviposition and egg hatch has been recommended in the past (Abraham, 1971; Abraham *et al.*, 1998). Recent studies carried out in Spain have shown that a single application of an insecticidal paint with 1.5% chlorpyrifos could prevent infestation for up to 6 months with a mean efficacy of 83.3% (Llácer *et al.*, 2010). Palms that are severely infested and beyond curative treatments have to be eradicated by shredding as burning may not result in complete destruction of the hidden pest stages deep inside the palm tissue. A recent report on the eradication of severely infested date palms estimated the annual loss was from USD $5-26 million in the Gulf countries of the Middle East at 1-5% infestation level, respectively of which 20% of the palms were removed and destroyed (El-Sabea *et al*., 2009).

Biological control : Several natural enemies such as insects, bacteria, entomopathogenic fungi (EPF), viruses, yeast, entomopathogenic nematodes (EPN) and birds have been reported attacking *R. ferrugineus* (Murphy and Briscoe, 1999; Faleiro, 2006). However, large scale classical biological control programs have yet to be developed to augment the management of this cryptic and lethal insect pest. Recent studies from Spain have shown that both EPN and EPF can provide an alternative to chemical control of *R. ferrugineus* (Dembilio and Jacas,2012). *Steinernema carpocapsae* (Weiser) (Nematoda: Steinernematidae) proved effective against *R. ferrugineus* in semi-field trials in both preventive and curative assays with efficacies of up to 98% and 80%, respectively (Llácer *et al*., 2009). Under field conditions, treatments using imidacloprid and *S. carpocapsae*, either alone or in combination were not significantly different from each other, with efficacies ranging from 73 to 95% (Dembilio *et al*., 2010a; Tapia *et al*., 2011). With regard to EPF, Dembilio *et al*. (2010b) recorded adult mortality under field conditions in Spain of up to 86% with a Spanish strain of *Beauveria bassiana.* More recently, Llácer *et al*. (2013) showed that the release of irradiated males infected with the same strain of *B. bassiana* should be considered as a practical way to improve the biological control of *R. ferrugineus* with EPF. Thus, both EPN and EPF have potential for large-scale deployment against *R. ferrugineus* in the field but would need standardization to withstand local conditions prevailing in different agro-ecosystems.

Autocidal control : Al-Aydeh and Rasool (2010) studied the influence of gamma radiation on the mating behavior and the efficacy of sterile insect technique (SIT) and found no adverse effects of gamma radiation on the mating behavior parameters of the weevil in the laboratory. During the mid 1970s, large scale field studies in India in a 400 ha coconut plantation using SIT revealed that 70% of the eggs laid by female weevils captured in log traps from the test plot were viable eggs indicating that SIT had almost no impact in curtailing the population build up (Rahalkar *et al*., 1977). The opportunity for females to mate with the natural male population before flying out of the brood (Faleiro, 2006) and expensive artificial rearing of *R. ferrugineus* (Dembilio and Jacas, 2012) to warrant repeated mass releases of male weevils would limit the success of any technique to manage this pest that depends on mating disruption or interference.

Plant resistance : Naturally occurring defense mechanisms against *R. ferrugineus* have not been fully understood and therefore not exploited to manage this lethal pest of palms. Barranco *et al*. (2000) and Dembilio *et al*. (2009) reported antibiotic and antixenotic mechanisms in some palm species.

Washingtonia filifera and *Chamaerops humilis* could not be naturally infested by *R. ferrugineus* adult females and antibiosis was found to be the main mechanism operating in *W. filifera* (Dembilio *et al.*, 2009). Studies carried out in China showed that *P. canariensis* and *W. filifera* were more suitable hosts while *P. sylvestris* was the least suitable host for *R. ferrugineus* based on population growth parameters (Ju*et al.*, 2011). In coconut, the cultivar Chowghat Green Dwarf was highly preferred for oviposition by *R. ferrugineus* while, the least number of eggs were laid in Malayan Yellow Dwarf (Faleiro and Rangnekar, 2001). Farazmand (2002) and Al-Ayedh (2008) found that date palm cultivars with high sugar content enhanced growth of *R. ferrugineus*. Further studies are needed to understand palm defense mechanisms against this pest.

Regulatory control : Local, regional and international shipments of palms for landscape gardening and farming have resulted in the quick spread of *R. ferrugineus* during the last three decades. The first report of an infestation of *R. ferrugineus* in the Middle-East, North African and Mediterranean countries by Faleiro *et al.* (2012) suggests that palms shipped for ornamental landscaping is the primary source of new infestation foci. The need to implement strict pre- and post-entry quarantine regimes to certify transport of planting material free of *R. ferrugineus* has been proposed previously (Faleiro, 2006). The European Union (EU) has put strict legal measures in place that govern the movement of palms including the demarcation of a 10-km buffer zone around infested areas, three monthly official inspection and annual crop declaration of plant nurseries, implementation of phytosanitary treatments, registration of planting material movement, and an EU plant passport for all palm shipments moving within the EU (EPPO, 2008). In Spain, fumigating palms at a dose of 1.14 g aluminium phosphide m^{-3} for three days was enough to kill all stages of *R. ferrugineus* in an infested *P. canariensis* palm (Llácer and Jacas, 2010) while in Saudi Arabia, it is recommended to dip date palm offshoots in 0.004% Fipronil for 30 minutes before transporting to ensure mortality of the hidden larval stages (Al-Shawaf *et al.*, 2013). These findings could serve as treatment protocols for certification prior to shipment.

Other IPM tactics : Besides the above mentioned IPM components for control of *R. ferrugineus*, identifying and treating hidden breeding sites, particularly in closed gardens and other areas difficult to access, is vital to curb the build-up of field populations and sustain success where the pest is controlled (Abraham *et al.*, 1998). Periodic training of key stakeholders including officials implementing IPM strategies, farmers, and palm nursery dealers is important to ensure capacity building through the technology transfer of the latest IPM strategies and also to gain cooperation of all concerned in implementing an areawide *R. ferrugineus* IPM program.

Periodic validation of Geographic Information Systems aided with spatial and temporal models (Massoud *et al.*, 2011) and trap capture data and infestation reports (Al-Shawaf *et al.*, 2012; Faleiro *et al.*, 2008, 2010) could enable pest managers to deploy the scarce resource of personnel and materials to areas where it is most needed in an areawide *R. ferrugineus* IPM operation.

16.2.2 Rhynchophorus palmarum

The American palm weevil (APW), *Rhynchophorus palmarum*, is one of the most important pests in tropical America and considered to have high invasive potential (Giblin-Davis, 1990; Esser, 1969). *Rhynchophorus palmarum* has been reported attacking 35 plant species of different families, but it is mostly found affecting species within the Arecaceae (Estévez and Cedeño, 2008; Briceño and Ramírez, 2000; Hernández *et al.*, 1992; Griffith, 1987). The main hosts within this family are *C. nucifera, E guineensis, Euterpe edulis* and *Bactris gasipaes*. It is also reported from *Saccharum officinarum* L., *Musa* spp., *Carica papaya* L., and *Annona* spp. *Rhynchophorus palmarum* affects not only healthy palms, but is also attracted to palms affected by pathogens that elicit palm fermentation. The female deposits its eggs inside the palm stem and developing larvae tunnel through the trunk weakening the palm; however, larval damage to the apical meristem is considered lethal (Griffith, 1987). In addition, *R. palmarum* adults are vectors of the nematode *Bursaphelenchus cocophilus* (BC), which is the causal agent of red ring disease, which affects Central and South American commercial plantations of oil palm, coconuts, as well as pejibaye, which is an important crop for heart of palm production (Pérez and Iannacone, 2006a; Locarno, *et al.*, 2005; Arroyo *et al.*,2004; Peña-Rojas, 2000). Previously, Lepesme (1947) and Hagley (1965a) maintained that APW is typically a secondary pest of palms, rarely observed on healthy trees, but always found in injured or diseased trees. Majaraj (1965) and Schuiling and van Dinther (1981) reported that the level of APW infestation in palm groves is positively correlated with the abundance of injured trees.

Biology and behavior : *Rhynchophorus palmarum* can be found attacking palms located at sea level up to 1200 m (Jaffe and Sanchez, 1990). Adults have a black, hard cuticle and measure 4-5 cm in length and 1.4 cm wide. Adults show sexual dimorphism; males have a conspicuous brush of hairs on the antero-central dorsal region of the rostrum. The adults are active during the day showing bimodal daily activity cycle. Hagley (1965a) reported activity peaks between 7 and 11 and between 17 and 19h. Field observations showed that adults fly at velocities of 6.01 m/sec (Hagley, 1965a).

The eggs are placed individually inside the palm tissue near the apical area of the palm and protected by a brown waxy secretion. Sánchez *et al.* (1993)

studied the biology of the insect and indicated that females deposit their eggs into holes in the plant made by the rostrum. Oviposition or attraction appears to be increased when the plant tissue is either damaged near or on the internodal area of the palm trunk next to the crown. The cream-white, legless early instars are 3-4 mm long and older larvae may reach 5-6 cm in length. The egg incubation period is 3.2±0.93 days and the larvae have 6-10 instars. The larval stage might last ca 50-60 days.The pupa is exarate and brown and inhabits a cylindrical cocoon built with plant fibres organized in a spiral configuration.

The longevity and high oviposition capacity of *R. palmarum* contribute to the importance of this pest. The long time span for the larval and adult stages can last between 40-80 days or 120 to 180 days depending on the report (Mexzón *et al.,*1994; Sánchez, *et al.,*1993; Gonzales and García, 1992; Restrepo *et al.,* 1982; Jiménez, 1969).The female can deposit an average of 100-250 eggs and a maximum of 900 during its life span (Mexzón *et al.,*1994; Hagley, 1963; 1965a). Another important factor is the larval boring behavior inside the trunk that makes it difficult to control with insecticidal spray applications. Lastly, thedispersion capability of adults being able to fly 6 m/sec and 1.6 km/day (Griffith, 1987;OEPP/ EPPO, 2005; Bech, 2011, 2012;) and its adaptability to different climatic regions in 29 countries in the Americas and the Caribbean allow for easy spread to new areas.

Damage and economic importance : The first symptoms of infested palms are a progressive yellowing of the foliar area, destruction of the emerging leaf and necrosis in the flowers. Leaves start to dry in ascending order in the crown; the apical leaf bends and eventually droops. If the palm trunk is dissected, the galleries and damage to leaf bases produced by the larvae are easily detected in heavily infested palms. Pupae and old larvae are frequently found when inspecting the crown of infested plants. Affected plant tissue turns foul, producing strong characteristic fermented odors (Jaffe *et al.,*1993; Chinchilla and Oehlschlager 1993; Sanchez and Jaffe, 1993).

The association *R. palmarum* with the nematode *Bursaphelenchus cocophilus* (BC) which is the causal agent of coconut red ring disease (RRD) (Ashby, 1921), has resulted in a complex plant health problem that can cause the total decline of coconut and African oil palm plantations. Esser and Meredith (1987) reported the incidence of APW in 92% of coconuts infected with BC. Fenwick (1967) and Griffith (1968) reported that populations of 30 larvae are sufficient to cause the death of an adult coconut palm. Coconut palms 3-10 years old infected with BC died during the first 2 months after incoculation (Griffith, 1967). Moreover, Thurston (1984) reported that affected palms take 23-28 days to show the symptoms of red-ring disease, and die 3-4 months after showing the first symptoms.

In Colombia, the complex of *R. palmarum* and BC is present in the Pacific coast. For example, during 2006-2009, budrot affected approximately 40,000 hectares of African oil palm where the infected palms became the focus of infestation of *R. palmarum* (Aldana *et al.,*2011). Later, Quintana (2012) reported that during 2010, a million US dollar loss to this crop due to APW and BC. A similar situation is commonly observed in other countries in Central and South America (Esser and Meredith, 1987).

Chemical control : Arango and Rizo (1977) suggested that management of *R. palmarum* should be aimed at adults instead of larval stages. Earlier, chemical control applied to the palm rachis or to the whole palm was the norm (Alfonso and Ramírez, 2004; Dean and Velis, 1976). At the same time, Arango and Rizo (1977) considered chemical control as inefficient and expensive and suggested that mass trapping was the best control tactic. The potential list of insecticides used against APW is extensive. The insecticides methomyl, triclorfon, pirimiphos-ethyl, fipronil, imidacloprid, abamectin, carbaryl and imidacloprid+deltametrin have been tested (Dean and Velis, 1976; Martínez *et al.,*2010), but control has not been satisfactory. Moura (1994) and Moura *et al.*(1995) suggested a preventive measure to brush the leaf rachis with a 0.2% solution of carbofuran or to make incisions to the trunk of a diseased palm and then apply carbofuran to the incisions in order to kill the insects attracted to the treated palm. Latest reports deal with the effect of botanical extracts (i.e., birthworth fruit) on *R. palmarum* larvae, demonstrating that this product caused 73% larval mortality. Perez and Iannacone (2006bb) and Perez *et al.*(2010) tested the toxicological effects of extracts of *Paulliniae clavigera* against APW and showed that the hydroalcoholic and decoction extract showed toxicity to this pest. While these data are encouraging, it must be emphasized that these are experimental data from laboratory experiments. Further tests need to be conducted to analyze the potential of botanical extracts under field conditions.

Trapping : *Rhynchophorus palmarum* trapping has been practiced since the 1970s (Mariau, 1968; Griffith, 1969). Earlier, trapping was based on the fact that adults are attracted to odors of fermenting palms for feeding and reproduction (Morin *et al.,*1986). The most commonly used traps before the discovery of a pheromone were insecticide-treated palm stems (Mariau, 1968, Morin *et al.,*1986, Chinchilla *et al.,*1990; Morales and Chinchilla, 1990; Dean and Velis, 1976) or using other plant tissue, *i.e.*, pieces of sugarcane, *Musa* pseudostems, pineapple and citrus fruits (Costa Miguens *et al.,*2011; Sumano *et al.,*2012; Cerda *et al.,*1994; Vera and Orellana, 1986; Raigosa, 1974). Hernandez *et al.*(1992) evaluated baits using tissues from *C. nucifera, E. guineensis, Jessenia polycarpa* Kartst, *Jessenia batatua* Burret, *Mauritia flexuosa* L., *Phenakospermium* sp., *Euterpe* sp. and *Antrocarium* sp. and reported that coconut, *Phenakospermium* sp., *Antrocarium* sp. and *J. batatua* attracted the highest number of adults.

Pheromones : Rochat *et al.*(1991a,b) and Oehlschlager *et al.*(1992) reported that APW males produce an aggregation pheromone (4S,5E)-2-methyl-5-hepten-4-ol or rhynchophorol. Racemic rhynchophorol synergistically enhances the attraction of plant material in field trapping (Oehlschlager *et al.*,1992). For example, Oehlschlager *et al.*(1992; 1993) demonstrated that more APW were captured using plastic bucket traps from which rhynchophorol was placed together with insecticide-laden sugarcane or palm stems. Moura *et al.*(1997) reported that capture of APW on oil palm treated with pheromone plus insecticide was 65 to 89% more than those treated with insecticide only. One of the problems of using plant material is that plant tissue generally loses attractiveness after 12-15 days, necessitating bait renewal (Oehlschlager *et al.*,1992, 1993; Jaffe *et al.*,1993).This fact corroborates the attraction of *R. palmarum* to plant kairomones in an effective trapping scheme.Rochat *et al.*(2000) reported that blends of "characteristic coconut" odor molecules were as efficient as sugarcane in synergizing rhynchophorol and luring *R. palmarum* into traps. Also, host plant volatiles from sugarcane, coconut, *Jacarantia digitata* and *E. guineensis* were tested. Oehlschlager *et al.* (2002) reported that pheromone-based trapping of *R. palmarum*using trap densities of 1 trap per 5 ha lowered RRD by over 80%. Pheromone lures last for 2 to 5 months depending on the season, but the necessary food component lasts only for a week or two with water evaporation and loss of attractancy as the two main problems. Oehlschlager (2010) reported that the key to trapping success was to make the food component of the trap more attractive as well as adding ingredients to the trap to extend the field life of the kairomones. This author found that emission of ethyl acetate from dispensers in pheromone/food traps increases captures compared to pheromone/food traps by 2-5 times. Also, addition of propylene glycol to traps extends the effective life of kairomones in traps. It has been suggested that replacement of food for attractive blends has not been effective in the field to date (Oehlschlager, 2010).

Oehlschager and Gonzalez (2001) reported that pheromone trapping and removal of red ring nematode infested palms are the only economically viable techniques used in the Americas to combat palm weevil problems. Trapping is made difficult by the requirement of replacement of water and food bait in the traps. Also, the right dose of the aggregation pheromone must be used in order to avoid repellency. Non-repellant additives can extend the effective life of trap food bait from 2 to 7 weeks (Oehlschlager and Gonzalez, 2001). While repellants reduce captures of *R. palmarum* in pheromone traps by more than 50%, they could make push-pull strategies to improve management of palm weevils possible (see *R. ferrugineus* for push and pull strategies).

Biological control : According to Costa Miguens *et al.*(2011) few studies have been conducted on natural enemies of *Rhynchophorus* species. During 1990-1991, Moura *et al.*(1993) found 51.1 % of *R. palmarum* parasitized by

Paratheresia menezesi Townsed (Diptera: Tachinidae). Ramirez (1998) reported that in Colombia the prepupae of *R. palmarum* was parasitized by an unidentified tachinid and preyed upon by *Hololepta* sp. However, the action of these natural enemies was considered insignificant. Mosquera and Viafara (2008) tested the effect of *Paratheresia claripalpis* (Wulp.) and *Metagonistylum minensis* (Townsed) in the laboratory against *R. plamarum*and found both species ineffective.Moura *et al.*(2006) reported that in the piassava palm in Brazil, *R. palmarum* larvae were parasitized by *Billaea rhynchophorae* (Blanchard) (Diptera: Tachinidae) with levels of parasitism ranging from 18-50%. Methods of mass production of these parasitoids are needed.

Microbial control : Four species of commensal nematodes in addition to BC have been reported from *R. palmarum*, but they are not pathogenic (Gerber and Giblin-Davis, 1990).The role of entomopathogenic fungi such as *B. bassiana* Balsamo Wuillemin and its effectiveness against *R. palmarum*is unclear (Costa Miguens *et al.,*2011).

16.2.3 Conclusions

The importance of the *R. palmarum* has increased due to the presence of budrot in oil palm plantations in Central and South America. Palms that are affected by budrot decompose fast, and the foul odors attract high weevil densities. Management of this insect-disease complex has become a key objective to maintain the African oil palm plantations in this region.The following steps are considered: 1) reduction of *R. palmarum* densities by use of mass trapping; 2) removal of red-ring affected palms.

16.3 Mites (Acari: Tenuipalpidae and Eriophyidae)

16.3.1 *Raoiella indica* (Acarina: Tenuipalpidae)

Distribution : *Raoiella indica* Hirst, is a tenuipalpid mite species that recently invaded the Western Hemisphere. Previous to its arrival in the New World, *R. indica* was found in tropical and subtropical areas of Africa and Asia including Oman and Egypt (Elwan, 2000), Israel (Gerson *et al.,*1983), Iran (Askari *et al.,*2002), Pakistan (Chaudri, 1974), Mauritius (Moutia, 1958), Phillipines (Gallego *et al.,*2003), India (Puttarudriah and Channabasavanna, 1956) and Benin and Tanzania (Zannou *et al.,*2010). In 2004, this mite was detected in Martinique (Flechtmann and Etienne, 2004) and subsequently spread to multiple islands of the Caribbean. It is now reported in Florida, U.S.A. (FDACS, 2007), Venezuela (Vásquez *et al.,*2008), Brazil (Marsaro Jr. *et al.,*2009), Colombia (Carrillo *et al.,*2011) and Mexico (NAPPO, 2009). Amaro and de Morais (2013) used a model to predict the potential geographical distribution of *R. indica* in South

America and detected suitable areas for *R. indica* in northern Colombia, central and northern Venezuela, Guyana, Suriname, east French Guiana and, the eastern Amazonas, northern Roraima state (Brazil) and the coastal zones from north eastern Brazil to north of Rio de Janeiro. Their results indicate the potential for significant *R. indica* related economic and social impacts in all of these countries.

Damage : *Raoiella indica* is a polyphagous species with a wide host plant range mostly within the Arecaceae (Carrillo *et al.,* 2012a ; Godim *et al.,*2013), but it also attacks some plants within the Pandanaceae, Musaceae, Heliconiaceae, Zingiberaceae and Strelitziaceae. Its major host is coconut, *C. nucifera*. *Raoiella indica* is highly destructive to coconut sometimes reaching 4,000 individuals per pinna (Peña *et al.,*2009). Infestations found on other palms (approximately 78 palm species) are lower in numbers but still problematic because of their importance as ornamental or native plants (Carrillo *et al.,*2012a; Godim *et al.,*2013). Damage caused by this species to most of its host plants has not yet been characterized. In coconut, *R. indica* feeding causes an initial bronzing of the leaves which later turns into necrotic tissues (Figure 16.1). Red palm mite feeding on coconut results in extensive chlorosis and yellowing of the leaves similar to damage caused by 'lethal yellowing' (LY) (Peña *et al.,*2012). However, LY disease differs in that it causes coconuts of all stages to drop from the tree, it disorts the emerging inflorescences and it causes the male flowers to become dark brown. The establishment of *R. indica* in the Caribbean has caused serious economic harm to coconut production with over 50% yield reductions at some locations (CARDI 2010).

Biology : *Raoiella indica* is usually found forming large multigenerational colonies on the abaxial surface of the leaflets (pinnae) of their host plants (Figure 16.2). Eggs are oblong, smooth, glossy, red in color, and measure 0.12 mm in length and 0.09 mm in breadth, with a stipe 0.15 long (Moutia, 1958) (Figures 16.2A and 16.2B). A single droplet of an unknown substance is sometimes found at the tip of the stipe. Upon hatching, larvae are slow in their movements and start feeding on leaf tissue near the site of emergence. Development of *R. indica* passes through two nymphal stages with a quiescent period before each molt. The immature stages of *R. indica* are red with blackish marks in the dorsum and develop near the site of oviposition. The duration of the immature development varies from approximately 22 to 34 days depending on the host plant and environmental conditions (Table 16.1).

Males are triangular in form, develop faster and are smaller than females. They actively seek out females and settle closely behind a quiescent deutonymph or adult female in a "guarding state" until mating (Figure 16.2B). The preoviposition period ranges from 3 to 7 days (Moutia, 1958; Flores-Galano *et al.,*2010). The type of reproduction in *R. indica* is unusual because mated females (sexual reproduction) produce progeny consisting solely of females and unfertilized females (asexual reproduction) produce only males (Nageshachandra and Channabasavanna,1984). The species is reported to have four and two chromosomes (females and males, respectively) which suggest that its reproduction is arrhenotokous (Helle *et al.*,1980). According to Nageshachandra and Channabasavanna (1984) mated females produce at an average of 22 ± 4.41 (females/female) and unmated females produce 18.4± 1.95 (males/female) during their adult life on coconuts. Moutia (1958) reported a similar average fecundity (approximately 28.1 eggs/female) but other studies with other hosts reported substantially lower fecundities (12.6 ±4.3) on coconut (Gonzales and Ramos, 2010), and 7.0 ± 3.5 eggs per female on Areca palms (Flores-Galano *et al.*, 2010). According to Nageshachandra and Channabasavanna (1984), longevity is greater in females (50.9 ± 11.4 days) than in males (21.6 ± 1.95 days), but this can also be influenced by the host plant (Gonzales and Ramos, 2010). Adults are more active and are responsible for dispersion within the leaf, to colonize new leaves or new plants.

Seasonality and influence of abiotic factors : Few studies have examined the effects of abiotic factors on *R. indica*. Moutia (1958) reported that population build-up was positively related with temperature and sunshine hours, but negatively related with rainfall and relative humidity. In agreement, Nageshachandra and Channabasavanna (1983) reported an increase of mite populations associated with periods of lower relative humidity, and Taylor *et al.*(2012) found that when conditions were hotter and drier, *R. indica* densities were significantly higher. Sakar and Somchoudhury (1989) reported that mite density had a significant positive relation with temperature and that *R. indica* densities declined with the onset of the rainy season in India. The available literature suggests that this mite adapts well to tropical climatic conditions; however, prolonged dry conditions could favor *R. indica* population growth.

Sampling : Roda *et al.* (2012) studied the intra tree spatial distribution of red palm mite populations on coconut and developed sampling methods to aid in the timely detection of the pest in a new area. Their studies revealed that the middle stratum of a palm had signiûcantly more mites than fronds from the upper or lower canopy; and mite populations did not vary within a frond. Roda *et al.* (2012) indicated that the most efficient way to detect infestations and estimate population densities is to sample one pinna per tree from as many trees as possible.

Chemical control : Rodriguez and Peña (2012) conducted several acaricide trials in Puerto Rico and Florida in order to provide chemical control alternatives to minimize the impact of this pest. Spiromesifen, dicofol and acequinocyl were effective in reducing the population of *R. indica* in coconut in Puerto Rico.

Spray treatments with etoxanole, abamectin, pyridaben, milbemectin and sulfur showed mite control in Florida.

Biological control : The host plant range and dispersal of *R. indica* throughout natural, agricultural, recreational and residential areas suggest that large scale mitigation programs are required for managing this species. Chemical control, host plant resistance and cultural control tactics could be used to manage local populations; however, the most promising approach is to find a practical, long-lasting solution in the form of biological control. One of the factors contributing to the aggressiveness of invasive species is that they often invade new areas without their specific natural enemies (Herren and Neuenschwander, 1991). Thus, one of the most common approaches to suppress invasive species is to search for natural enemies in their site of origin and reunite pest and natural enemies through importation of the latter (*i.e.* Classical biological control) (Van Driesche and Bellows, 1996). However, in many cases indigenous natural enemies may provide some suppression of the invasive pest. In addition, biological control might be found among species that have not experienced close prior relationship with the target organism (Hokanen and Pimentel, 1984).

Twenty-eight species of predatory arthropods, including mites and insects, have been reported in association with *R. Indica* in Asia, Africa and the Neotropics (Carrillo *et al.,*2012c). In addition, pathogenic fungi associated with *R. indica* in the Caribbean have been reported. The available literature indicates that each site has a different natural enemy complex with only one predator species, *Amblyseius largoensis (Muma)* (Acari: Phytoseiidae), present in all the geographical areas.This predatory mite species has shown a conspicuous association with *R. indica* and it represents the most important biotic factor known to be acting over *R. indica* in the different places where this pest is present.The interaction between *A. largoensis* and *R. indica* in Florida has been investigated. An initial study demonstrated that *A. largoensis* was able to feed, develop and reproduce on a diet consisting solely of *R. indica*. The intrinsic rate of natural increase of the predator was significantly higher when fed on *R. indica* than on other arthropods inhabiting coconut, including *Tetranychus gloveri* Banks (Acari: Tetranychidae), *Aonidiella orientalis* (Newstead) (Hemiptera: Diaspididae) and *Nipaecoccus nipae* (Maskell) (Hemiptera: Pseudococcidae) (Carrillo *et al.,*2010). Additional studies showed that *A. largoensis* had a marked preference for *R. indica* eggs over other stages of the pest. In addition, the predator showed a Type II functional response and a positive numerical response to increasing densities of *R. indica* which could explain the population increase observed in areas of recent invasion (Carrillo and Peña, 2012). Moreover, *A. largoensis* may be responding to the invasion by *R. indica* either by learning or evolving genetically to be better predators of this pest (Carrillo *et al.,*2012b). *Amblyseius largoensis* populations with a long time association with *R. indica* showed increased prey consumption and higher reproductive rates compared to populations with short time exposure to this invasive pest (Carrillo *et al.*,2012b; Domingos *et al.*,2013). Based upon previous studies, researchers have identified *A. largoensis* populations that are superior predators of *R. indica* and could be used as classical biological control agents (Domingos *et al.*,2012). Moreover, de Assis *et al.* (2013) identified acaricides that are selective for *A. largoensis* such as fenpyroximate and spirodiclofen, which could be used to integrate chemical and biological control strategies against *R. indica*. Additional studies using exclusion and release techniques suggested that *A. largoensis* could be efficient at regulating low prey population densities. However, it's likely that other mortality factors will be required to effectively suppress the populations of this invasive pest. Search for effective natural enemies should be intensified and ways to improve the levels of control by the existing natural enemies (i.e. provision of alternative food sources) should be further investigated.

Host plant interactions : Host plants could have an important influence in the population dynamics of *R. indica*. However, very little is known about the influence of plant characteristics on their quality as hosts for *R. indica*. Sakar and Somchoudhury (1989) studied possible resistance mechanisms to *R. indica* in 8 coconut varieties based on morphological (pinnae length and width, leaf thickness, depth of midrib groove, and intervienal distance) and biochemical (crude protein, moisture, nitrogen and mineral content) characteristics. The study concluded that there was no relationship between mite populations and the physical (morphological) characteristics of the coconut varieties but the chemical characteristics had an influence on mite densities. Varieties which contained higher amounts of nitrogen and crude protein showed higher incidence and higher mite population densities.

Raoiella indica was reported as the first mite species observed feeding through the stomata of their host plants (Kane *et al.*,2005). Later it was shown that feeding via stomatal aperture occurred among several *Raoiella* species (Ochoa *et al.*,2011). Further studies are needed to understand the implications of the feeding habits of *R. indica*. Through this specialized feeding habit, *R. indica* probably interferes with the photosynthesis and respiration processes of their host plants. Moreover, the type of stomata and their densities could determine the suitability of a plant as host for *R. indica*. For instance, coconut varieties can be divided in two main groups, tall and dwarf varieties. Tall palms may attain heights of 30 m (approximately) while dwarfs have shorter internodes and thus are much shorter (Gomes and Prado, 2007). Tall varieties are greatly influenced by wind currents that remove the humidity from the leaf surface and thus have more efficient osmoregulation mechanisms than dwarf varieties (Gomes and Prado, 2007). Dwarf varieties have a greater number of stomata per leaf area and lower wax content on the leaf surface compared to the tall varieties (Rajagopal *et al.*, 1990). These differences among coconut palm varieties could be influential on their suitability as hosts for *R. indica* which merits investigation.

Other IPM tactics : *Raoiella indica* is a quarantine pest for some countries (EPPO, 2009). Quarantine-mandatory acaricide sprays before shipping *R. indica* hosts have been adopted in order to prevent *R. indica*'s rapid dissemination (TDA, 2008;FDACS, 2011).More research is needed to mitigate the potential impact of this damaging invasive species.

16.3.2 Aceria guerreronis (Acarina: Eriophyidae)

Distribution : The coconut mite, *Aceria guerreronis* Keifer, is perhaps the most destructive mite affecting coconuts, *Cocos nucifera*. The species was described by Keifer (1965) from specimens from Guerrero, Mexico. Navia *et al.*(2005) show records from as early as 1948 with a wide distribution throughout the neotropical region (Colombia, Brazil, Mexico, Venezuela, islands of the Caribbean, Costa Rica) and the subtropics of the U.S.A. The same report shows that almost simultaneously with its original description, the mite was reported in Africa and in 1966 in the Gulf of Guinea Islands, in 1967 Benin, and in the 1980s in Tanzania in the east of the African continent. The most recent records of the coconut mite are from India and Sri-Lanka, where the species was unknown until the end of 1990s. Results from DNA sequence data from two mitochondrial and one nuclear loci obtained from samples of 29 populations from the Americas, Africa and the Indo-ocean region, support the hypotheses of an American origin of the mite corroborating the idea that the original host plant for the mite was not coconut, but an American host (Navia *et al.*,2005).

Host plants : Besides coconuts, *A. guerreronis* has been reported from other palm hosts (Navia *et al.*,2005). Two of them are palms of South American origin used as ornamentals, *Lytocaryum weddellianum* (H. Wedl.) in Brazil (Fletchman, 1989) and *S. romanzoffiana*, in southern California, U.S.A (Ansaloni and Perring, 2004). In both cases, the mite was found only in nurseries and not in natural areas. However, Ramaraju and Rabindra (2002) reported the species from *Borassus flabellifer* from India (Navia *et al.*,2007). Besides the species mentioned above, *A. guerreronis* has not been found in other Arecaceae in Brazil (Navia *et al.*,2007), thus, the original plant host remains as an open question.

Biology : The adult female mite is between 30-52 µm in width and 205-255 µm in length. Ansaloni and Perring (2004) determined that on *Syagrus romanzoffiana*, *A. guerreronis* development from egg to adult required 30.5, 16, 11.5, 8.1 and 6.8 days at 15, 20, 25,30 and 35°C, respectively and the thermal maximum was estimated to be close to 40°C. Optimal immature development was observed at 33.6°C and lower temperature threshold at 9.36°C. Using these values, these authors found that 172.71 degree days were required for mites to develop from egg to the adult stage. Fertilized females laid more eggs and lived longer than non fertilized females, with a maximum of 51 eggs produced during 43 days. Unfertilized females laid only male offspring, indicating that the species is arrhenotokous. According to Mariau (1977) the mite life cycle in coconut is completed in 7-10 days. In India, population peaks occur during the summer months of April to May (Nair, 2000).

Damage : The mites infest the abaxial surfaces of perianth and that part of the fruit surface that is covered by the perianth. They are able to penetrate between the tepals of the perianth and fruit surface a month after the fruit begins its development; prior to this the tepals are too tightly appressed to allow entry of the mites (Moore and Howard, 1996). Mite feeding causes necrosis and suberization of tepals and the uneven growth often results in distortion and stunting of coconuts, leading to reductions of copra yield up to 30% (Moore and Howard, 1996). The fruits are susceptible to mite attack through most of the development of the fruit, but fewer mites are found on fruits that are 10-13 months or older (Moore and Alexander, 1987). Yield losses are compounded because the compacted fibers of the mesocarp increase the labor requirement for de-husking. In addition to damaging the fruit, *A. guerreronis* can kill coconut seedlings by feeding on their meristematic tissues (Arruda, 1974). According to Nampoothiri *et al.*(2002), infestation by the mite results in button shedding and formation of malformed or undersized nuts. In India, pest outbreaks during the late 1990s resulted in 85-90% of malformed nuts. In Kerala, copra losses can reach 31% while total husk production can be reduced in 41% (Nampoothiri *et al.*,2002). Copra dealers in North Malabar, India, estimated that copra production dropped from 18–20 to 10–12 kg per 100 nuts after the coconut mite upsurge at the end of the 1990s. Reduction in copra weight ranged from 6.8 to 31.6 from low to high infestation level. However, as indicated in the categorized plot, the values were more consistent for low and medium infested groups of nuts (Maq, 2011).

Sampling : Proper management of *A. guerreronis* requires an assessment of its population density on coconuts. The high mobility,microscopic size and hidden lifestyle of colonies aggregating in inner and outer bracts (tepals) and under the bracts of developing coconuts, make an effective sampling plan time consuming, tedious and often inaccurate (Siriwardena *et al.*,2005). Howard *et al.*(1990) developed a method where bracts of infested coconuts were removed, and the mites were counted in 1 mm^2 of a 25 mm^2 area of the colonies on the nut surface. Moore and Alexander (1987) scored the number of mites on the coconut surface in a log scale (e.g., 0 = 0 mites, 1 = 1–10 mites, 2 = 11–100 mites, 3 = 101–1000 mites, *etc.*). Both assessments were considered time consuming and difficult. Siriardena *et al.* (2005) developed a method where the coconut mites were removed by washing the bracts and surface of an infested coconut with 30 ml of a detergent solution. Shaking the wash for 5 seconds allowed the mites to distribute uniformly. The number of mites in the first 1 ml of the first wash (X) yielded a very accurate predictor of the total number of mites on a coconut (Y): Y = 30.1X ($R^2 = 0.99$; $p < 0.0001$), also confirming that the wash was homogeneous. The present study envisaged developing a new technique to estimate the actual population density of *A. guerreronis* on a coconut by washing off the mites into a solution, distributing the mites homogeneously in the wash by shaking, and counting the mites in a subsample for estimation of the population. Fernando *et al.* (2003) reported that coconut mites were present on 88 and 75% of the nuts at two sampled sites in Sri Lanka, but no more than three-quarters of the nuts showed damage symptoms.Their results indicated also that assessment of infestation levels by damage symptoms alone is not reliable. Because the mean number of coconut mites increased until the nut bunches were 5-months-old, these authors suggested that sampling could be improved if nuts from 6-month-old bunches were used.

From this we conclude that sampling for *A. guerreronis* will be based on availability of susceptible organs on coconuts, pest prevalence, etc.

Economic thresholds : Determining economic thresholds for *A. guerreronis* is a difficult task due to the pest behavior, difficulty to accurately determine the pest population and its correlation with economic damage. For instance, experiments conducted in India showed that nut fall was observed in infested and uninfested nuts, making unclear if the pattern of nut fall was completely related to the presence of *A. guerreronis* (Maq, 2012). In order to ascertain the degree of the impact by mite feeding, total surface area on the damaged area of the nut was studied statistically through regression analysis and ANOVA by Maq (2011). This author showed a significant quadratic relation between these parameters, indicating a relatively strong reduction of damaged area with total nut surface area. While his results suggest that when infestation is delayed, the nuts become larger and mite damage is reduced, the very low value of R^2 (0.15) indicated that little of the variation in the damaged area could be explained by the variation of the total nut area (Maq, 2011).

Chemical control : In the 1970s to 1980s, chemical control of the coconut mite was considered possible but not practical (Moore and Howard, 1996). The insecticides (i.e, dicrotophos, monocrotophos) were either applied as a spray to bunches of developing fruit or injected, but the results varied from satisfactory to ineffective. Muthiah and Bhaskaran (1999) reported that spraying methyl demeton at 4 ml/litre effectively reduced nut damage and suggested that drenching the roots with monocrotophos or methyl demeton would result in less environmental problems. According to Nair (2000) control can be achieved by spraying monocrotophos, dicofol and methyl demeton and 2% neem mixed with garlic.

Pushpa (2006) reported that among different treatments, fenazaquin 10 EC at 2 ml/l was significantly superior in reducing both egg and active stages of mite population and was on par with monocrotophos 36 SL at 4 ml/l and dicofol 18.5 EC at 4 ml/l. The efficiency of treatments continued upto 21 days, but later the mite population again increased. Throughout the experimental period these treatments were found superior in recording minimum mite population. With regard to per cent damaged nuts, fenazaquin produced a lower percentage of damaged nuts (30.97), but monocrotophos was superior with the lowest proportion of damaged nuts (24.80). When acaricides were applied, more undamaged nuts were obtained compared to the untreated control.

Spraying botanicals three times at an interval of 3 months against *A.guerreronis* indicated that among the different botanicals neem seed kernel extact (NSKE) 5 percent was found to be effective and significantly superior at different intervals of observations in reducing the mite and egg population. NSKE continued to be effective in suppressing the population of the mite upto 21 days after treatment. Neem oil at 3 percent and neemazal at 5 ml/l were proved to be the next best treatments. The standard check dicofol at 4 ml/l was superior over all treatments in minimizing the mite population, whereas sweet flag and biocare throughout the experiment maintained their inferiority in reducing mite populations.

With regard to their efficacy on nut damage NSKE 5 percent was significantly superior in recording the least percentage of damaged nuts (57.58 %). Maximum healthy nuts (36) and total nuts (89) were recorded in NSKE 5 percent which was followed by neem oil and neemazal. Dicofol, NSKE 5 percent, neem oil and neemazal sprayed palms recorded less damaged grade nuts. Among the different schedules, the treatment consisting of root feeding 6 times per year which was scheduled at two months interval during January, March, May, July, September and November was significantly superior in reducing the mite (both active stages and egg) population and was also superior in recording a minimum number of damaged nuts. The treatment consisting of spraying 2 times per year scheduled at a 6 months interval *i.e.* during April and October was significantly inferior in reducing the mite population.

Biological control : During a survey for acarine fauna associated with *A. guerreronis* in Brazil, Lawson-Balagbo *et al.* (2008) found the phytoseiids *Neoseiulus paspalivorus* DeLeon, *N. baraki* Athias-Henriot and *Amblyseius largoensis* Muma as well as the ascidid *Proctolaelaps bickleyi* Bram, *Proctolaelaps* sp. Nov and *Lasioseius subterraneus* Chant associated with *A. guerreronis*. The predators, *N. baraki, N. paspalivorous* and *P. bickelyi,* are commonly found associated with the mite in American, African and Asian coconut regions reported by Howard *et al.* (1990), Fernando *et al.* (2003) and by Moraes *et al.* (2004). Earlier, Moore (2000) supported the idea that no natural enemy appears to be successful as a classical biological control agent. However, Fernando *et al.* (2010) showed that inundative releases of *N. baraki* resulted in a highly significant reduction of *A. guerreronis* populations during 6 months of the releases.

Microbial control : Pathogens have potential to produce effective control and a programme to develop an effective myco-acaricide with the most promising *Hirsutella*spp. candidates should begin immediately (Moore, 2000). Fernando *et al.*(2007) reported that 4 Sri Lankan isolates of *Hirsutella thompsoni* Fisher obtained from *A.guerreronis* were tested as biopesticides resulting in mite reduction during 4 weeks after treatment, but the efficacy was reduced when the population was evaluated 30 weeks after treatment. A second application of the isolate IMI 391722 and IMI 390486 resulted in dead mites throughout the study.

Plant resistance : Most coconut varieties are susceptible to *A. guerreronis* including most subpopulations of the East African Tall (EAT) and dwarf varieties. However, some introduced varieties including Polynesian Tall (PYT), Malayan Red Dwarf (MRD), Rennel Tall (RLT), Cameroon Red Dwarf (CRD) and Equatorial Green Dwarf (EGD) have nuts which show more tolerance to mite attack (Seguni, 2000). During a 4-year screening, Sujatha *et al.*(2010) recorded the lowest damage in Laccadive ordinary (LO) followed by Andaman Ordinary (AO). The same authors reported that the hybrids ECT x GB (Godavari Ganga-1.87, LO x COD-1.95, YHC-I (ECT x MGD) -2.08 recorded the lowest mite damage whereas LM x GB -2.68, Java x GB-2.83 and Fiji x GB-2.75 recorded the highest damage among crosses.

16.4 Aphids, Scales, Mealybugs, and Whiteflies (Hemiptera:Sternorrhyncha and Fulgoromorpha)

Howard *et al.*(2001) recognized Hemiptera as one of six orders of insects that contain notable palm pests worldwide.Hemiptera is comprised of 3 suborders: Heteroptera (true bugs), Auchenorrhyncha (free-living hemipterans), and Sternorrhyncha (plant-parasitic hemipterans).Hemipteran insects can be recognized by the specialized morphology of the mouthparts which are modified into concentric stylets forming food and salivary channels (Forero, 2008).

In general, pests in the Suborder Heteroptera that feed on palms attack fronds of specific palm species and generally fall into 4 families; Miridae (plant bugs), Pentatomidae (stink bugs), Tingidae (lace bugs) and Thaumastocoridae (flat bugs). Pests in the Suborder Auchenorrhyncha are typically not damaging to palms through their feeding, generally occur in low to moderate populations, and can be rather inconspicuous. Of most concern is the potential of some of these species in this suborder to vector palm diseases. The Sternorrhyncha are probably the most represented of the 3 suborders on palms. Most species on palms feed on the foliage but may also be found on stems, fruits and sometimes roots. Within the Sternorrhyncha are Aphididae (aphids), Aleyrodidae (whiteflies), and the Superfamily Coccoidea (scales and mealybugs). In this chapter, we will highlight key pests of Sternorrhyncha and Auchenorrhyncha as vectors of plant diseases.

16.4.1 Vectors of Palm Diseases (Fulgoroidea)

In the US, lethal yellowing of coconut and other palms is apparently caused by mycoplasma like organisms (Howard, 1983; Howard *et al.,*1983). This disease has also been known from the Caribbean Islands for at least 100 years and from West Africa for at least 50 years (Howard, 1983). The species considered to be the most important vectors of palm diseases are from the families Cixiidae and Derbidae. Derbids are a large family of planthoppers with approximately 1,700 species worldwide. Many Derbid species have been reported from palms (Howard, 2001) although the immature stages are associated with fungi in rotten wood. A well known example of a disease vector of palms is *Myndus crudus* (*Haplaxius crudus*) (Heteroptera: Cixiidae), the vector of lethal yellowing in the Americas and the Caribbean Islands (Eden-Green 1978; Howard *et al.,*1983). Investigations in Vanuatu indicated that *Myndus taffini* Bonfils is the disease vector for coconuts on that island (Julias, 1982).The nymphs are found in the hollows where leaflets join the frond rachis (Julias, 1982) or on heaps of decaying wood of *Hibiscus tiliaceaus* L. (Julias, 1982). Nymphs of *M. crudus* feed on the roots of grasses, such as *Stenotaphrum secundatum* (Walt.), *Axenopus compresus* Beauv.-Gram. and *Cynodon* spp. (Howard, 1990; Eden-Green, 1978) while some grasses such as *Brachiaria brizantha* Hocht., *Chloris gayana* Kunthe and *Hemarthria altissima* (Poir) appear to be poor hosts (Howard, 1990). In Kerala, *Proutista moesta* (Westwood) (Hemiptera: Derbidae) has been identified as the vector of Kerala wilt disease of coconuts (Edwin and Mohankummar, 2007). However, other authors also identify the lacewing, *Stephanitis typicus* Distant (Tingidae), as a vector of Kerala wilt of coconuts (Srinivasan *et al.,*2000).

IPM of disease vectors : Until recently, management of plant diseases caused by phytoplasmas was focused on controlling the vector with insecticides (Weintraub and Beanland, 2006). Removal of alternative vector host plants and/ or reservoirs of phytoplasma-infected crop plants and weeds by roguing or by treating infected trees with insecticide before roguing, can reduce the incidence of disease spread. Weintraub and Beanland (2006) maintain that chemical control of vectors likely will continue for the foreseeable future, but vector management or management of phytoplasma spread within the plant is now slowly shifting to habitat management and the use of genetically modified crops. Reports of effectiveness of natural enemies of these vectors are scarce. For instance, *Halictophagous* sp. (Strepsiptera: Halictophagidae), has been reported parasitizing *P. moesta* in India (Ponnamma and Biju, 1998). Habitat management can reduce pest incidence. The type of mulching materials used around coconut trees influences the abundance of the planthopper vector of lethal yellowing, *H. crudus*. Fewer nymphs are found around trees mulched with coarse materials such as pine bark nuggets (Howard and Oropeza, 1998). Although some parasitoids of leafhoppers have been identified, no studies have investigated the use of thesenatural enemies to effectively manage pest species. Unfortunately, the vegetation that can increase the incidence and abundance of natural enemies of vectors can also be favorable to those taxa that transmit phytoplasmas. More effort should be made to determine those elements of the cropping environment that enhance the survival of natural enemies but do not increase vector numbers (Weintraub and Beanland, 2006).

16.4.2 Aphididae (Aphids)

Distribution and biology : Aphids are small, generally soft bodied insects with long, spindly legs. Aphids are phloem-feeders and often act as vectors of numerous plant pathogenic viruses (Buss, 2010). In palms, however, there are few virus or viroid diseases (Elliott *et al.*,2004). Aphids are commonly found on numerous host plants, living in dense aggregations. The only aphids known to commonly occur on palms are the palm aphids, *Cerataphis* species, which are native to Asia, and do not resemble "typical" aphids. They are more similar in appearance to some whitefly immature stages or scale insects with a dark circular, somewhat convex shape with a fringe of white waxy filaments around the body. They have short, hidden legs making them appear to be legless and sessile.The two species of palm aphids, *C. brasiliensis* (Hempel) and *C. lataniae* Boisduval, are only differentiated by the presence of dagger-like setae occurring in *C. brasiliensis*. As such, there is some question if they are actually two species. Another morphologically similar species, *C. orchidearum* (Westwood), which occurs on orchids, may also be confused with the palm aphids. Both *C. brasiliensis* and *C. lataniae* have been widely distributed and are common on palms throughout the humid tropical regions of the world.

Damage and management : Palm aphids have been reported on numerous palms including coconut. They generally occur in dense aggregations on young, unopened fronds and occasionally on flowers and young fruit. High populations of palm aphids occasionally become severe in production and the landscape. The most severe damage is seen on young coconut palms. Typical symptoms include yellowing and loss of plant vigor. Excessive honeydew production by the aphids promotes sooty mold growth which can interfere with photosynthesis when dense. Heavy infestations can cause stunted growth. Unless populations are excessive or host plants are young, management is not necessary. Unlike other types of aphids, palm aphids have limited mobility and are, therefore, more prone to attack by predators. Syrphid and coccinellid larvae have been reported feeding on palm aphids (Sumalde and Calilung, 1982). Howard (2001) reported two coccinellid predators, *Cycloneda sanguinea* L.and *Hippodamia convergens* (Guërin-Mëneville) preying on palm aphids in Florida.Although parasitoids have been commonly reported on other types of aphids, they have not been commonly reported attacking palm aphids. Horticultural oils and soaps can be used to manage palm aphids particularly on smaller plants where good coverage can be obtained with frequent foliar sprays. Insecticides may be warranted in production systems with heavy infestations. There are numerous insecticides available for aphid management; however, there have been few tests conducted on the efficacy against palm aphids. It is probable that many of these insecticides would also provide some control against palm aphids.In more recent years, the neonicotinoids which are systemic have been commonly used for aphid control. Availability of insecticide products will vary around the world.

16.4.3 Coccoidea (Scales and mealybugs)

Distribution and biology : The superfamily Coccoidea contains nearly 8,000 species of plant-feeding hemipterans comprising up to 32 families and are more diverse in species richness and morphology than either the aphids or whiteflies (Gullan and Cook, 2007). Most scale species are small and cryptic in habit, often resembling part of their host plant. Most species produce a waxy secretion that covers the body as a structure detached from the body or as a secretion that sticks to the cuticle. Adult females are typically elongate, elliptical or circular and are flat to convex with little distinction between the head, thorax and abdomen and no visible legs or antennae. Adult females of mealybugs have functional legs. Although there is some variation among scales, they have a mobile crawler stage, with females becoming less mobile as the mature. In most groups, all other stages are sessile and remain in one location to feed. Adult males, if they exist, look distinctly different from females and immature stages. Males are rarely seen or noticed and resemble winged aphids. In some families, all stages are mobile (more primitive) and in other families only the crawler stage and male are mobile (more advanced). The more advanced families which contain more sessile stages are more highly protected by their wax secretions (Howard *et al.,*2001).

All species contain modified mouth parts that form a sucking stylet from which they feed on plant juices. All immature and female stages feed. Unlike the aphids and whiteflies, not all families produce honeydew. The largest family, Diaspididae (armored scales), do not produce honeydew. Important species on palms reside in approximately 10 families. The three biggest families are the mealybugs (Pseudococcidae), soft scales (Coccidae) and the armored scales (Diaspididae).

There are approximately 2,000 mealybug species worldwide. The bodies of most mealybugs are covered with a white, waxy substance. The bodies are typically elongate or oval in shape, pink or yellowish, and often have wax filaments. Mealybugs typically produce an ovisac from their waxy secretions in which they lay their eggs. Mealybugs lay eggs or first instars, called crawlers, which walk around the plant to locate a suitable feeding site. Although all mealybug instars (in most species) have legs, mealybugs are primarily sedentary. Mealybugs feed on a wide variety of palms. Most species aggregate in protected sites such as folds and crevices of leaves and other above-ground plant parts; some feed on roots. Mealybugs can produce excessive amounts of honeydew and are often associated with ants. Most mealybugs are not host specific and as of 2001, fifty-two known species have been reported on palms (Howard *et al.,*2001). The most commonly reported mealybugs on palms also feed on numerous other types of plants. Examples include pineapple mealybug, *Dysmicoccus brevipes* Cockerell, coconut mealybug, *Nipaecoccus nipae* (Maskell), and longtailed mealybug, *Pseudococcus longispinus* Targioni-Tozetti.Pineapple mealybug is known from all zoogeographic regions of the world and is commonly intercepted at U.S. ports of entry. Although reported on palms, it is of more concern on pineapple and other bromeliads. In southern Florida, pigmy date palms grown in containers that were root bound were often infested with pineapple mealybug on the roots (Mannion, personal observation).

The coconut mealybug is a common pest of palms. It is widespread worldwide; found outdoors in warmer parts of America, Europe and Africa and in greenhouses in more northern latitudes. This mealybug can be economically damaging to palms. It is widely distributed in North, Central and South America, Europe, Asia, Oceania and Africa (Miller *et al.,*2007). It can be found feeding on above-ground plant parts and roots of containerized palms. Longtailed mealybug is most easily identified by the extremely long caudal wax filaments. This mealybug is a pest of many ornamental plants including palms.

There are a few mealybugs restricted to palms. One of these, palm mealybug, *Palmicultor palmarum*, primarily occurs on the foliage. It is commonly found at U.S. ports of entry on coconut and occasionally other palms. It has been recorded from nearly all areas of the world where coconuts are grown except it is not known from Africa. It does little damage to adult palms but can damage or kill seedlings.

The soft scales, Coccidae, comprise 1,148 species in 173 genera worldwide (Miller and Ben-Dov, 2015). In appearance, they are elliptical to pyriform and flat to convex but can vary greatly in color. The soft scales that are of most concern on palms are those scales that are cosmopolitan and polyphagous. Howard *et al.* (2001) reported 40 species of soft scales from palms.Of those he reported, 6 have only been collected from palms. The three most commonly reported soft scales on palms are brown soft scale, *Coccus hesperidum* L., tessellated scale, *Eucalymnatus tessellatus* (Signoret), and (wax scales) *Ceroplastes* species. Brown soft scale attacks a wide variety of plants including palms and has cosmopolitan distribution. It is more likely to be a pest of palms in greenhouses than in outdoor production or landscapes. Damage may result from direct feeding but also indirect damage is from the honeydew production and growth of sooty mold. Tessellated scale is believed to be native to South America but is widely distributed in Africa, Australia, North, Central, and South America, the Caribbean, Asia and Europe. The female scale is flat and brown with a distinctive mosaic pattern. Although it is a very common scale in tropical countries on palms and other plants, and it has been reported to be a serious pest of palms outdoors, it can be unusually sparse. However, in a dense planting of Malayan dwarf coconut palms in Florida, a serious infestation with up to 200 mature females per pinna had been observed.Dense populations of this scale may result from interruptions of natural enemies through pesticide spraying. Wax scales, which are in the genus, *Ceroplastes*, have also been commonly reported on palms. Wax scales are easily identifiable due to their globular shape and heavy layer of beige, pinkish, whitish or grayish wax. From the top view, they appear rectangular, oval or lobed. The Florida wax scale, *Ceroplastes floridensis* Comstock, is one of the most commonly encountered soft scales (Sharma and Buss, 2013).

The armored scales, Diaspididae, are the largest family of Coccoidea with almost 2,500 species in approximately 400 genera worldwide. Female scales secrete a waxy coating along with remnants of cast skins of earlier instars, called a scale, which is detached but covers the body. They may be circular to pyriform to elongate in shape and white, yellow, purple, red, or orange in color. All stages except the crawler stage are legless. The crawlers are very small, oval and flattened with functional legs. In contrast to the soft scales, mealybugs, aphids and whiteflies, armored scales do not produce honeydew. Feeding occurs at almost the same site throughout the life of the scale. Under tropical conditions, most armored scales develop from egg to adult in a couple weeks and feed on above ground plant parts. Most armored scales reported on palms are not restricted to palms but are polyphagous. However, armored scales are the most consistently encountered hemipterous pest of cultivated palms and are considered a major pest of palm production and in the landscape (Howard *et al.,*2001).

Armored scales are unusually adapted for long-range dispersal on their host plants due to the highly protective wax covering, and they are often difficult to detect due to their small size and cryptic habits. One of the economically important armored scales of palms and one that is reported only on palms is *Parlatoria blanchardi* Targioni Torzetti, a serious pest of date palm. The mature female is reddish pink to reddish purple, pyriform, and commonly infests leaf bases. As populations increase, they feed on older, then younger foliage and finally fruits. A discolored area of injured tissue develops where individuals settle and feed. Significant damage to the leaves results in the pinnae withering and eventually dying. This scale is presently recorded from most date-growing countries in the world except the U.S. (Benassy, 1990). In 1920, it killed about 100,000 date palms at one oasis in Algeria (Rosen, 1990).

The coconut scale, *Aspidiotus destructor* Signoret, is one of the most widespread pests of coconut. It is highly polyphagous and been recorded on 75 genera in 45 families of plants worldwide (Borchsenius, 1966) but is common on many palm species (Bondar, 1922; Howard, 1999). The scale is small, flat, and whitish with a semitransparent waxy cover. The scale is found on the lower plant surface which is marked by a yellow, discolored spot. Heavy infestations on coconut will stunt new leaves and can stop production. Coconut scale is commonly found with other armored scales on palms such as Florida red scale, *Chrysomphalus aonidum* (L.),oriental scale, *Aonidiella orientalis* (Newstead), and black thread scale, *Ischnaspis longirostris* Signoret.Lever (1979) reported that palms in neglected coconut groves were more susceptible to attack by this scale. Other accounts of stressed plants being more susceptible to this scale have been reported (Howard *et al.*, 2001). The black thread scale is another frequently encountered armored scale. It can easily be distinguished from other armored scales because the female has a shiny, black, long and narrow, covering. This scale can be found on fronds, petioles, and the fruits of palms. It is considered one of the 43 most serious worldwide armored scale pests (Miller and Davidson, 2005). It is most commonly found in tropical and subtropical areas throughout the world. It has been reported on over 50 plant families including numerous palm species. It is most commonly found on parlor and oil palms.

There are several examples of scale pests on palms outside the three largest families discussed above. One example, the red date scale, *Phoenicococcus marlatti* Cockerell, in the Family Phoenicococcidae, is also a frequent pest of date palm. It is native to North America and the Middle East but has been found almost everywhere date palms are grown (Miller *et al.*,2007). Within the U. S., it has been limited to Florida but intercepted and eradicated in Arizona, California and Texas. The female is spherical, red to reddish-brown, and may be embedded in a mass of white, cottony wax on plant tissue. Its appearance can be deceiving and at first glance appear to be fungi. This scale typically is found on the stem behind the overlapping leaf bases but can also infest the bases of inflorescences or roots. Heavy infestations can cause premature leaf aging, drying of the fruit and disruption of normal plant functions and plant death (Zaid *et al.,*2002).

Other scale families that include members that have been reported on palms are Margarodidae, Eriococcidae, Conchaspididae, Asterolecaniidae, Halimococcidae, and Beesoniidae. For most of these families, there are relatively few scales reported as economic pests of palms. There are several species in Halimococcidae associated with palms which tend to infest young, unfolded fronds.These scales generally go unnoticed. The palmetto scale, *Comstockiella sabalis*, may be native to the southern U.S. This scale has been placed in the Diaspididae but appears to be unique and not closely related to other members of Diaspididae. It is distributed throughout the southern United States, Mexico, and the Caribbean and has been most commonly reported on *Sabal* species as well as a few other palms. Although it is common, infestations are not conspicuous. Females are light pink to reddish brown with an almost circular shape (Espinosa *et al.,*2012).

Damage and management : Scales and mealybugs can be serious pests of numerous types of plants including palms. Damage by these pests is not always evident until populations are high. Heavily infested plants can look unhealthy and produce little new growth. Leaf yellowing, premature leaf drop, branch dieback and sometimes plant death can result from infestations of scales and mealybugs. Infestations on palms, however, are not always initially obvious. Detection of these pests is the most critical step in management because many of these pests are cryptic, difficult to see, and can sometimes appear to be other types of disorders or diseases. In some cases, direct damage from feeding is minor, but indirect damage from excessive production of honeydew and resulting growth of sooty mold is major. Scales and mealybugs on palms often thrive in warm, humid conditions, so increase of air flow or decrease of plant density may make conditions less conducive.

Numerous predators and parasites have been observed feeding on scales and mealybugs, and there are many cases in which natural enemies have been released for management. For example,the use of *Pseudaphycus utilis* Timberlake, a parasitic wasp, successfully controlled coconut mealybug in Hawaii and Puerto Rico (Watson, 2007). For longtailed mealybug, population resurgence after insecticide spraying in Australia suggested that natural enemies are an important control factor (Furness, 1976). There are over 30 known natural enemies of the soft brown scale worldwide and populations are normally controlled by natural enemies. *Metaphycus stanleyi* Compere (Hymenoptera: Encryptidae) is a common natural enemy of tessellated scale. A fungus,

Lecanicillium lecanii (Zimmerman) Viegas, was reported attacking tessellated scale in the Seychelles (Vesey-FitzGerald, 1940). This is a cosmopolitan fungus that attacks many kinds of insects and has been developed as a biocide. Three parasitic wasps have been reported from Florida wax scale (Sharma and Buss, 2013). Two different biological control strategies for *P. blanchardi* were discussed by Benassy (1990) which involved enhancement of local natural enemies and the introduction of a predator that had successfully controlled it in several localities. More than 40 predators and parasitoids were reported to attack *A. destructor* on coconut (Beardsley, 1970). It is under natural control in some areas by a coccinellid predator.In date palms, a new sub species of the predator, *Chilocorus bispustulatus* var *iranensis* is reported from Iran preying on *P. blanchardii* (Iperti *et al.,*1970).

There are numerous contact and systemic insecticides labeled for control of scales and mealybugs. Most contact insecticides cannot penetrate the waxy covering, so the crawler stage is usually the best target. Failure of contact sprays often is a result of poor timing or poor coverage. Horticultural oils can kill all stages of scales, but there must be sufficient exposure to get this result. Neonicotinoid insecticides are systemic insecticides that are commonly used for mealybug and scale control.

One of the problems in managing scales and mealybugs is determining when they are dead. For example, scales often remain on the plant long after death. This is particularly true of the armored scales. They can be physically checked to see if they are dry (and likely dead) or moist (and likely live). Additional insecticide applications should never be made without determination the result of previous applications.

16.4.4 Aleyrodidae (Whiteflies)

Distribution and biology : Whiteflies are widespread in the tropics and warm areas of the world with more than 1,550 described species (Martin, 2004). Adults are small, typically 2-3 mm, soft-bodied insects that resemble very small moths (Hodges and Evans, 2005). Adults typically appear white due to a white, waxy secretion. Males and females are similar in appearance. Eggs are typically laid on the underside of leaves in circles, spirals or, loose clumps. The first instar, crawler stage, has functional legs and is mobile. Subsequent immature stages have vestigial legs and are sessile, superficially resembling scale insects. The immature stages are typically oval, flat to convex, and may have ornate waxy secretions in the later instars.

Whiteflies are phloem feeders typically associated with dicotyledonous woody perennial plants. All species are phytophagous and some are capable of transmitting viruses which has made them one of the most serious crop protection problems (Byrne *et al.,* 1990). Most of the virus diseases transmitted by whiteflies affect annual crop plants; very few are known in woody perennials and none are known to affect palms. The impact of whiteflies on palms is through feeding and indirectly, by production of honeydew which supports the growth of sooty mold. Most whiteflies that infest palms, also infest dicotyledonous hosts as well. Forty-three species of Aleyrodidae were listed on palms in 1978 (Mound and Halsey). The largest percentage of whiteflies have been reported from coconut which may be more due to the amount of coconut grown in the landscape and in production in the tropics. Also coconuts are commonly inspected for palm pests.

A few noteworthy whitefly species on palms that have been reported as pests of coconut are all in the genus *Aleurodicus*. These species are larger than the typical whitefly, white, and lay their eggs in spirals covered in waxy patches. These whiteflies produce a thick flocculant of wax which can cover the plant surface they are feeding on. These species often produce excessive amounts of honeydew leading to excessive growth of sooty mold. At least 47 species of Aleyrodidae attack *C. nucifera* (Howard*et al.*, 2001), among them *Aleurodicus cocois* (Curtis), *A. pulvinatus, A. destructor* and *Aleuroctanthus destructor* (Mackie). An unexpected increase of the spiraling whitefly, *Aleurodicus dispersus* Russell co-existing with *Lecanoideus floccissimus* Martin, were affecting canary palms in Spain (Carnero *et al.,* 1999). *Aleurodicus dispersus*, is probably the most widely spread whitefly that infests palms around the world. Although commonly reported on coconut, it is also considered an occasional pest of bananas and many tropical dicotyledonous crop plants. Cherry (1980) found the leaves and fronds of seagrape and palms are very favorable for establishment of this whitefly, however, host plant preference seems to differ with locality and conditions. *Aleurodicus dispersus* is native to the Caribbean region and Central America, where it is known from a very wide number of host plants. Its current geographical distribution is cosmopolitan (Carnero *et al.,* 1999). Both whiteflies cause direct damage and heavy infestations might kill palms. The sooty mold that develops from the heavy honeydew produced by both whiteflies interferes with photosyntheses (Carnero *et al.,* 1999). *Aleurodicus pseudugessii* Martin has recently been found infesting coconut in Brazil (Mendonca de Omena *et al.,* 2012) weakening palms and causing unusual sooty molds on the fronds. Mendonca de Omena *et al.* (2012) also report that infestation is first characterized by the fronds in the palm crown acquiring a silver tone. Increase of damage to coconuts and other palm species by *Aleurotrachelus atratus* Hempel and *Paraleyrodes bondari* Perachi has been reported (Streito *et al.,* 2004; Borowiec *et al.,* 2010).

More recently, specimens of a new, exotic whitefly, *Aleurodicus rugioperculatus* Martin were reported in Florida in 2009 and subsequently given the common name rugose spiraling whitefly (previously referred to as the gumbo limbo spiraling whitefly) (Stocks and Hodges, 2012). It was described in 2004 by Martin and is thought to be native to Central America.It appears to have a wide host range but most seriously affects gumbo limbo and palms, especially coconuts. Rugose spiraling whitefly is relatively large in size, docile and lays its eggs in a spiral pattern typical of this genus (Mannion, 2010). These whiteflies are spreading throughout southern and central Florida. The extent of damage from direct feeding from this whitefly is unclear, however, these whiteflies are extremely pestiferous due to the excessive wax and honeydew produced and subsequent sooty mold growth.

Damage and management : Although whiteflies are relatively common on coconut and other palm species, they are not always considered damaging. They often occur in small aggregations on the underside of fronds and on occasion reach dense populations. They often are not listed in the top pest problems of palms in the coconut-growing regions of the world. Yet, when there are heavy populations that do require management, they are not always easy to manage. Often, whitefly populations are kept in check by natural enemies, but in situations in which the natural balance has been disturbed as in production and landscape systems, high and damaging populations may occur (Hodges and Evans, 2005). Feeding damage can reduce plant vigor and in some cases with young or stressed plants can ultimately cause plant death. Fronds may turn yellow and drop from extensive whitefly feeding.

Numerous predators and parasites have been observed feeding on whiteflies and there are many cases in which natural enemies have been released for management of whitefly. For example, in the Caribbean, two predatory coccinellid beetles were recorded feeding on *A. cocoiș* (Mound and Halsey, 1978) and have been attributed to keeping populations of the whitefly under control. Five natural enemies were introduced into Hawaii from the Caribbean to control *A. dispersus* which had become s major economic pest by 1979 (Martin Kessing and Mau, 1993). Introductions and establishment of the coccinellid, *Nephaspis ammnicola* Wingo and the aphelinid *Encarsia haitiensis* Dozier have resulted in successful control of *A. dispersus* in Hawaii (Kumashiro *et al.,* 1983), American Samoa and Guam (Carnero *et al.,*1999). One of three coccinellid beetles, *Nephaspis oculatus*, has proven to be effective when populations of whitefly are high (Kumashiro *et al.,*1983). Two parasitic wasps, *Encarsia haitiensis* Dozier and *Encarsia* sp., were also introduced and found to be effective (Kumashiro *et al.,*1983). Whitefly populations were ultimately reduced 79 to 98 percent.

Aleurodicus dispersus was reported for the first time in Mauritius in July 2000 and initial measures to control it with insecticides were found difficult due to its wide host range and high infestations (Gungah *et al.,*2005). A predator, *Nephaspis bicolor* Gordon, was imported from the Caribbean and Trinidad in 2003 and is credited to suppressing the pest populations (Gungah *et al.,*2005).

In Florida in the late 1970's, no parasitoids were found attacking *A. dispersus* (Cherry, 1980). By the early 1980s, populations of *A. dispersus* subsided in Florida. It was believed that a parasitoid, *Euderomphale vittata* Dozier, was believed to have caused the decline (Bennett and Noyes, 1989). *Aleurotrachelus atratus* Hempel is found throughout the Caribbean, northern South America and southern Florida infesting various palms. Currently in Florida, populations are low and parasitoid exit holes are commonly observed in final instars. Current efforts to manage the rugose spiraling whitefly in Florida as resulted in the development of a website, www.flwhitefly.org, which presents current information and management on this and other invasive whiteflies. Several predators and parasitoids have been identified attacking rugose spiraling whitefly in Florida. In Florida, the natural enemies that appear to have the most promise for reducing the population are the parasitoid, *Encarsia noyesi,* and the predator, *Nephaspis oculatus* (Mannion, unpublished data). In other regions, the species *Eretmocerus cocois* Delvare and *Cales noacki* Howard, *Encarsia cubensis* Gahan and *E. hispida* De santis appear to be the most important natural enemies of this speciesin Reunion (Borowiec *et al*., 2010).

In the case of whiteflies on palms and the abundance of natural enemies, many times management is not warranted. In other situations, where whiteflies are causing an economic reduction in production or impacting a landscape, insecticides may be necessary. Management of whiteflies is often still heavily reliant on insecticides. Spraying with soaps and contact insecticides has proved ineffective for control of *L. floccissimus* (Carnero *et al*., 1999). As with aphid management, there are numerous options of contact and systemic insecticides but will vary on availability in different areas. A combination of cultural, biological and chemical controls can be effective in managing whiteflies and reducing the overall impact. Some cultural strategies may not be useful with palms as they are with other types of plants. However, conservation of natural enemies helps to prevent outbreaks or reduce the impact when they are present. The systemic insecticides (neonicotinoids) are heavily relied upon in many parts of the world for many different types of insects. In Florida, these insecticides are used in both the landscape and production for the rugose spiraling whitefly. Although natural enemies have been identified and are spreading, the impact of this whitefly requires some insecticide intervention. Tremendous efforts are being placed on education on how to incorporate the use of insecticides without harming the natural enemies.

16.5 Conclusions

The recent movement of *Rhynchophorus* weevils, and the mites *Aceria guerreronis* and *Raoiella indica*from their areas of origin to other continents and their establishment in palm growing regions of the world have necessitated the development of palm IPM programs in areas where the pest invasion have taken place. Standard IPM methodologies, involving a number of control alternatives used alone or in combination (such as early detection, sanitation, monitoring, mass-trapping, and infested host removal) have achieved some success in controlling *Rhynchophorus* weevils. These tactics need to be implemented in a coordinated manner, otherwise their effectiveness is compromised. The interaction of *R. palmarum* with bud rot of oil palm increases the economic importance of this weevil not only in the Americas, but in other areas of the world. Much more research is neededto achive better levels of control with respect to *R. ferrugineus* particularly with regard to; i) detection of infested palms in the early stages of attack, ii) testing potential insect repellents and their deployment in area-wide management programs through push-pull strategies and also to protectinfestation sites on individual palms, iii) eliminationof trap servicing currently associated with food-baited pheromone traps through attract and kill technology, iv) isolation of virulent strains of biological control agents especially EPNs and EPFs in *R. ferrugineus* native habitatsin addition to developing sound biological control strategies for large scale deployment of these organisims in the field and, v) identifying factors of resistance and incorporating these into palms as the first line of defense against *R. ferrugineus*.

The mite *Aceria guerreronis* is now a serious pest in India and other countries in Asia where coconut is of great importance. Here we have summarized basic studies and IPM tactics (chemical, natural enemies, microbial control). From these studies it appears that monitoring techniques for this pest are difficult and are in need of more studies in order to predict economic damage in a more accurate manner; knowledge gained on predaceous mites and the result of inundative releases as well as results of microbial control show promise for a better IPM for this pest. *Raoiella indica* invasion of the neotropics and its damage to coconut predominantly in the Caribbean islands, has called for research that deals with sampling, chemical and biological control. Still, IPM of this pest is in its infancy and much more research is needed to achieve greater pest suppression through augmenting the effect of biotic factors that are already acting on *R. indica*. Considering that the natural enemy complexes of *R. indica* observed in areas of recent invasion are composed of only generalist predators it will be desirable to discover more specific natural enemies. Searching for natural enemies of this pest, particularly in the area originwhich has yet to be clearly identified, should be intensified. The high diversity of *Raoiella* species in the Middle East region suggests the species could have originated in this region and, theoretically, its most effective natural enemies could be found there. Recent studies suggest that along time association between *A. largoensis* and *R. indica* could result in more aggressive strains of this predator. The possible existence of more aggressive strains, biotypes or cryptic species of *A. largoensis* from

Table 16.1 : Developmental times of immature stages of *R. indica* females reported by various studies using variations of the leaf flotation method under laboratory conditions.

Host Plant	Temp. (ºC)	Rel. Hum. (%)	Develop. Time (d)	Reference
Cocos nucifera	24.2	*	22	Moutia 1958
Cocos nucifera	17.9	*	33	Moutia 1958
Phoenix dactylifera	26.1	57.90	21.4	Zaher et al. 1969
Cocos nucifera	23.9-25.7	59.85	24.5± 1.92	NageshaChandra and Channabasavanna 1984
Cocos nucifera	26.3± 1.26	74.8± 4.3	31.4± 3.31	Gonzales and Ramos 2010
Areca catechu	25.4± 1.20	57.5± 6.5	31.0± 4.11	Flores-Galano et al. 2010

*information not provided

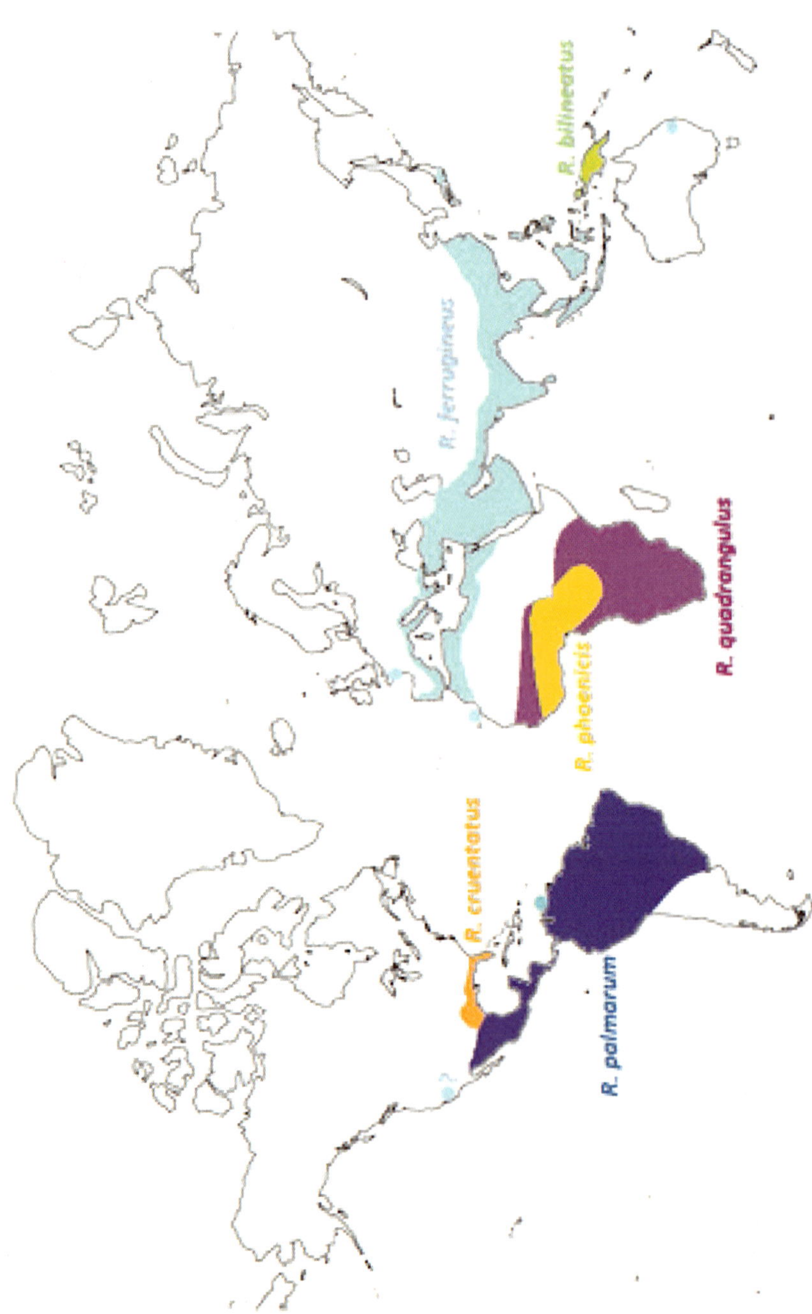

Fig. 16.1 : World distribution of *Rhynchophorus* species

Fig. 16.2 : *Raoiella indica* on coconut. A. Oviposition by *R. indica*. B. *R. indica* colony, the arrow shows a male closely behind a female deutonymph in a "guarding state". C. *R. indica* colonies on the underside of coconut leaves. D. Damage on coconut leaves caused by *R. indica*.

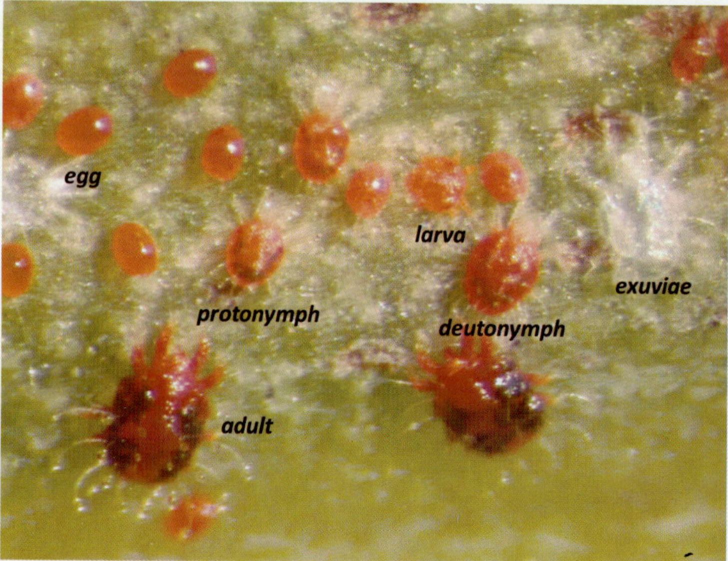

Fig. 16.3 : *Raoiella indica* forming multigenerational colonies on the abaxial surface of coconut leaflets (pinnae).

Fig. 16.4 : Palm aphid *Cerataphis* sp.

Fig. 5 : Pineapple mealybug *Dysmicoccus brevipus* on Phoenix roots

Fig. 6 : Coconut mealybug *Nipaecoccus* nipae

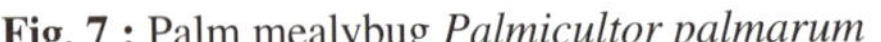

Fig. 7 : Palm mealybug *Palmicultor palmarum*

Fig. 8 : Wax scale *Ceroplastes* sp. on coconut

Fig. 9 : Red dae scale *Phoenicococcus marlatti* **Fig. 10 :** Rugose spiraling white fly *Aleurodicus rugioperculatus*

Fig. 11 : Rugose spiraling whitefly *Aleurodicus rugioperculatus* on coconut

other parts of the world is currently being investigated. Moreover, the recent report of acaropathogenic fungi attacking *R. indica* in Puerto Rico is promising and deserves special attention. The tropical conditions and high *R. indica* populations found in areas of recent invasion could favor epizootics caused by acaropathogenic fungi which could cause a decline in *R. indica* densities and possibly enhance the regulatory capacity of *A. largoensis* and other predators. In addition, cultural practices such as the effects of different fertilization regimes on the developmet of *R. indica* should be explored More control alternatives are needed to implement an IPM program to suppress *R. indica* populations and mitigate the adverse effects caused by this invasive mite.

The hemipteran pest problems, *i.e.*, whiteflies, armored scales appear to be increasing in plantation palms as well as in the landscape. However, vectors of palm diseases remain as the most important within this order. Chemical control of vectors as mentioned eaerlier, likely will continue for the foreseeable future, but vector management or management of phytoplasma spread within the plant is now slowly shifting to habitat management and the use of genetically modified crops.

References

Abdallah, S., Al Abbad, A.H., Al-Dandan, A.M., Abdallah, A.B. and Faleiro, J.R. (2008). Enhancing trapping efficiency of red palm weevil pheromone traps with ethyl acetate. *Indian Journal of Plant Protection* 36, 310-311.

Abe, K., Hata, K. and Sone, K. (2009) Life History of the Red Palm Weevil, *Rhynchophorus ferrugineus* (Coleoptera: Dryophtoridae) in Southern Japan,*Florida Entomologist* 92, 421-425.

Abozuhairah, R.A., Vidyasagar, P.S.P.V. and Abraham, V.A. (1996) Integrated management of red palm weevil, *Rhynchophorus ferrugineus* F. in date palm plantations of the Kingdom of Saudi Arabia. XX *International Congress of Entomology*, Florence, Italy, August 25-31, 1996. *Tropical Entomology* (Sect.). Paper 17-033.

Abraham, V.A. (1971) Prevention of red palm weevil entry into coconut palms through wounds. *Mysore Journal of Agricultural Science* 5, 121-122.

Abraham, V.A., Abdulla Koya, K.M. and Kurian, C. (1989) Integrated management of red palm weevil (*Rhynchophorus ferrugineus* Oliv.) in coconut gardens. *Journal of Plantation Crops* 16 (Supplement), 159-162.

Abraham, V.A., Al Shuaibi, M.A., Faleiro, J.R., Abozuhairah, R.A. and Vidyasagar, P.S.P.V. (1998) An integrated management approach for red palm weevil, *Rhynchophorus ferrugineus* Oliv., a key pest of date palm in the Middle East. *Journal of Agricultural and Marine Sciences* 3, 77-84.

Abraham, V. A., Faleiro, J. R., Al Shuaibi, M. A. and Saad Al Abdan. (2001) Status of pheromone trap captured female red palm weevils from date gardens in Saudi Arabia. *Journal of Tropical Agriculture* 39, 197-199.

Abuagla, A. M. and Al-Deeb, M. A. (2012) Effect of bait quantity and trap color on the trapping efficacy of the pheromone trap for the red palm weevil, *Rhynchophorus ferrugineus*. *Journal of Insect Science* 12,120. Available online:http://www.insectscience.org/12.120.

Al-Ayedh, H. (2008) Evaluation of date palm cultivars for rearing the red date palm weevil, *Rhynchophorus ferrugineus* (Coleoptera: Curculionidae). *Florida Entomologist* 91, 353-358.

Al-Aydeh, H.Y. and Rasool, K.G. (2010) Determination of the optimum sterilizing radiation dose for control of the red date palm weevil *Rhynchophorusferrugineus* Oliv.(Coleoptera: Curculionidae). *Crop Protection* 29, 1377-1380.

Aldana, R.,Aldana, J. and Moya,O. (2011) Manejo del picudo *Rhynchophorus palmarum* L. (Coleoptera: Curculionidae). *Boletín Instituto Colombiano Agropecuario – ICA Ola Invernal.* 47 p.

Aldryhim, Y. and Al- Bukiri S. (2003) Effect of irrigation on within – grove distribution of red palm weevil*Rhynchophorus ferrugineus*.*Sultan Qaboos University Journal for Scientific Research, Agricultural and Marine Sciences* 8, 47-49.

Aldryhim,Y. and Khalil, A. (2003) Effect of humidity and soil type on survival and behaviour of red palm weevil *Rhynchophorus ferrugineus* (Oliv.) adults.*Sultan Qaboos University Journal for Scientific Research, Agricultural and Marine Sciences* 8, 87-90.

Alfonso, J. A. andRamirez, T. (2004) Manual técnico del cultivo del cocotero (*Cocos nucifera* L.). La Lima, Cortés, Honduras, C. A.

Al-Jboory, I.J. (2007) Survey and indentification of the biotic factors in date palm environment and its application for designing IPM-Program of date palm pests in Iraq.*University of Aden Journal of Natural and Applied Sciences* 11, 423-457.

Al-Saoud, A.H. (2011) Comparative effectiveness of four food baits in aggregation pheromone traps on red palm weevil, *Rhynchophorous ferrugineus,* Olivier. *Arab Journal of Plant Protection* 29, 83-89.

Al-Shawaf, A. M. A., Al-Abdan,S., Al-Abbad A. H., Ben Abdallah, A. and FaleiroJ.R. (2012)Validating area-wide management of *Rhynchophorus ferrugineus* (Olivier) (Coleoptera: Curculionidae) in date plantations of Al-Hassa, Saudi Arabia. *Indian Journal of Plant Protection* 40, 255-259.

Al-Shawaf, A. M., Al-Shagag A.,Al-Bagshi M., Al-Saroj S., Al-Bather S., Al-Dandan A. M., Ben Abdallah A. and Faleiro J. R.(2013) A quarantine protocol against red palm weevil *Rhynchophorus ferrugineus* (Olivier) (Coleptera: Curculiondae) in date palm. *Journal of Plant Protection Research* 53(4), 409-415.

Amaro, G. and de Morais, E.G.F. (2013) Potential geographical distribution of the red palm mite in South America.Exp Appl Acarol DOI 10.1007/s10493-012-9651-9.

Anonymous (2013) Save Algarve palms. http://www.savealgarvepalms.com/en/weevil-facts/host-palm-trees (accessed on 24th March, 2013).

Ansaloni, T., and Perring, T.M. (2004) Biology of *Aceria guerreronis* (Acari: Eriophyidae) on queen palm, *Syagrus romanzoffiana*(Arecacea). *International Journal of Acarology* 30, 63-70.

Arango, G., and Rizo, G. (1977) Algunas consideraciones sobre el comportamiento de *Rhynchophorus palmarum* y *Metamasius hemipterus* en caña de azúcar.*Revista Colombiana de Entomología.* 3, 23 – 28.

Arroyo, C.,Mexzon, J.and Moura,U. (2004) Insectos fitófagos en pejibaye (*Bactris gasipaes* K.) para Palmito.*Agronomía Mesoamericana* 15, 201–208.

Arruda de., P.G. (1974) Fenologia do *Eriophyies guerreronis* (Keifer, 1965) (Acarina, Eriophyidae), em Pernambuco *Boletin Tecnico Instituto de Pesquisas Agronomicas*. 66, 1-56.

Ashby, S. F. (1921) Some recent observations on red ring disease of the coconut.*Agricultural News* 20,350-351.

Askari, M., Arbabi, M., and Golmohammad, Z.N. (2002) Plant mite fauna of Sistan-Baluchestan and Hormozgan provinces.*Journal of the Entomolgical Society of Iran* 22, 87–88.

Avand Faghih, A. (1996) The biology of red palm weevil, *Rhynchophorus ferrugineus* Oliv. (Coleoptera: Curculionidae) in Saravan region (Sistan and Balouchistan Province, Iran), *Applied Entomology and Phytopathology* 63, 16-18.

Barranco, P., Pena, J. A. De la, Martin, M.M. and Cabello, T. (1998). Efficacy of chemical control of the new palm pest *Rhynchophorus ferrugineus* (Olivier 1790) (Curculionidae: Coleoptera). *Boletin de Sanidad Vegetal, Plagas*. 24, 301-306.

Barranco, P., De La Pena, J.A., Martin, M.M. and Cabello, T. (2000) Rango de hospedantes de *Rhynchophorusferrugineus* (Olivier, 1790) y diametro de la palmera hospedante. (Coleoptera, Curculionidae). *Boletín de Sanidad Vegetalde Plagas* 26, 73-78.

Beardsley, J. W. (1970) *Aspidiotus destructor* Signoret, an armored scale pest new to the Hawaiian Islands. *Proceedings Hawaiian Entomological Society* 20,505-508.

Bech, R. (2011). Detection of South American Palm Weevil (*Rhynchophorus palmarum)* in California. http://www.aphis.usda.gov/plant_health/plant_pest_info/palmweevil/index.shtml.

Bech, R. (2012) Detection of South American Palm Weevil (*Rhynchophorus palmarum)* in Texas. pp. Downloaded as : http://www.aphis.usda.gov/plant_health/plant_pest_info/palmweevil/ index.shtml.

Benassy, C. (1990) Date palm. In Rosen, D. [ed.] *Armored Scale Insects, their Biology, Natural Enemies and Control*. Vol. 4B.World Crop Pests. Elsevier, Amsterdam, the Netherlands. pp 8-26.

Bennett, F. D. and Noyes, J.S. (1989) Three chalcidoid parasites of diaspines and whiteflies occurring in Florida and the Caribbean. *Florida Entomologist* 72, 370-373.

Blumberg, D. (2008) Date palm arthropod pests and their management in Israel.*Phytoparasitica* 35, 411-448.

Boheman, C. H. (1845) Genus 598-*Rhynchophorus* of *familia* Cuculionides; In C. J. Schoenherr's Genera *et* Species. *Curculionidum* 8, 216-219.

Bondar, G. (1922) Insect Pests and Diseases of the coconut in Brazil.113 pp. http:// www.cabidirect.org/abstracts/19230500347.html;jsessionid=33EE8EB08A11F78C58.(June 12, 2013).

Borchsenius, N. S. (1966) A Catalogue of the Armored Scle Insects *(Diaspidoidea)* of the World. Nauka, Moscow. 449 pp.

Borden, J.H. (1985) Aggregation pheromones, in G.A. Kerkut and L.I. Gilbert (eds). Comprehensive Insect Physiology, Biochemistry and Pharmacology, vol. 9.pp. 257-285, Pergamon Press, Oxford.

Borowiec, N., Quilici, S., Martin, J., Issimaila, M. A., Chadhouuliati, M., Youssoufa, A., Beaudoin-Olivier, L., Delvare, G., and Reynaud, B. (2010) Increasing distribution and damge to palms by the neotropical whitefly *Aleurotrachelus atratus* (Hemiptera: Aleyrodidae). *Journal of Applied Entomology* 134, 498-510.

Briceño, A., and Ramírez, W. (2000). Diagnóstico de insectos coleoptera asociados a las plantaciones de plátano en el sur del lago de Maracaibo – Venezuela. *RevistaForestalVenezolana* 44, 93-99.

Buss, E. (2010). Aphids on Landscape Plants.University of Florida IFAS Extension.http:// edis.ifas.ufl.edu/mg002

Byrne, D. N., Bellows, T.S. and Parrella, M. P. (1990) Whiteflies in agricultural systems, pp. 227-261 In D. Gerling [ed.] *Whiteflies: Their Bionomics, Pest Status and Management.* Intercept LTD, United Kingdom, 348pp.

Cabello, T. (2006) Biology and population dynamics of red palm weevil in Spain., *I Jornada Internacional sobre el Picudo Rojo de las Palmeras* 180pp. ISBN: 84-690-1742-X.

CARDI (2010) Caribbean Agricultural Research and Development Institute.Natural resource management,invasive species. DIALOG.http://www.cardi.org/default.asp?id=46. Accessed July 2010.

Carnero, A., Hernadez-Suarez, E., Hernandez-Gracia, M., Torres, R., and Palacios, I. (1999) Peststatus of the spiraling whiteflies affecting some species of the Arecaceae and Musaceae families in the Canary Islands.Proceedings of the 2nd International Symposiumon Ornamental Palms and Other monocots from the tropics.Eds., Caballero-Ruano,M. *Acta Horticulturae*, 486, 159-164.

Carrillo, D., Amalin, D., Hosein, F., Roda, A., Duncan, R., and Peña, J.E. (2012a) Host plant range of *Raoiella indica* Hirst (Acari: Tenuipalpidae) in areas of invasion of the new world. *Experimental and Applied Acarology* 57, 271-289.

Carrillo, D., de Coss, M.E., Hoy, M.A., and Peña, J.E. (2012b) Variability in response of four populations of *Amblyseius largoensis* (Acari: Phytoseiidae) to *Raoiella indica* (Acari: Tenuipalpidae) and *Tetranychus gloveri* (Acari: Tetranychidae) eggs and larvae. *Biological Control* 60, 39-45.

Carrillo, D., Frank, J.H., Rodrigues, J.C.V., and Peña, J.E. (2012c) A review on the natural enemies of the red palm mite, *Raoiella indica* (Acari: Tenuipalpidae). *Experimental and Applied Acarology* 57, 347-360.

Carrillo, D., Navia, D., Ferragut, F., Peña, J.E. (2011) First report of *Raoiella indica* HIRST (Acari: Tenuipalpidae) in Colombia. *Florida Entomologist* 94,370–371.

Carrillo, D. and Peña, J.E. (2012) Prey-stage preference, functional and numerical responses of *Amblyseius largoensis* (Acari: Phytoseiidae) to *Raoiella indica* (Acari: Tenuipalpidae). *Experimental and applied Acarology* 57, 361-372.

Carrillo, D., Peña, J.E., Hoy, M.A., Frank, J.H. (2010) Development and reproduction of *Amblyseius largoensis* (Acari: Phytoseiidae) feeding on pollen, *Raoiella indica* (Acari: Tenuipalpidae), and other microarthropods inhabiting coconuts in Florida, USA. *Experimental and Applied Acarology* 52,119–129.

Cerda, H., Hernández, J.,Jaffé, J., Martínez, R., Sánchez, P. (1994) Estudio olfatométrico de a atracción del picudo del cocotero *Rhynchophorus palmarum* (L) a volátiles de tejidos vegetales. *Agronomía Tropical* 44, 203- 214.

Chaudhri, W.M., Akbar, S., Rasol, A. (1974) Taxonomic studies of the mites belonging to the families Tenuipalpidae, Tetranychidae, Tuckerellidae, Caligonellidae, Stigmaeidae and Phytoseiidae. University of Agriculture, Lyallpur, Pakistan, PL–480 Project on Mites, pp 250.

Cherry, R. H. (1980) Host plant preference of the whitefly, *Aleurodicus dispersus* Russell.*Florida Entomologist* 63, 222-235.

Chinchilla, C., and Oehlschlager, A. C. (1992) Captures of *Rhynchophorus palmarum* in traps baited with the male-produced pheromone .*A. S. D. Oil Palm Papers* 5, 1-8.

Chinchilla, C., and Oehlschlager, A. C. (1993) Traps for capture of adult *Rhynchophorus palmarum,* using the aggregation pheromone produced by the male .*Manejo Integrado de Plagas (Costa Rica)* 29, 28-35 (in Spanish).

Chinchilla, C., Menjívar, R.,Arias, E. (1990) Picudo de la palma y enfermedad del anillo rojo/hoja pequeña en una plantación comercial en Honduras.*Turrialba* 40, 471 – 477.

Costa Miguens, F., Scaramento, J. A., Amorin, L. Rossi, V., Coustour, N., Lummerzheim, M., lacerda, J. I. Motta, R. (2011) Mass trapping and classical biological control of *Rhynchophorus palmarum* L (Coleoptera; Curculionidae). A hypothesis based in morphological evidences. *EntomoBrasilis* 4, 49-55.

Dean, C. G. and Velis, M. (1976) Differences in the effects of red ring disease on coconut palms in Central America and the Caribbean and its control.*Oléagineux* 31, 321 – 324.

De Assis, C.P.O., De Morais, E.G.F. and Gondim, M.G.C.Jr. (2012) Toxicity of acaricides to *Raoiella indica* and their selectivity for its predator, *Amblyseius largoensis* (Acari: Tenuipalpidae: Phytoseiidae) *Experimental and Applied Acarology* DOI 10.1007/s10493-012-9647-5.

Dembilio, O., Jacas, J.A. and Llácer, E. (2009) Are the palms *Washingtonia filifera* and *Chamaeropshumilis* suitable hosts for the red palm weevil, *Rhynchophorus ferrugineus* (Coleoptera: Curculionidae).*Journal of Applied Entomology*.133, 565-567.

Dembilio, O., Llacer, E., Martinez de Altube, M.M. and Jacas, J.A. (2010a) Field efficacy of imidacloprid and *Steinernema carpocapsae* in a chitosan formulation against the red palm weevil *Rhynchophorusferrugineus* (Coleoptera: Curculionidae) in *Phoenix canariensis*. *Pest Management Science* 66, 365-370.

Dembilio, O., Quesada-Moraga, E., Santiago-Alvarez, C. and Jacas, J.A. (2010b) Biocontrol potential of an indigenous strain of the entomopathogenic fungus *Beauveria bassiana* (Ascomycota; Hypocreales) against the red palm weevil, *Rhynchophorus ferrugineus* (Coleoptera: Curculionidae). *Journal of Invertebrate Pathology* 104, 214- 221.

Dembilio, O. and Jacas, J.A. (2011) Basic bio-ecological parameters of the invasive Red Palm Weevil, *Rhynchophorus ferrugineus* (Coleoptera: Curculionidae), in *Phoenix canariensis* under Mediterranean climate. *Bulletin of Entomological Research* 101, 153-163.

Dembilio, O., Tapia, G.V., Téllez, M.M., and Jacas, J.A. (2012) Lower temperature thresholds for oviposition and egg hatching for the red palm weevil, *Rhynchophorus ferrugineus* (Coleoptera: Curculionidae), in a Mediterranean climate. *Bulletin of Entomological Research* 102, 97-102.

Dembilio O and Jacas J.A. (2012) Bio-ecology and integrated management of the red palmweevil, *Rhynchophorus ferrugineus* (Coleoptera: Curculionidae),in the region of Valencia (Spain). *Hellenic Plant Protection Journal* 5, 1-12.

Dhileepan, K (1991) Insects associated with oil palm in India. *FAO Plant Protection Bullettin* 39, 94-99.

Domingos, C.A., Oliveira, L.O., de Morais, E.G.F., Navia, D., de Moraes, G.J., and Gondim, M.G.C.Jr. (2012) Comparison of two populations of the pantropical predator *Amblyseius largoensis* (Acari: Phytoseiidae) for biological control of *Raoiella indica* (Acari:Tenuipalpidae). *Exp Appl Acarol*—DOI 10.1007/s10493-012-9625-y.

Eden-Green, S. J. (1978) Rearing and transmission techniques for *Haplaxius* sp. (Horn:Cixiidae), a suspected vector of lethal yellowing disease of coconuts.*Annals of Applied Biology* 89, 173-176.

Edwin, B.T., and Mohankumar, C. (2007) Kerala wilt disease phytoplasma analyisis and identification of a vector, *Proutista moesta*. *Physiological and Molecular Plant Pathology* 71, 41-47.

El-Garhy, M.E. (1996) Field evaluation of the aggregation pheromone of *Rhynchophorus ferrugineus* in Egypt. Brighton Crop Protection Conference: Pests and Diseases (3) *Proceedings of an International Conference*. Brighton, U.K. pp 18-21.

Elliott, M. L., Broschat, T. K Uchida, J. Y. and Simone, G. W. (2004) Diseases caused by viroids and viruses. *Compendium of Ornamental Palm Diseases and Disorders*. APS Press, Minnesota. Pp 40-42.

Elwan, A. (2000) A survey of the insect and mite pests associated with date palm trees in Al-Dakhliya region, Sultanate of Oman. *Egyptian Journal of Agricultural Research* 78,653–664.

El-Sabea A. M. R., Faleiro J. R. and Abo El Saad M. M. (2009) The threat of red palm weevil *Rhynchophorus ferrugineus* to date plantations of the Gulf region of the Middle East: an economic perspective. *Outlook on Pest Management* 20, 131-134.

El-Shafie, H.A.F., Faleiro, J.R., Al-Abbad, A.H., Stoltman, L. and Mafra-Neto, A. (2011). Bait-free attract and kill technology (Hook[tm] RPW) to suppress red palm weevil *Rhynchophorus ferrugineus* (Coleoptera: Curculionidae) in date palm. *Florida Entomologist* 94, 774-778.

EPPO (European and Mediterranean Plant Protection Organization) (2008) Data sheets on quarantine pests. *Rhynchophorus ferrugineus. EPPOBulletin* 38, 55-59.

EPPO (2009) *Raoiella indica* (Acari: Tenuipalpidae) *In: European and Mediterranean plant protection organization.* DIALOG.http://www.eppo.org/QUARANTINE/Alert_List/insects/raoiella_indica.htmAccessed Sept 2009.

Espinosa, A., Hodges,A.,and Hodges, G. (2012) Palmetto scale, Comstockiella sabalis Comstock.University of Florida IFAS ExtensionEENY 465. http://edis.ifas.ufl.edu/in835.

Esser, R. (1969) *Rhadinaphelencus cocophilus* a potential threat to Florida palms. *Nematology Circular* 9, Florida Department of Agriculture and Consumer Service, Division of Plant Industry, 2 p.

Esser, R., and Meredith, J. (1987) Red ring nematode. *Nematology Circular of the Florida Department of Agriculture* no. 141, Gainesville, FL (US).

Estévez, E and Cedeño, R. (2008) Susceptibilidad de *Rhynchophorus palmarum* (L.)(Coleoptera: Curculionidae) a cepas de hongos entomopatogenos. Proyecto de graduacion, Ingeniero Agronomo, Universidad Earth, Guassimo, Limon, Costa Rica downloaded as:
http://usi.earth.ac.cr/glas/sp/ColeccionVirtual/pdf/PG01-2008 EstevesE CedenoR%5B1%5D.pdf; March 28, 2013.

Faleiro J. R. and Rangnekar P.A. (2001) Ovipositional preference of red palm weevil *Rhynchophorus ferrugineus* Oliv. to coconut cultivars. *Indian Coconut Journal* 32, 22-23.

Faleiro J. R., Ashok Kumar J. and Rangnekar P. A. (2002) Spatial distribution of red palm weevil Rhynchophorus ferrugineus Oliv. (Coleoptera: Cuculionidae) in coconut plantations. *Crop Protection* 21, 171–176.

Faleiro, J. R., Rangnekar, P. A. and Satarkar, V. R. (2003) Age and fecundity of female red palm weevils *Rhynchophorus ferrugineus* (Olivier) (Coleoptera : Rhynchophoridae) captured by pheromone traps in coconut plantations of India. *Crop Protection* 22, 999-1002.

Faleiro, J.R. (2005) Pheromone technology for the management of red palm weevil, *Rhynchophorus ferrugineus* (Olivier) (Coleoptera : Rhynchophoridae) –A key pest of coconut, Technical Bulletin No.4, ICAR Research Complex for Goa, India. 40 pp.

Faleiro, J.R. (2006) A review of the issues and management of the red palm weevil *Rhynchophorusferrugineus* (Coleoptera: Rhynchophoridae) in coconut and date palm during the last one hundred years. *International Journal of TropicalInsect Science* 26, 135-154.

Faleiro J. R. and Ashok Kumar J. (2008) A rapid decision sampling plan for implementing area-wide management of red palm weevil, *Rhynchophorus ferrugineus*, in coconut plantations of India. *Journal of Insect Science* 8, 15 (available at: insectscience.org/8.15).

Faleiro, J.R., Abdallah, B.A. and Ashok, K.J. (2010) Sequential sampling plan for area-wide management of *Rhynchophorus ferrugineus* (Olivier) in date palm plantations of Saudi Arabia.*International Journal of Tropical Insect Science*30, 145-153.

Faleiro, J.R., El-Saad, M.A. and Al-Abbad, A.H. (2011) Pheromone trap density to mass trap *Rhynchophorus ferrugineus* (Coleoptera: Curculionidae/Rhynchophoridae/Dryophoridae) in date plantations of Saudi Arabia. *International Journal of Tropical Insect Science* 31, 75-77.

Faleiro, J.R., Ben Abdullah, A., El-Bellaj, M., Al Ajlan, A.M. and Oihabi, A. (2012) Threat of red palm weevil, *Rhynchophorus ferrugineus* (Olivier) to date palm plantations in North Africa.*Arab Journal of Plant Protection* 30, 274-280.

Farazmand, H. (2002) Investigation on the reasons of food preference of red palm weevil, *Rhynchophorus ferrugineus Oliv. Applied Entomology and Phytopathology* 70, 11-12.

FDACS (2007 and 2011) Red palm mite infestation identified in Palm Gardens. In: Florida Department of Agriculture and Consumer Services. DIALOG.http://www.doacs.state.fl.us/press/2007/12052007_2.html. Accessed Sept 2009.

Fenwick, D (1967)The effect of weevil control on the incidence of red ring disease. *Journal of the Agricultural Soceity of Trinidad and Tobago* 67, 231-244.

Fernando, L.C.P., Aratchige, N.S., and T.S. G. Ferris.(2003) Distribution patterns of coconut mite, *Aceria guerreronis*, and its predator *Neoseiulus aff paspalivorus* in coconut palms.*Experimental and Applied Acarology* 31, 71-78.

Fernando, L.C.P., Manoj, P., Hapuarachchi, D.C.I., and Edginton, S. (2007) Evaluation of four isolates of Hirsutella thompsoni against coconut mite (*Aceria guerreronis*) in Sri Lanka. *Crop Protection* 26, 1062-1066.

Fernando, L.C.P., Waidyaraithne, K.P., Perera, K.F.G., and De Silva, P.H.P.R. (2010) Evidence for suppressing coconut mite *Aceria guerreronis* by inudnative release of the predatory mite, *Neoseiulus baraki. Biological Control* 53, 108-111.

Fiaboe, K.K.M., Peterson, A.T., Kairo, M.T.K. and Roda, A.L. (2012) Predicting the potential worldwide distribution of the red palm weevil *Rhynchophorus ferrugineus* (Olivier) (Coleoptera: Curculionidae) using ecological niche modeling. *Florida Entomologist* 95, 559-673.

Fletchman, C.H. W. (1998) *Coccus wendelliana* H. Wendl (Palmae: Arecaceae), a new host plant for *Eriophyies guerreronis*(Keifer, 1965) (Acari: Eriophyidae) in Brazil. *International Journal of Acarology* 15, 241.

Flechtmann, C.H.W., and Etienne, J. (2004) The red palm mite, *Raoiella indica* Hirst, a threat to palms in the Americas (Acari: Prostigmata: Tenuipalpidae). *Systematic and Applied Acarology* 9,109–110.

Flores-Galano, G., Montoya and A., Rodriguez, H. (2010) Biologia de *Raoiella indica* Hirst (Acari: Tenuipalpidae) sobre *Areca catechu*L. *Revista deProteccion Vegetal* 25,11–16.

Forero, D. (2008) The systematic of the Hemiptera. *Revista Colombiana de Entomologia* 34, 1-21.

Foster, W.A., Snaddon, J.L., Turner, E.C., Fayle, T.M., Cockerill, T., Farnon Ellwood, M.D., Broad, G., Chung, A.Y., Eggleton, P., Vun Khen, C., and Yusah, K.M. (2011) Establishing the evidence base for maintaining biodiversity and ecosystem function in the oil plam landscapes of South East Asia. *Phylosophical transactions of the Royal Society B: Biological Sciences* 366, 3277-3291.

Furness, G. O. (1976) The dispersal, age-structure and natural enemies of the long-tailed mealybug, *Pseudococcus longispinus* (Targioni-Tozzetti), in relation to sampling and control. *Australian Journal of Zoology* 24, 237-47.

Gallego C. E, Aterrado E. D and Batomalaque C. G (2003) Biology of the false spider mite, *Rarosiella cocosae* Rimando, infesting coconut palms in Camiguin, northern Mindanao (Philippines). *Philippine Entomology* 17, 187.

Gassouma SM (2005) Pests of date palm (Phoenix dactilifera). Downloaded as :http:// www.uae.gov.ae/uaeagricent/palmtree2/chap8.stm. Accessed 25 July 2009

Gerber, K., and Giblin-Davis, R. (1990) Association of the red ring nematode and other nematode species with the palm weevil, *Rhynchophorus palmarum.The Journal of Nematology* 22, 143-149.

Gerson, O., Venezian, A., and Blumberg, D. (1983) Phytophagous mites and date palms in Israel *Fruits* 38, 133-135.

Ghosh, C.C. (1912) Life-histories of Indian insects, III.The rhinoceros beetle (*Oryctes rhinoceros*) and the red or palm weevil (*Rhynchophorus ferrugineus*).*Memoires of the Department of Agriculture of India* 2, 193-217.

Giblin-Davis, R. (1990) The red ring nematode and its vectors.Nematology Circular No. 181. Florida Department of Agriculture and Consumer Service, Division of Plant Industry.4 p.

Giblin-Davis, R. M., Faleiro, J. R., Jacas, J. A., Peña, J. E., and Vidyasagar, P.S.P.V. (2013) Coleoptera: Biology and management of the red palm weevil, *Rhynchophorus ferrugineus*. Pp. 1-34. *In* J. E. Peña [ed.], Potential Invasive Pests of Agricultural Crop Species.CABI Wallingford, UK.

Godim, M.G.C.Jr., Castro, T.M.M., Masaro, A.L.Jr., Navia, D., Melo, J.W.S., Demite, P.R., and de Moraes, G.J. (2013) Can the red palm mite threaten the Amazon vegetation? *Systematics and Biodiversity* 10, 527-535.

Gomes, F.P., and Prado, C.H.B.A. (2007) Ecophysiology of coconut palm under water stress. *Brazilian Journal of Plant Physiology* 19,377–391.

Gonzales, A.I., and Ramos, M. (2010) Desarrollo y reproduccion de *Raoiella indica* Hirst (Acari: Tenuipalpidae) en laboratorio. *Revista deProteccion Vegetal* 25, 7–10.

Gonzalez, P., and García, U. (1992) Ciclo biologico de *Rhynchophorus palmarum* (Col: Curculionidae) sobre *Washingtonia robusta* en laboratorio. *Revista Peruana deEntomologia* 35, 60 – 62.

Griffith, R. (1968)The relationship between the red ring nematodes and the palm weevil *Journal of Agricultural Society of Trinidad and Tobago* 17, 1-3.

Griffith, R. (1987) Red ring disease of coconut palm. *Plant Disease* 71, 193-196.

Guarino S., Peri, E., Bue, P. L., Germanà, M. P., Colazza, S., Anshelevich, L., Ravid U. and Soroker V. (2013) Assessment of synthetic chemicals for disruption of *Rhynchophorus ferrugineus* response to attractant-baited traps in an urban environment. Phytoparasitica, 41: 79-88.

Gullan, P. J. and L. G. Cook.(2007) Phylogeny and higher classification of the scale insects (Hemiptera: Sternorrhyncha: Coccoidea). *Zootaxa* 1668, 413-425.

Gungah, B., Seewooruthun, S. I Nundloll, P and Rambhunjun, M. (2005) *Biological control of the spiraling whitefly, Aleurodicus dispersus*.MAS.Food and Agricultural Research Council, Reduit, Mauritius.307-311.

Hagley, E. (1965a) On the life history and habits of the palm weevil *Rhynchophorus palmarum* L. *Annals of the Entomological Society of America* 58, 22-28.

Hagley, E. (1965b) Test of attractant for the palm weevil. *Journal of Economic Entomology* 58, 1002-1003.

Hagley, E.(1963) The role of the palm weevil, *Rhynchophorus palmarum* as a vector of the red ring disease of coconuts. I. Results of preliminary investigations. *Journal of Economic Entomology* 56,375-380.

Hallett, R.H., Gries, G., Gries, R., Borden, J.H., Czyzewska, E., Oehlschlager, A.C., Pierce, Jr., H.D., Angerilli, N.P.D. and Rauf, A. (1993) Aggregation Pheromones of Two Asian Palm Weevils *Rhynchophorus ferrugineus* and *R. vulneratus.Naturwissenschaften* 80, 328-331.

Hallett, R. H., Oehlschlager, A.C. and Borden, J.H. (1999) Pheromone trapping protocols for the Asian palm weevil, *Rhynchophorus ferrugineus* (Coleoptera: Curculionidae). *International Journal of Pest Management* 45, 231-237.

Hallett, R.H., Crespi, B.J. and Borden, J.H. (2004) Synonymy of *Rhynchophorus ferrugineus* (Oliver), 1790 and *R. vulneratus* (Panzer), 1798 (Coleoptera: Curculionidae, Rhynchophorinae). *Journal of Natural History* 38, 2863-2882.

Haq, M. A. (2011) Coconut destiny after the invasion of *Aceria guerreronis* (Acari: Eriophyidae) in India *In:* Moraes, G.J. de & Proctor, H. (eds) Acarology XIII: Proceedings of the International Congress.*Zoosymposia* 6, 1–304.

Hartley, C.W.S. (1967) The oil palm, 706 pp.

Hassan, E. (1972) Problems of applied entomology in Papua New Guinea.*Anzeiger fur Schalingskunde und Pflanzenschuts* 45, 129-134.

Helle, W., Bolland, H.R., and Haitmans, W.R.B. (1980) Chromosomes and types of parthenogenesis in the false spider mites (Acari: Tenuipalpidae). *Genetica* 54, 45–50.

Hernández, J. V., Cerda, H., Jaffe, K., and Sánchez, P.(1992) Localización del hospedero, actividad diaria y optimización de las capturas del picudo del cocotero *Rhynchophorus palmarum* L. (Coleoptera: Curculionidae) mediante trampas inocuas. *Agronomía Tropical* 42, 211 – 226.

Herren, H.R., and Neuenschwander, P. (1991) Biological control of cassava pests in Africa.*AnnualReview of Entomology* 36, 257–83.

Hodges, G. S. and Evans, G. A. (2005) An identification guide to the whiteflies (Hemiptera: Aleyrodidae) of the southeastern United States. *Florida Entomologist* 88, 518-534.

Hokkanen, H., and Pimentel, D. (1984) New approach for selecting biological control agents. *Canadian Entomologist* 116, 1109–1121.

Hoong, H.W., and Hoh, C.K.Y. (1992) Major pests of oil palm and their occurrence in Sabah. *Planter* 68, 193-210.

Howard, F. W (1983) World distribution and possible geographic origin of palm lethal yellowing disease and its vectors. Food and Agricultural Organization, Rome 31, 101-113.

Howard, F. W. (1987) *Myndus crudus*, a vector of lethal yellowing of palms. *In:* Wilson, M. R. and Nault, L. R. (eds.) 2nd International Workshop on Leafhoppers and Planthoppers of Economic Importance held at Brigham Young University, Provo, Utah, USA. CAB International Institute of Entomology. London, pp 117-129.

Howard, F.W. (1990) Evaluation of grasses for cultural control of *Myndus crudus,* a vector of lethal yellowing of palms. *Entomologia Experimentalis et Applicata* 56, 131-137.

Howard F.W., Abreu-Rodriguez E. and Denmark H.A. (1990) Geographical and seasonal distribution of the coconut mite, *Aceria guerreronis* (Acari: Eriophyidae), in Puerto Rico and Florida, USA. *Journal ofAgriculture University of Puerto Rico* 74, 237–251.

Howard, F. W. (1999) An introduction to insect pests of palms. Proceedings of the 2nd International Symposium on Ornamental Palms and Other monocots from the tropics. Ed., Caballero-Ruano,M. *Acta Horticulturae*, 486, 133-139.

Howard, F.W. (2001) Sap feeders on palms.In : Howard, F.W., Moore, D., Giblin-Davis, R., and Abad, R. G. (eds) Insects onPalms. CABI Publishing Wallingford, United Kingdom, pp.199-232.

Howard, F. W., Norris, R.C., and Thomas, D.L. (1983) Evidence of transmission of palm lethal yellowing agent by a planthopper, *Myndus crudus* (Homoptera: Cixiidae). *Tropical Agriculture* 60, 168-171.

Howard F. W,andOropeza C. (1998) Organic mulch as a factor in the nymphal habitat of *Myndus crudus* (Hemiptera: Uchenorrhyncha: Cixiidae). *Florida Entomologist* 8, 92–97.

Huger, A. M. (2005) The Orcytes virus: its detection, identification, and implementation in biological control of the coconut palm rhinocerus beetle, *Orcytes rhinoceros* (Coleoptera: Scrabaeidae). *Journal of Invertebrate Pathology* 89, 78-84.

Iperti, G., Laudeho, Y., Brun, J., Janvry, E. C., de (1970) The natural enemies of Parlatoria blanchardi Targ. In the palm groves of the Adrar región of Mauritania. III. Introduction, acclimatization and effectiveness of a new predaceous coccinellid, *Chilocorus bipustulatlus* L, variety iranensis (new variety). *Annals de Zoologie Ecologie Animale* 2, 617-638.

Jaffé, K. and Sanchez, P. (1990) *Final report of the Project on Ethological Study ofRhynchophorus palmarum* Universidad Simon Bolivar –FONAIAP Caracas (VE).

Jaffe, K., Sanchez, P., Cerda, H., Hernandez, J. V., Jaffe, R., Urdaneta, N., Guerra, G., Martinez, R., and B. Miras. (1993) Chemical ecology of the palm weevil *Rhynchophorus palmarum* (L.) (Coeloptera: Curculionidae): attraction to host plants and to a male-produced aggregation pheromone. *Journal of Chemical Ecology* 19, 1703-1720.

Jiménez O. D. (1969) Biología y hábitos del *Rhynchophorus palmarum* L. Universidad Nacional de Colombia Medellín. Facultad de Agronomía e Instituto Forestal. Tesis Ingeniero Agrónomo. 48 p

Ju, R.T., Wang, F., Wan, F.H. and Li, B. (2011) Effect of host plants on development and reproduction of *Rhynchophorus ferrugineus* (Olivier) (Coleoptera: Curculionidae). *Journal of Pest Science* 84, 33-39.

Julias, J. F. (1982) *Myndus taffini* (Homoptera: Cixiidae), a vector of foliar decay of coconut in Vanuatu. *Oleaginex* 37,409-414.

Kaakeh, W. (2006) Toxicity of imidacloprid to developmental stages of *Rhynchophorus ferrugineus* (Curculionidae: Coleoptera): Laboratory and field tests. *Crop Protection* 25, 432-439.

Kaakeh, W., El-Ezaby, F., Abu Al-Nour, M.M. and Khamis, A.A. (2001) Management of the Red Palm Weevil by a Pheromone / Food-Based Trapping System, UAE Second Int. Conf. On Date Palms, Al-Ain, UAE, March 25-27, p 325-343.Available on line http://wwwpubhort.org/datepalm2/datepalm2-38.pdf

Kalshoven, L.G.E. (1950) The pests of crops in Indonesia. Revised and translated by P.A. van der Laan, 1981, with G.H.L. Rothschild, Jakarta: P.T. Ichtiar Barn - van Hoeve, 701 p.

Kane, E.C., Ochoa, R., Mathurin, G., Erbe, E.F. (2005) *Raoiella indica* Hirst (Acari:Tenuipalpidae): An island-hopping mite pest in the Caribbean. Available via DIALOG. http://www.sel.barc.usda.gov/acari/PDF/Raoiella indica-Kane et al.pdf. Accessed July 2010.

Keifer, H. H. (1965) *Eriophyid studies B-14*. Sacaramento, California, Department of Agriculture, Bureau of Entomology.

Kumashiro, B.R., Lai, p.Y., Funasaki, G.Y., Teramoto, K.K. 91983) Efficacy of *Nephaspis amnicola* and *Encarsia hatiensis*in controlling Aleurodicus dispersus in Hawaii. *Proceedings of the Hawaiian Entomological Society* 24, 261-269.

Kurian, C., Sathiamma, B., Sukumaran, A.S., and Ponnamma, K.N. (1979) Role of attractants and repellents in coconut pest control in India. Paper presented at the 5th session of the Food and Agriculture Organization Technical Working party, Manila. Philippines.

Lawson-Balagbo, L.M., Gondim, M.G. C., de Moraes, G.J., Hanna, R., and Schausberger, P. (2008) Exploration of the acarine fauna on coconut palm in Brazil with emphasis on *Aceria guerreronis* (Acari: Eriophyidae) and its natural enemies. *Bulletin of Entomological Research* 98, 83-96.

Lever, R. J.(1969) Pests of the coconut palm.Food and Agricultural Organization, Rome, 190 pp.

Lever, R. J. A. W. (1979) Pests of the Coconut Palm. FAO Agricultural Studies No. 77, Food and Agriculture Organization of the United Nations, Rome.

Llácer, E., Martinez, J. and Jacas, J.A. (2009) Evaluation of the efficacy of *Steinernema carpocapsae* in a chitosan formulation against the red palm weevil, *Rhynchophorus ferrugineus*, in *Phoenixcanariensis*. *BioControl* 54, 559-565.

Llácer, E., Dembilio, O. and Jacas, J.A. (2010) Evaluation of the Efficacy of an insecticidal paint based on chlorpyrifos and pyriproxyfen in a microencapsulated formulation against *Rhynchophorusferrugineus* (Coleoptera: Curculionidae). *Journalof Economic Entomology* 103, 402-408.

Llácer, El., Negre, M. and Jacas, J.A. (2012) Evaluation of an oil dispersion formulation of imidacloprid against *Rhynchophorus ferrugineus* (Coleoptera, Curculionidae) in young palm trees.*Pest ManagementScience*.DOI: 10.1002/ps.3245.

Llácer, E., Santiago-Álvarez, C., and Jacas, J.A. (2013) Could sterile males be used to vector a microbiological control agent? The case of *Rhynchophorus ferrugineus* and *Beauveria bassiana*. *Bulletin of Entomological Research* 103, 241–250.

Locarno, L., Constantino, L., Agudelo, R., Alarcon, A., Caicedo, V. (2005) Observaciones sobre el gualapán (Coleoptera: Chrysomelidae: Hispinae) y otras limitantes entomológicas en cultivos de chontaduro en el Bajo Anchicayá. *Acta Agronómica* 54, 25 – 30.

Maharaj, S (1965) A new trap design of trap for collecting the palm weevil, *Rhynchophorus palmarum*(L.) *Tropical Agriculture*, 42, 373-375.

Mannion.C. (2010) Rugose spiraling whitefly: a new whitefly in south Florida. Fact Sheet http://trec.ifas.ufl.edu/mannion/pdfs/Rugose%20spiraling%20whitefly.pdf

Mariau, D. (1982) Phyllophagous oil palm and coconut pests. Importance of entomopathegic parasites for population regulation. *Oleagineux* 37, 3-7.

Mariau, D (1977) *Aceria (Eriophyies) guerreronis* : un important ravageur des cocoteraies africainnes et americaines. *Oleagineux* 32: 101-111.

Mariau, D. (1968) Methods de lutter contre le Rhynchophore. *Oléagineux* 23, 443-446.

Marsaro, A.L.Jr., Navia, D., Gondim, M.G.Jr., Silva, F.R., and de Moraes, G.J. (2009) Chegou ao Brasil-o ácaro vermelho das palmeiras *Raoiella indica*. *Cultivar, Hortaliças e Frutas* 57,31.

Martin, J. H. (2004) Whiteflies of Belize (Hemiptera: Aleyrodidae). Part 1 – Introduction and Account of the Subfamily Aleurodicinae Quaintance & Baker. *Zootaxa* 681:1-119.

Martin Kessing, J. L. and Mau, R. F. L. (1993).*Aleurodicus dispersus* (Russell). Crop Knowledge Master. http://www.extento.hawaii.edu/kbase/crop/Type/a_disper.htm.

Martínez, L., Moya O. and Aldana, R. (2010) Evaluacion de insecticidas para prevenir el ataque de *Rhynchophorus palmarum* (L) (Coleoptera: Curculionidae) en palma de aceite afectada por pudricion del cogollo. In: *Resúmenes XXXVI Congreso Sociedad Colombiana de Entomomología. Julio 29, 30 y 31, Medellín, Colombia.*

Martin-Molina, M.M. (2004) Biologia y ecologia del curculionido rojo de las palmeras, *Rhynchophorusferrugineus* (Olivier, 1790) (Coleoptera: Dryophthoridae). *Unpublished PhD Thesis*. Universidad de Almeria, 203 pp.

Massoud A.M., Faleiro J.R., El-Saad M.A andSultan E. (2011) Geographic information system used for assessing the red palm weevil*Rhynchophorus ferrugineus* (Olivier)in date palm oasis of Al-Hassa, Saudi Arabia. *Journal of Plant Protection Research* 51,234-239.

Mendonca de Omena, R.P. , Guzzo, E C., Santos Ferreira, J M., Cavalcante de Mendonca, F A., de Lima, F., Racca-Filho, F., Goulart Santana, A E (2012) First report on the whitefly *Aleurodicus pseudugesii* on the coconut palm, *Cocos nucifera* in Brazil. *The Journal of Insect Science* 12, 26-30.

Menon, K.P. V., and Pandalai, K.M. (1958) The coconut palm. A monograph, 384 pp Downloaded as: http://www.cabdirect.org/abstracts/19601604549.html?freeview=true(June 20, 2013)

Mexzon, R., Chinchilla, C., Castillo, G., and. Salamanca, D. (1994) Biología y hábitos de *Rhynchophorus palmarum* L. asociado a la palma aceitera en Costa Rica. ASD *Oil Palm Papers*. 8, 14 – 21.

Miller, D. R. and Davidson, J. A. (2005) Armored Scale Insect Pests of Trees and Shrubs (Hemiptera: Diaspididae). Cornell University Press. Ithaca, NY. 456 pp.

Miller, D. and Y. Ben-Dov. 2015. Coccidae. http://www.sel.barc.usda.gov/scalenet/Scale Net scalenet.htm (last updated 13 April 2015

Miller, D. R., Rung, A. Venable, G. L,. Gill, R. J.(2007) Scale insects: Identification tools, images and diagnostic information for species of quarantine significance. *Systematic Entomology* USDA-ARS. http://www.sel.barc.usda.gov/scalekeys/ScaleInsectsHome/ScaleInsects Home.html.

Monteiro, V.B., Lima, D.B., Gondim, Jr., M. G. C., and Siqueira, H. A. (2012) Residual bioassay to assess the toxicity of acaricides against *Aceria guerreronis* (Acari: Eriophyidae) under laboratory conditions. *Journal Economic Entomology* 105, 1410-1425.

Moore, D. (2000) Non-chemical control of *Aceria guerreronis* on coconuts. In, Fernanado, L.C.P., de Moraes, G. J., and Wickramananda, I.R. *Proceedings of the international workshop on coconut mite (Aceria guerreronis)*, Coconut Research Institute, Sri Lanka, January 2000, pp.63-70.

Moore, D., andAlexander, L. (1987) Aspects of migration and colonization of the coconut palm by the coconut mite *Eriophyies guerreronis* (Keifer) (Acari: Eriophyidae). *Bulletin of Entomological Research* 77, 641-650.

Moore, D., and Howard, F. W.(1996) Coconuts. In: Eriophyoid mites: their biology, natural enemies and control, E. Lindquist, M.W. Sabelis and J. Bruin (eds.) Elsevier Science, New York. pp. 561-570.

Moraes, G. J., McMuretry, J.A., Denmark, H. A., and Campos, C.B. (2004) A revised catalogue of the family Phytoseiidae. *Zootaxa* 434,1-494.

Morales, J., and Chinchilla, C (1990) Picudo de la palma y enfermedad del anillo rojo/hoja pequeña en una plantación comercial en Costa Rica. *Turrialba,*. 40, 478 – 485.

Morin, J. P., F. Lucchini, F., de Araujo, J., Ferreira, J., and Fraga, L.(1986) *Rhynchophorus palmarum* control using traps made from oil palm cubes. *Oléagineux,* 41, 60 – 61.

Mosquera, M. andViafara,J. D. (2008) Evaluation of *Paratheresia claripalpis* and *Metagonistylum minense* you as possible parasitoides of larvas of *Rhynchophorus palmarum*, under conditions of laboratory in the municipality of Buenaventura, Valle del Cauca, Colombia *Bioetnia* 5, 110-114.

Mound, L.A. and Halsey, S. H. (1978) Whiteflys of the World: a Systemic Catalogue of the Aleyrodidae (Homoptera) with Host Plant and Natural Enemy Data.British Museum and John Wiley & Sons, Chichester.

Moura, J. I.L. (1994) Management of pests of coconut palms in the state of Bahia, Brazil. *Acta Horticulturae* 360, 223-229.

Moura, J. I.L., Bento, J.,de Souza, J. and Vilela, E. (1997) Field trapping of *Rhynchophorus palmarum* (L.) using trap plants treated with aggregation pheromone and insecticide.*Anais da sociedade Entomologica do Brasil* 26, 69-73.

Moura, J.I.L., Mariau, D., and Delabie, J.H.C. (1993) Efficacy of *Paratheresia meneszi* Townsend (Diptera: Tachnidae) for natural biological control of *Rhynchophorus palmarum* (L.) (Coleoptera: Curculionidae) *Oleagineaux* 48, 219-223.

Moura, J.I.L., Resende, M. L. B., Vilela, E.F. (1995) Integrated pest management of *Rhynchophorus palmarum* (L.)(Coleoptera: Curculionidae) in oil plams in Bahia.*Anais da Sociedade entomomologica do Brasil* 24, 501-506.

Moura, J.I., Toma, I., Sgrillo, R.B., Delabie, J.H.C. (2006) Natural efficiency of parasitism by *Billaea rhynchophorae* (Blanchard) (Diptera:Tachinidae) for the control of *Rhynchophorus palamarum* (L.) (Coleoptera: Curculionidae) *Neotropical Entomology* 35, 273-274.

Moutia, L.A. (1958) Contribution to study of some phytophagous Acarina and their predators inMauritius. *Bulletin ofEntomological Research* 49,59–75.

Murphy, S.T. and Briscoe, B.R. (1999) The red palm weevil as an alien invasive: biology and prospects for biological control as a component of IPM. *BioControl* 20, 35-45.

Muthiah, C., and Bhaskaran, R (1999) Screening of coconut genotypes and management of eriophyid mite, Aceria guerreronis (Eriophyidae: Acari) in Tamil Nadu. *Indian Coconut Journal* 30, 10-11.

Nageshachandra, B.K. and Channabasavanna, G.P. (1983) Studies on seasonal fluctuation of the population of *Raoiella indica* Hirst (Acari: Tenuipalpidae) on coconut with reference to weather parameters. *Indian Journal ofAcarology* 8,104–111.

Nageshachandra, B.K., and Channabasavanna, G.P. (1984) Development and ecology of *Raoiella indica* Hirst (Acari: Tenuipalpidae) on coconut. In: Griffiths DA, Bowman CE (eds), Acarology VI. Ellis Horwood Publishers, Chicester, UK, pp 785–790.

Nair, C.P. R. (2000) Status of coconut eriophyid mite *Aceria guerreronis* Keifer in India. In, Fernando, L.C.P., de Moraes, G. J., and Wickramananda, I.R. *Proceedings of the international workshop on coconut mite (Aceria guerreronis), Coconut Research Institute, Sri Lanka,* January 2000, pp. 9-12.

Nampoothiri, K.U.K., Nair, C. P.R., Kannaiyan, S., Doraisamy, S., Saradamma, K., Naseema Beevi, S. and Sreerama Kumar, P. (2002) Coconut eriophyid mite (*Aceria guerreronis* Keifer) an update.*Proceedings of Placrosym* 15, 487-496.

NAPPO (2009) North American Plant Protection Organization.Detection of the red palm mite (*Raoiella indica*) in Cancun and Isla Mujeres, Quintana Roo, Mexico.Available via DIALOG:http://www.pestalert.org/oprDetail.cfm?oprID=406.Accessed 30 December 2010.

Navia, D., de Moraes, G., Roderick, G., and Navajas, M. (2005) The invasive coconut mite *Aceria guerreronis* (Acari: Eriophyidae): origin and invasion sources inferred from mitochondrial (16S) and nuclear (ITS) sequences. *Bulletin of Entomological Research* 95, 505-516.

Navia, D., Gondim, M. G. C., de Moraes, G. (2007) Eriophyid mites (Acari: Eriophyoidea) associated with palm trees. *Zootaxa* 1389, 1-30.

Nirula, K.K. (1956) Investigations on the pests of coconut palm. Part .*Rhynchophorus ferrugineus. Indian Coconut Journal*, 9: 229-247.

Ochoa, R., Beard, J.J., Bauchan, G.R., Kane, E.C., Dowling, A.P.G., and Erbe, E.F. (2011) Herbivore exploits chink in armor of Host. *American Entomologist* 57, 26–29.

Oehlschlager, A.C. (1998) Trapping of the date palm weevil. *Food and Agriculture Organization workshop on date palm weevil (Rhynchophorus ferrugineus) and its control.* Cairo, Egypt, December 15-17.

Oehlschlager, A.C. (2006) Mass trapping as a strategy for management of *Rhynchophorus* palm weevils, *I Jornada Internacional sobre el Picudo Rojo de las Palmeras* 143-180 pp. ISBN: 84-690-1742-X.

Oehlschlager, A.C. (2010) Efficiency and Longevity of Food Baits in Palm Weevil Traps.Proceedings. 4th International Date Palm Conference Eds.: A. Zaid and G.A. Alhadrami. *Acta Horticulturae*. 882, 399-406.

Oehlschlager, A.C., Chinchilla, C., Gonzalez, L.M.(1992) Management of red ring disease in oil palm through pheromone-based trapping of *Rhynchophorus palamarum.International seminar on coconut research*, Kingston, Jamaica, 16 pp.

Oehlschlager, A.C., Chinchilla, C., Gonzalez, L.M., Jiron, L.F., Mexzon, R., and Morgan, B. (1993) Development of a pheromone-based trap for the American palm weevil, *Rhynchophorus palmarum* (L).*Journal of Economic Entomology* 86, 1381-1392.

Oehlschlager, A.C., Chinchilla, C., Castillo, G., and Gonzalez, L. (2002) Control of the red ring disease by mass trapping of *Rhynchophorus palmarum* (Coleoptera: Curculionidae).*Florida Entomologist* 85, 507-513.

Oehlschlager, A. C., and Gonzalez, L. (2001) Advances in trapping and repellency of palm weevils. In:*Proceedings of the Second International Conference on Date Palms* (Al-Ain, UAE). www.pubhort.org/datepalm/datepalm2/datepalm2_40.pdf.

OEPP/EPPO, (2005) *Rhynchophorus palmarum.*European and Mediterranean Plant Protection Organization. Organisation Européenne et Méditerranéenne pour la Protection des Plantes. *Bulletin OEPP/EPPO Bulletin* 35. Pp. 468–471.

Peña, J.E., Bruin, J. and Sabelis, M.W. (2012) Biology and control of the red palm mite, *Raoiella indica*: an introduction. *Experimental and Applied Acarology* 57, 211-213.

Peña, J.E., Rodrigues, J.C.V., Roda, A., Carrillo, D., and Osborne, L.S. (2009) Predator-prey dynamics and strategies for control of the red palm mite (*Raoiella indica*) (Acari: Tenuipalpidae) in areas of invasion in the Neotropics. *Proceedings of the 2nd Meeting of IOBC/WPRS, Work group integrated control of plant feeding mites*. Florence, Italy 9–12 March 2009, pp 69–79.

Peña-Rojas, E. (2000) Plagas y enfermedades del chontaduro. In: Reyes,R., E. Peña y J.Soto(Eds) El cultivo del chontaduro para palmito. *Corporación Colombiana de Investigación Agropecuaria- CORPOICA Manual Técnico* 4, 140 p.

Pérez, D., and Iannacone, J. (2006a) Aspectos de la bioecología de *Rhynchophorus palmarum* / Linnaeus) (Coleoptera: Curculionidae) en el pijuayo (*Bactris gasipaes* H. B. K.) (Arecaceae), en la Amazonia peruana. *Revista Peruana de Entomología* 45, 138 – 140.

Pérez, D. J., and Iannacone, J. (2006b) Effectiveness of botanical extracts from ten plants on mortaility and larval repellency of *Rhynchophorus palmarum* L., an insect pest of the peach palm Bactris gasipaes Kunth in Amazonian Peru. *Agricultura Tecnica* 66, 21-30.

Pérez, D., Iannacone, J., and Pinedo, H. (2010) Toxicological effect from the stem cortex of the amazonic plant soapberry *Paullinia clavigera* (Spindaceae) upon three arthropods. *Ciencia e Investigacion Agraria* 37, 133-143.

Pikkheng, H., Bong, C.F. J., Jugah, K., Rajan, A. (2009) Evaluation of *Metarrhizium anisopliae* var. *anisopliae* (Deuteromycotina: Hyphomycete) isolates and their effects on subterranean termite Coptotermes curvignathus (Isoptera: Rhinotermitidae). *American Journal of Agricultural and Biological Science*4, 289-297.

Pillai, G. B. (1987) Integrated pest management in plantation crops. *Journal of Coffee Research*, 17, 150- 153.

Ponnamma, K. N., and Biju, B. (1998) Record of Halictophagous sp (Strepsiptera: Halictophagidae) a parasitoid of Proutista moesta (Westwood) (Homoptera: Derbidae), avector of phytoplasma diseases of palms. *Planter* 74, 37-40.

Pushpa, V. (2006) Management of coconut perianth mite, *Aceria guerreronis* Keifer. Department of Entomology, College of Agriculture, Dharwad, University of Agricultural Sciences, Dharwad, India, Master Thesis, 76 pp.

Puttarudriah, M., and Channabasavanna, G.P. (1956) Some beneficial coccinellids of Mysore. *Journal Bombay Natural History Society* 54,156–159.

Queiroz Cysne, A., Araujo Cruz, B., Viera da Cunha, R. N., Carvalho da Rocha, R.N. (2013) Population dynamics of *Rhynchophorus palmarum* (L.) (Coleoptera: Curculionidae) on oil palms in the Amazonas State.*Acta Amazonica* 43,9 p. http://www.scielo/br/scielo.php?pid=S0044-59672013000200010&script=sci_arttext on 06/14/2013. (not listed in text)

Quintana, C. (2012) Cadena nacional de coco de Colombia. Acuerdo de Competitividad 2012. C. A. Quintana Jiménez (ed), Actualización Octubre de 2012. 43 p.

Rahalkar, G. W., Harwalkar, M.R., Rananvare, H.D., Kurian, C., Abrham, V.A. and Abdulla Koya, K.M. (1977) Preliminary field studies on the control of the red palm weevil, *Rhynchophorus ferrugineus* using radio sterilized males. *Journal of Nuclear Agriculture Biology*. 6, 65-68.

Raigoza, J. (1974) Nuevos diseños de trampas para control de plagas en caña de azúcar (*Saccharum officnarum*). En: *Memorias Segundo Congreso de la Sociedad Colombiana de Enomología (Socolen)*. Cali, Colombia. Pp. 5 – 24.

Rajagopal, V., Kasturi-Bai, K.V., and Voleti, S.R. (1990) Screening of coconut genotypes for drought tolerance. *Oleagineux* 45,215–223.

Ramaraju, K. and Rabindram, R.J. (2002) Palmyra, *Borassus flabellifer* L. (Palmae); a host of the coconut eriophyid mite*Aceria guerreronis* Keifer. *Pest Management Horticultural Ecosystems* 7,149-151.

Ramirez, F. (1998) Reconocimiento de enemigos naturales y studio de la fluctuacion poblacional de *Rhynchophorus palmarum* L., y *Metamasius hemipterus* (Coleoptera: Curculionidae) Acacias, Meta.Universidad Nacional de Colombia, Thesis, Bogota, Colombia 84 p.

Restrepo, L., Rivera, F., and Raigosa, J. (1982) Ciclo de vida, hábitos y morfometría de *Metamasius hemipterus* Oliver y *Rhynchophorus palmarum* L. Publication E P R.

Rochat, D., Gonzalez, V.A., Mariau, D., Villanueva, G.A., and zagati, P. (1991a) Evidence for male-produced aggregation pheromone in American palm weevil, *Rhynchophorus palmarum*(L.) (Coleoptera: Curculionidae). *Journal of Chemical Ecology* 17, 1221-1230.

Rochat, D., Malosse, C., Lettere, M., Ducrot, P.H., Zagatti, P., Renou, M., and Descoins, C. (1991b) Male-produced aggregation pheromone of the American palm weevil, *Ehynchophorus palmarum* (L.) (Coeloptera: Curculionidae) : Collection, indentification, electophysiological activity, and laboratory bioassay. *Journal of Chemical Ecology* 17, 2127-2141.

Rochat, D., Nagnan-Le, P., Esteban-Duran, J. R., Malosse, C., Perthuis, B., Morin, J.P., and Descoins, C. (2000) Identification of pheromone synergists in American palm weevil, *Rhynchophorus palmarum,* and attraction of related *Dynamis borassi. Journal of Chemical Ecology* 26, 155-187.

Roda, A., Nachman, G., Hosein, F., Rodrigues, J.C.V., and Peña, J.E. (2012) Spatial distributions of the red palm mite, *Raoiella indica* (Acari: Tenuipalpidae) on coconut and their implications for development of efficient sampling plans. *Experimental and Applied Acarology* 57,291-308.

Rodrigues, J.C.V., and Peña, J.E. (2012) Chemical control of the red palm mite, *Raoiella indica* (Acari: Tenuipalpidae) in banana and coconut. *Experimental and Applied Acarology* 57, 317-329.

Rosen, D. (1990) Biological control: selected case histories. In D. Rosen [ed.], Armored Scale Insects, their Biology, Natural Enemies and Control.Vol.4B.World Crop Pests. Elsevier, Amsterdam, the Netherlands. pp 497-505.

Salama, H., Hamdy, M. and Magd El-Din, M. (2002) The thermal constant for timing the emergencenof the red palm weevil, *Rhynchophorus ferrugineus* (Oliv.), (Coleoptera, Curculionidae). *Journal of Pest Science*, 75, 26–29.

Sallam, A.A., El-Shafie, H.A.F. and Al-Abdan, S. (2012) Influence of farming practices on infestation by red palm weevil *Rhynchophorus ferrugineus* (Olivier) in date palm: A case study. *International Research Journal of Agriculture Science and Soil Science* 2, 370-376.

Sánchez, P., Jaffé, K. Hernández, J., and Cerda, H.(1993). Biología y comportamiento del picudo del cocotero *Rhynchophorus palmarum* L. (Coleoptera: Curculionidae). *Boletin de Entomologia Venezolana* 8, 83 – 93.

Sarkar, P.K., and Somchoudhury, A.K. (1989) Influence of major abiotic factors on the seasonal incidence of *Raoiella indica* and *Tetranychus fijiensis* on coconut. In: Channabasavanna GP, Viraktamath CA (eds) Progress in Acarology. Oxford and IBH, New Delhi, India, pp 59–65.

Schuiling.M. and van Dinther, J.B. M. (1981) Red ring disease in the Paricatuba oil palm estate, Para, Brazil. *Zeitschrift für Angewandte Entomologie* 91, 154-169.

Sebay, Y. (2003) Ecological Studies on the red palm weevil, *Rhynchophorus ferrugineus* Oliv (Coleoptera: Curculionidae) in Egypt. *Egyptian Journal of Agricultural Research* 81, 523-529.

Seguni, Z. (2000) Incidence, distribution and economic importance of the coconut eriophyid mite, *Aceria guerreronis* Keifer in Tanzanian coconut based cropping systems. In, Fernanado, L.C.P., de Moraes, G. J., and Wickramananda, I.R. *Proceedings of the international workshop on coconut mite (Aceria guerreronis), Coconut Research Institute, Sri Lanka*, January 2000, pp. 54-57.

Sharma, S. and Buss, E. (2013) Florida wax scale. Featured Creatures, University of Florida, IFAS, Entomology and Nematology.EENY-510. http://entomology.ifas.ufl.edu/creatures/orn/scales/florida_wax_scale.htm.

Siriwardena, P.H.A.P., Fernando, L.C.P., and Peiris, T.S.G. (2005) A new method to estimate the population size of coconut mite, *Aceria guerreronis*, ona coconut.*Experimental and Applied Acarology* 37, 123-129.

Soroker, V., Blumberg, D., Haberman, A., Hamburger-Rishard, M., Reneh, S., Talebaev, S., Anshelevich, L., and Harari, A.R. (2005) Current status of RPW in date palm plantations in Israel. *Phytoparasitica* 33, 97-106.

Srinivasan, N., Koshy, P.K., Amma, P. G., Sasikala, M., Gunasekaran, M., and Solomon, J. J. (2000) Appraisal of the distribution of coconut root wilt and heavy incidence of the disease in Cumbum Valley of Tamil Nadu. *Indian Coconut Journal* 31, 1-5.

Stocks, I. C. and Hodges, G. (2012) The rugose spiraling whitefly, Aleurodicus rugioperculatus Martin, a new exotic whitefly in south Florida (Hemiptera: Aleyrodidae). Florida Department of Agriculture and Consumer Service, Division of Plant Industry Pest Alert DACS-P-01745. http://www.freshfromflorida.com/pi/pest-alerts/pdf/aleurodicus-rugioperculatus-pest-alert.pdf.

Streito, J.C., Olivier, J., and Beaudoin-Olivier, L. (2004) Two new pests of Cocos nucifera L., from Comoro islands: *Aleurotrachelus atratus* Hempel, 1922 and *Paraylerodes bondari*Perachchi , 1971 (Hemiptera: Aleyrodidae). *Bulletin de la Société Entomologique de France* 109, 67-72.

Sudtharto, P.S., Sipayung, A., and De Chenon, R.D. (1990) Leaf eating caterpillars of secondary importance on oil palm, á potential risk in plantations.International conference on plant protection in the tropics. Genting Highlands, Pahang (Malaysia) 20-23 March 1990. Summary 1 page. Downloaded as: http://agris.fao.org/agris-search/search/search/display.do?f= 2012%2FOV%2FOV2012062090062

Sujatha, A., Chalam, M.S.V., Kalpana, M. (2010) Screening of coconut germplasm against coconut eriophyid mite, *Aceria guerreronis* Keifer in Andhra Pradesh.*Journal of Plantation Crops* 38, 53-56.

Sumalde, A. C. and Calilung. V. J. (1982) Life history of *Cerataphis palmae* Ghesquiére (Pempigidae, Homoptera) on coconut *Philippine Entomologist* 5, 273-290.

Sumano Lopez, D., Sánchez Soto, S., Romeron Napoles, J., and Sol Sanchez, A. (2012) Efficiency in capturing *Rhynchophorus palmarum* L., (Coleoptera: Dryophthoridae) with different traps designs in Tabasco, Mexico. *Fitosanidad* 16, 43-48.

Takhouk, A.S. (1991) on the management of the date palm and its arthropods enemies in the Arabian Peninsula. *Journal of Applied Entomology* 111, 514-520.

Tapia, G., Ruiz, M.A. and Tellez M.M. (2011) Recommendations for a preventive strategy to control red palm weevil (*Rhynchophorus ferrugineus*, Olivier) based on the use ofi nsecticides and entomopathogenic nematodes. *Bulletin OEPP/EPPOBulletin* 41, 136-141.

Taylor, B., Rahman, P.M., Murphy, S.T., and Sudheendrakumar, V.V. (2012) Population dynamics of the red palm mite (*Raoiella indica)* and phytoseiid predators on two host palm species in southwest India. *Experimental and Applied Acarology* 57, 331-345.

TDA (2008) Texas Department of Agriculture.Red Palm Mite, *Raoiella indica*Hirst. DIALOG.http://www.agr.state.tx.us/agr/main_render/0,1968,1848_27454_0_0,00.html?channelId=27454. Accessed May 2011.

Tiong, R.H.C., Heong, K.L., Lee, B.S., Lim, T.M., Teoh, C., H., and Ibrahim, Y. (1982) Oil palm pests in Sarawark and the use of natural enemies to control them. In:Henog, KL., Lee., B.S., Lim, T.M., and Ibrahim, Y. (eds.) *Proceedings of the international conference on plant protection in the Tropics*. 1-4 March, Kuala Lumpur, Malaysia, pp. 363-372.

Thurston, H (1984) Red ring disease of coconut . In: *Tropical Plant Disease*, pp. 161-164 American Phytopathological Society, St Paul (US).

Tuck, H.C. (1998) Safe and efficient management systems for plantation pests and diseases. *Planter* 74, 369-385.

Van Driesche, R.G., and Bellows, T.S.Jr. (1996) Biological control. Kluwer Academic Publishers, Massachusetts, USA, pp 539.

Vásquez, C., Quirós, G.M., Aponte, O., and Sandoval, D.M.F. (2008) First report of *Raoiella indica* Hirst (Acari: Tenuipalpidae) in South America. *Neotropical Entomology* 37,739–740.

Vera, H., and Orellana, F.(1986) Evaluación de atrayentes vegetales para la captura de adultos de "gualpa" (*Rhynchophorus palmarum*), insecto plaga de palma africana y cocotero. Estación Esperimental Santo Domingo. INIAP – Instituto Nacional de Investigaciones Agropecuarias. Quito, Ecuador. *Boletín técnico* 63,10 p.

Vesey-FitzGerald D. (1940) The control of Coccidae on coconut in the Seychelles. *Bulletin of Entomological Research* 31, 253-286.

Vidyasagar, P.S.P.V., Hagi, M., Abozuhairah, R. A., Al-Mohanna, O.E. and Al-Saihati, A.A. (2000) Impact of mass pheromone trapping on red palm weevil adult population and infestation level in date palm gardens of Saudi Arabia. *Planter* 76, 347-355.

Watson G. (2007) Invasive Insect Report 2004.06. Coconut mealybug *Nipaecoccus nipae* (Homoptera: Pseudococcidae). *University of Guam Cooperative Extension Service Knowledgebase Wiki.*http://guaminsects.net/uogces/kbwiki/index.php?title=IRR2004.06.

Wattanapongsiri, A. (1966) A revision of the genera *Rhynchophorus* and *Dynamis* (Coleoptera : Cuculionidae). Bangkok, Thailand. *Department of Agriculture Science Bulletin* 1, 328 pp.

Weintraub, P.G., and Beanland, L. (2006) Insect vectors of phytoplasms. *Annual Review of Entomology* 51, 91-111.

Wood, B.J., Liau, S.S. and Knecht, J.C. (1974) Trunk injection of systemic insecticides against the bagworm, *Metisa planta* (Lepidoptera: Pyralidae) on oil palm. *Oleagineux* 29, 499-505.

Wood, B.J. Piggot, C.J., Phang Sew and Nesbit, D. P. (1973) Aerial application in oil palms. *Oleagineux* 28, 275-282.

Zaher, M.A., Wafa, A.K., and Yousef, A.A. (1969) Biological studies on *Raoiella indica* Hirst and *Phyllotetranychus aegyptiacus* Sayed infesting date palm trees in U.A.R. (Acarina: Tenuipalpidae). *Z Angew Entomol* 63,406–411.

Zaid, A., P. F. de Wet, Djerbi, M. and A. Oihabi, A.(2002) *Diseases and pests of date palms.*In Date Palm Cultivation. Zaid A [ed.] FAO. Rome-Italy.

Zannou, I.D., Negloh, K., Hanna, R., Houadakpode, S., and Sabelis, M.W. (2010) Mite diversity in coconut habitat in West and East Africa. *XIII International Congress of Acarology*, Recife, Brazil, Abstract Book, p 295.

Integrated Pest Management in the Tropics; pp. 499-575

New India Publishing Agency, New Delhi (India)

CHAPTER - 17

Integrated Pest Management in Cotton

A K Dhawan
Former Professor and Head, Division of Entomology, Punjab Agricultural University, Ludhiana-141001, Punjab, India

17.1 Introduction

Cotton is a major fibre crop of global importance and has high commercial value. It is grown commercially in the temperate and tropical regions of more than 100 countries. Specific areas of production include countries such as China, USA, India, Pakistan, Uzbekistan, Turkey, Australia, Greece, Brazil, Egypt *etc.*, where climatic conditions suit the natural growth requirements of cotton. These include periods of hot and dry weather and adequate moisture obtained through irrigation. Cotton crop is considered as one of the most important cash crops that influenced the industrial and agricultural economy of the different cotton growing countries of world. Thus, cotton is an immensely important crop for the sustainable economy of these countries and livelihood of the farming community. Cotton crop is considered as one of the most important cash crops that influenced the industrial and agricultural economy of the different cotton growing countries of world. Cotton cultivation and textile industry provide livelihood to million persons, therefore any failure of cotton crop significantly influence the economic growth of cotton belt and textile sector. Out of about 50 species of cotton plants in the world, only four species *viz.*, *Gossypium arboreum, G. herbaceum, G.hirsutum and G. barbadense* along with intra- and interspecific hybrids, have been domestically cultivated for cotton fibres under diverse agro-climatic conditions along the diverse agro climatic conditions, varying from 8-32° N latitude and 70-80° longitude. *G. hirsutum* is the most commonly cultivated species of cotton in the world accounting for more than 90% of world fibre production. The area under cotton production in the world is estimated at around 30-31 million hectares. India has the largest area under cotton production but China is the largest producer of cotton in the world, whereas India is the second largest. Among the six major cotton growing countries in the world Brazil (2027 kg/ha) hold the highest productivity followed by China (1311kg/ha), USA (945 kg/ha), Uzbekistan (859 kg/ha), Pakistan (648 kg/ha) and India (478 kg/ha). India ranked first in terms of cultivated area followed by China, USA and Pakistan.

17.2 Insect Pest Complex

Cotton ecosystem provides home to about 1326 species of insects from sowing to maturity in different cotton growing areas of the world(Hargreaves, 1948). A relatively large number of pest species that are not susceptible to the *Bt* toxins expressed in transgenic cottons affect cotton production worldwide. The sucking pests, including the cotton aphid, thrips, whitefly, leafhopper, and spider mites, are the major non-target pests in *Bt*cotton fields, which are not susceptible to the *Bt* proteins currently used (Wu and Guo, 2005; Arshad and Suhail, 2010; Mann *et al.,* 2010). In general, most of these species exhibit the same pest status and continue to be managed identically in *Bt* and conventional cotton systems. However, due to the reduced use of insecticides for bollworms and the change of pest management regimes in *Bt* cotton fields, these secondary pest populations have increased and gradually evolved into key pests during the season in the USA, India, China, Australia, and other countries (Gouse *et al.* 2004; Williams, 2006; Wilson *et al.,* 2006; Lu *et al.,* 2008; Li *et al.,* 2010; Zhao *et al.,* 2011; Dhawan, 2011, Dhawan *et al.,* 2011, 2012). For example, in Australia, the green mirid (*Creontiades dilutus*), green vegetable bug (*Nezara viridula*), leaf hopper (*Austroasca viridigrisea* and *Amrasca terraereginae*), and thrips (*Thrips tabaci*, *Frankliniella schultzei*, and *Frankliniella occidentalis*) have become more prominent (Lei *et al.,* 2003; Wilson *et al.,* 2006). In India, the reduction in insecticide use during flowering phase increased the incidence of sucking and other pests, such as the mirid bug, mealy bug, thrips and leaf-eating caterpillar (Karihaloo and Kumar, 2009; Nagrare *et al*., 2009; Dhawan, 2011). Field surveys conducted over 10 years in six major cotton-growing provinces in northern China showed that mirid bugs have progressively increased and acquired pest status in *Bt* cotton fields (Lu *et al.,* 2010). In addition, it was also found that spider mites have been observed at higher levels in *Bt* cotton during the drought season (Wu and Guo, 2005). These emergent pests have forced Chinese farmers to use chemical pesticides, however, the increase in insecticide use for the control of these secondary insects is lower than the reduction in total insecticide use due to *Bt* cotton adoption (Wang *et al*., 2009).

In India, cotton crop is inhabited by 166 species of insect species and it is generally recognized that various arthropod predator species also play an important role in regulating pest populations (Khan and Rao, 1960). Before the introduction of *Btcotton*, eight were the key pests which cause major damage to the crop and pest status varies in different climatic zones due to different factors. Among these, cotton jassid [*Amrasca biguttula* (Ishida)], whitefly [*Bemisia tabaci* (Gennadius)], pink bollworm [*Pectinophora gossypiella* (Saunders)], spotted bollworms [*Earias vittella* (Fabricius) and *E. insulana* (Boisdual)], American bollworm [*Helicoverpa armigera* (Hubner)] are pest of national importance. The thrips [*Thrips tabaci* Lindman] cause moderate to heavy damage in Central India. Similarly, the phytophagous mites [*Tetranychus cinnabarinus* (Boisdual) and *T. neocaledonicus* Andre] are of prime importance in Central and South India. Stem weevil [*Pempherulus affinis* (Fabricius)], tobacco caterpillar [*Spodoptera litura* Fabricius] and thrips [*Scirtothrips dorsalis* Hood] are the key pests in South zone. Besides this, shoot weevil *Alcidodes affaber* Aurivillus has also established as serious pest in Karnataka. However, with introduction of BT cotton pest status has changed. Arthropod fauna was observed in cotton agro ecosystem in cotton belt of Punjab indicated 134 species of arthropods, including Hymenoptera (23.9%), Hemiptera (19.4%), Coleoptera (16.4%), Lepidoptera (14.2%), Orthoptera (8.2%), Diptera and spiders (4.5% each), Odonata (2.9%), Dictyoptera, Isoptera and mites (1.5% each) and Neuroptera and Thysanoptera (0.7% each) were found to be associated with the cotton crop (Bal and Dhawan, 2009b,2009c, Dhawan *et al*., 2011) (Table 17.1). Fifty-four species of herbivorous insects and mites were recorded from the cotton crop. Natural enemies accounted for one third of the total arthropod fauna recorded on cotton crop. With introduction of *Btcotton*, bollworms are not serious pests as *Bt*cotton provides effective control of this group of pests. The jassid damage has increased and new sucking pests like mealy bugs, and mirid bugs damaged the cotton crop in recent past (Dhawan *et al.* 2007, 2011, 2012). The tobacco caterpillar is another new pest that is has gained importance (Arora *et al.,* 2007). But with introduction of stacked Bollgard II, the status has declined to minor pest.

In India, pest outbreaks are of common occurrence in cotton ecosystem. The first outbreak on cotton crop was of whitefly in 1932-33 in undivided Punjab (Hussain and Trehan, 1933) and thereafter in 1990's in Andhra Pradesh and Punjab, Harayana, Rajasthan and Gujarat. Outbreaks of *Helicoverpa armigera* were observed in Maharashtra, Gujarat, Andhra Pradesh, Punjab, Haryana, Rajasthan and Madhya Pradesh in recent years. Similarly, outbreaks of *Spodoptera* in recent years in Central (Gujarat, Maharashtra) and Southern zone (Andhra Pradesh) and of whitefly in north zone (Punjab, Haryana and

Rajasthan), central zone (Gujarat, Maharashtra) and south zone (Andhra Pradesh) were observed as result of deleterious effect of insecticides. Similarly, there was early season outbreak of leaf miner in northern Karnataka and some parts of Tamil Nadu. The mealy bug (*Phenococcus solenopsis* Tinsley) in whole cotton belt and mirid bugs in South zone has gained importance as pest. *Bt*cotton is effective against bollworm complex and also provides effective control of *Spodoptera*. The incidence of jassid has increased and requires management during flowering phase. The whitefly population declined after introduction of *Bt*cotton and now again is serious concern for cultivation of *Bt* cotton. Mealy bug population has declined and mirid bug population is building up in some area. The population of aphid late in season and thrips early in season is showing upward trend. The sporadic pests like semilooper and leaf folder has also declined but cotton stainer population has increased (Dhawan, 2012).

Bt cotton increased the diversity of arthropod communities (Men *et al.* 2003) but the biological control function of natural enemies in *Bt* cotton fields did not change compared with conventional cotton (Narnajo 2005, Wolfenbarger *et al,* 2008). Research in Australia examined the diversity and species richness of invertebrate communities present in unsprayed *Bt* and conventional crops. Across a number of specific invertebrate categories, Fitt and Wilson (2002) found no significant impacts of *Bt* cotton. The analysis of whole community indicated that there were small differences between the communities in *Bt*cotton and conventional cotton, but these differences peaked at different times during the season, and the timing of differences was not consistent between years. *Bt* increases the diversity and stability of cotton agro ecosystem Many workers observed no difference in species richness, evenness and diversity in *Bt* and conventional cotton under unsprayed conditions (Dhawan and Bal 2007, Bal and Dhawan 2009a, 2009b, 2009c, 2009d, 2010).

Many insect pests cause severe crop losses, the cotton bollworm *Helicoverpa armigera,* bollweevil *(Anthonomis grandis),*whitefly *(Bemisia tabaci)* has been the main focus of attention because of potential to cause heavy losses. Cotton bollworm and whitefly has been recorded from different plant species including cotton. It was found to cause yield losses of up to US$290–350million in India annually (King 1994). There have been several estimates of crop losses in various parts of the world. Fitt (1994) reported cotton crop loss of 7.7% in Queensland, Australia despite an expenditure of US $ 4.2 million for its control. In Punjab (India) in 1998-99, cotton production was $US 45 million against use of insecticides of $US 50 million. The infestation of mealybug caused 42.21 per cent reduction in yield with monetary loss of $US 4595-7483/ ha in different *Bt* hybrid (Dhawan *et al.,* 2007). Several efforts have been made all over the world to minimize crop losses due to major pests of cotton with focus on sustainable control of the bollworms.

17.3 Historical Perspective of Cotton IPM

Insect pests have always limiting factor in cotton production and also governed the socio-economic status of cotton growers. Economy of many cotton growing countries is based on the success of cotton crops. This is only source of livelihood for small and medium cotton growers in developing and underdeveloped countries. Cotton being a commercial crop was always subjected to intense human interventions to protect it from damage of pests which damage the crop from sowing to harvest in differ tropical cotton growing. Pests are very serious in tropical counties due to better environment for survival and development of different pests. The cotton crop has rich diversity of insects in each cotton agroecosystem in Africa, Asia, and America and there are more than 100 insect species, but only a dozen of them are of economic importance. To manage the pests intervention start from sowing to harvest to manipulate the environment to reduce the damage. However, for the last six decades the dependence is on use of insecticides to reduce the losses. The pest spectrum on cotton before 1980 comprised mainly of the cotton jassid, pink bollworm, spotted bollworm as the major insect pests. The whitefly, *Spodoptera litura* were pests of less economic importance. The American bollworm *Helicoverpa armigera* was mentioned as 'not a regular or a serious pest' of cotton in India.

The term Integrated Pest Management is now more or less universally understood for sustaining productivity in cotton. The use of different strategies to manage the pests in cotton and reduced the use of pesticides was explained by citing some well documented historical cases before the introduction of concept of integrated pest management. The main reliance on the use of pesticides resulted in simplification of cotton agroecosystem and this resulted in appearance newer pest problems and more frequent outbreaks of pests in the cotton crop. Due to lack of resistant cultivars, non-adoption of cultural control measures, and non-availability of effective biocontrol agents, the indiscriminate use of insecticides resulted in development of resistance to insecticides in cotton pests such as American bollworm [*Helicoverpa armigera* (Hubner)], cotton jassid (*Amrasca biguttulla* (Ishida)), resurgence of pests such as spider mites (*Tetranychus* spp.), cotton jassid [*Amrasca biguttulla* (Ishida)] and whitefly (*Bemisia tabaci* (Gennadius) and destruction of natural enemies, which ultimately led to crop failures in some countries. Such failures in cotton production systems were documented in Latin America (Can˜ete Valley, Peru), Sudan and other places even before the term IPM was coined. Can˜ete Valley, Peru had been a successful cotton growing area with progressive farmers. In 1939, the tobacco bud worm [*Heliothis virescens* (Fabricius)] appeared in cotton crops. The spraying of arsenical insecticides and nicotine sulphate resulted in build-up of cotton aphid [*Aphis gossypii* (Glover)] and worsening of the tobacco bud worm problem. By 1949, cotton yields (lint) dropped from about 500 kg ha^{-1} to 365 kg ha^{-1} as natural enemies disappeared due to insecticide applications resulting pest populations to resurge after sprays were applied. The failure of cotton in undivided India due to whitefly is well documented (Husssain and Threhan 1933). Cotton production in Sudan also suffered due to over reliance on insecticides. DDT induced outbreaks of cotton whitefly, *Bemisia tabaci* (Gennadius) and the use of parathion against this pest increased the occurrence of cotton bollworm [*Heliothis armigera* (Hubner)] which resulted in reduction in yields (Joyce and Roberts 1959). A key feature in the history of IPM is that the concept was first articulated by scientists from the Entomology Department at the University of California, USA. In the 1950s scientists initiated the development of a new pest management strategy which brought applied ecologists and bio-control experts together (Perkins 2002). Up to this time, applied entomology in the USA had largely been taken over by a toxicology mind-set to find the right poison. This concept aimed to integrate the use of biological control with chemical control and was the beginning of IPM in the USA (Smith and Allen 1954; Perkins 2002). This early concept was based on the premise that pesticides could have a minimum impact on the natural enemies of the pest if applied at the correct time and under the correct conditions. Economic threshold, another important concept in IPM, was introduced at that time (Stern *et al.* 1959) and was the first attempt at providing a rational basis for deciding if a pest population based control, based on the value of expected loss from damage and the cost of control. Their major insight in creating IPM in fact rested upon their realization that the best suppression practices lay in preserving natural enemies and using the new insecticides only when needed to supplement the suppressive effects of natural enemies. In other words, they developed "integrated control" to use chemicals only if needed and in ways that did not decimate populations of natural enemies. This judicious use of insecticide also helped avoid the problems of resistance. The focus of IPM began to shift to non-pesticidal tactics in the 1980s, including expanded use of cultural control, introduction of resistant variety and biological control. In Asia, the Farmer Field School (FFS) approach for disseminating the IPM technology was adopted and FFS has become a preferred extension methodology for implementing IPM programs in Africa, Latin America, Caribbean and Eastern Europe. The concept of integrated control was introduced including judicious the use of synthetic organic pesticides, the reintroduction of beneficial insects, crop diversification schemes, planting of early maturing varieties and the destruction of cotton crop residues. Pest problems subsequently declined dramatically and pest control costs were substantially reduced in tropical and subtropical cotton growing areas. The cotton cultivation has passed to different phases of pest management during last eight decades in India.

17.3.1 Subsistence phase (1930-1960)

Before the discovery of DDT, the pest management stagy was based on of traditional agronomic practices such as use of pest tolerance varieties, timely sowing, destruction for infested plants/plant residues, low use of fertilizers, use of plants products etc. The pest spectrum comprised mainly the cotton jassid and pink bollworm during vegetative and reproductive phase of cotton crop, respectively. The spotted bollworm and whitefly were minor pests. *Spodoptera litura* and American bollworm (*Helicoverpa armigera*)presence in cotton agroecosystem was negligible.

17.3.2 Exploitation phase (1960-1985)

With advent of DDT followed by chlordane, aldrin, dieldrin, heptachlor the era of insecticide use in cotton dominated the pest management options. The impact of DDT on pest management was unmatched with other synthetic products. This was followed by organophosphates (monocrotophos, fenitrothion, quinalphos, chlorpyriphos, profenophos, dimethoate, formothion, phophamidon, phosalone, oxy demeton methyl, acephate, methyl parathion etc.), carbamates (carbosulfan, carbaryl, thiodicarb, methomyl) pyrethroids (cypermethrin, deltamethrin, fenvalerate, permethrin, λ-cyhalothrin, fluvalinate, fenpropathrin, bifenthrin, etc.) and formamidines (chlordimeform and amitraz). The sprays were based on fixed calendar at 15 and thereafter at 10 day interval. The economic threshold were defined but were not adopted. The concept of Integrated Pest Management was introduced to encourage the judicious use of insecticide, but major emphasis was on the use of insecticides. The synthetic pyrethroids provided excellent control of bollworm complex and the yield potential of varieties increased by 50 % with better control of pests on cotton.

17.3.3 Crisis phase (1985-1995)

The excessive and indiscriminate insecticide use representing 'exploitation phase' was followed by 'crisis' phase in cotton. , The use of high yielding and pest susceptible varieties along with higher use of energy subsides like fertilizer and water initially provided better return but thereafter leading to problems of insecticide resistance, pest resurgence, accumulation of harmful residues and toxicity to non-target organisms. Cotton IPM programmes have been built around cultural control, biological control and biopesticide interventions in many parts of the world but did not provided effective control of pests due to failure of insecticide as component of IPM.

17.3.4 Disastrous phase (1995-2000)

The decade 1995-2000 was dark era for cotton pest management. Excessive use of insecticides, especially synthetic pyrethroids, led to problems of insecticide resistance in *Helicoverpa armigera* and *Spodoptera litura,* which further resulted the repeated application of insecticides mostly as combination of insecticides. The resurgence of whitefly further complicated the pest management options in cotton. There was shift in pest scenario as American bollworm emerged as serious pest and was difficult to managed due to high level resistance to all group of insecticides(Mehrotra and Phokela 1992, Armes *et al.*, 1996; Sekhar *et al.*, 1996,Kranthi *et al.*, 2001a, 2001b, 2002a and 2002b). The *H. armigera* resistance to insecticides had emerged as a great challenge to cotton pest management in Asia and Australia. Subsequently non-insecticidal alternative methods of eco-friendly pest management were developed. Thus cotton IPM programmes were built around cultural control, biological control and biopesticide interventions in many parts of the world.

17.3.5 Reviving cotton cultivation (2000 onwards)

After heavy losses due to pests, a number of IPM programmes, based on a combination of cultural control methods, management of natural enemies and applications of biopesticides/bioagents interspersed with need based application of insecticides, were initiated all over the world in all cotton growing countries to ensure effective bollworm management. The non availability of quality biopesticides and biological control organisms, coupled with sub-optimal efficacy under field conditions was major constraints in adoption. The cotton cultivators had to depend on insecticides. Since, American bollworm resistance to insecticides had emerged as a major threat to pest control programmes, thereby rendering insect pest control ineffective, inefficient and unsustainable. IPM packages were refined to include IRM (Insecticide Resistance Management) as a major component. IRM was more relevant to the management of crisis caused by insect resistance to insecticides. Therefore specific IRM programmes were designed for regions affected by severe resistance problem.

After failure of cotton in major cotton growing countries of world like India, Pakistan and China the pest management changed drastically after 2000 all over the world under pressure from farmers and policy planners. The various IPM option were validated, demonstrated and many training programme were initiated to transfer the technology to farmers and build confidence of cotton growers in cultivation of cotton. In India and Pakistan, area under cotton has no suitable alternate crop and has to depend on cotton. They suffered huge losses and this shatter the economic condition of cotton belts. With the introduction of novel eco-friendly insecticides highly effective on bollworms, *Helicoverpa armigera*, era of reviving the cotton cultivation started. Insecticides such as spinosad and indoxacarb were highly effective against *H.armigera* and insecticides like novaluron and emamectin benzoate ensured effective control of *Spodoptera* and moderate control of *H.armigera.* All these insecticides were_less toxic to beneficial and non target insects in the cotton ecosystem. Unlike synthetic pyrethroids, none of these insecticides caused the resurgence of whitefly and subsequently the whitefly which emerged as serious pest did not pose any threat to cotton cultivation. The *H. armigera* infestation ceased significantly in cotton ecosystems since after 2000 due to various reasons. Additionally, the chloronicotinyl (imidacloprid, acetamiprid and thiomethoxam) replaced the dimethoate, oxydemeton methyl and phophamidon for control of cotton jassid, which are selectively more effective on the sucking pests and less toxic to beneficial insects as compared to all the conventional insecticides added to sustainable pest management. *Bt* cotton was introduced as component of IPM in different cotton growing areas of world from 1996-2006, which provided effective control of bollworm complex. The chloronicotinyl insecticides were used as_foliar sprays against cotton jassid before the introduction of *Bt* cotton. Thereafter, the Bt cotton seed marketed by various seed companies was treated with chloronicotinyl. In North India and also in other part of cotton growing, the jassid damaged the crop after 50-60 days of sowing when chemical lost the effectiveness. However, the growth promoting effect of these insecticides helped the farmers to have 8-10 percent higher yield. This resulted in development of resistance to these chemicals against jassid and farmers are not getting effective control even with double the approved dosages. In some cases sucking pests such as jassid continued their damage despite_seed treatment. By the late nineties insect resistant GM (genetically modified) *Bt*-cotton was introduced for bollworm control in major cotton growing countries such as the_USA, Australia, and China. *Bt*-cotton was introduced in India in 2002. The technology was so potent that within 3-4 years of its introduction, the area under *Bt*-cotton increased to more than 50% in almost all the countries. After the introduction of *Bt*-cotton, due to the reduction of insecticide_sprays, especially during flowering and boll formation phase, some minor pests(*Spodoptera litura*, mealybugs, miridbugs, thrips, jassid, weevils *etc.*), which are not_susceptible to Cry1Ac showed resurgence in many parts of the world. Resistance management strategies have been revised from time to time in light of the_introduction of *Bt* cotton and new insecticides.

17.4 Integrated Pest Management Options

17.4.1 Host plant resistance

Host plant resistance is key component of IPM. Several efforts have been made to identify pest resistant genotypes that could be used in plant breeding programmes to develop resistant cultivars. Attempts have been to have cultivars resistant to early season pests to delay the insecticide application and conserve the natural enemies. The cotton grown in the world today is American uplands or their derivatives. In reviewing the history of cotton variety development in the United Sates, indicated that there are 17 sources which provided the parental germplasm of the modern-day American Uplands. Of the four principal Upland types cultivated worldwide, the Coker, Deltapine, and Stoneville types have a common ancestor in the Bohemian variety which dates back to 1860. The Acala cottons have a different ancestry, tracing back to direct introductions from Mexico into the United States in 1907. Certain varieties of the Acalas developed in recent years represent some infusion of germplasm from "eastern" types. Despite the apparently narrow germplasm base of the Uplands, they retain a surprising degree of genetic plasticity. Most of the current improvement efforts with the Uplands involve direct selection or intervarietal crossing, and little of the available exotic materials have been incorporated successfully into practical breeding programme. Yet there is increasing interest by breeders in utilizing a broader range of germplasm in improvement programme, especially pest resistance is major objective. In old world India, Central and East Africa, Iran and Afghanistan and in the New World, Southern Mexico and Central America are considered the centers of variability for *G. hirsutum*; northern South America, the West Indies and Central America are centers of variability for *G. barbadense*.

Research on insect resistance in cotton is limited to breeding of varieties resistant to sucking pests particularly the cotton jassid and whitefly. The resistant sources are available for developing jassid and whitefly resistant varieties. However, the resistant sources for cotton bollworms are not available except for few wild species which show small degree of tolerance to bollworms (Table 17.2). Wild species of cotton namely *G. davidosnii* (aphids), *G. rainmodii* and *G. tomentosum* (thrips), *G. armourianum* (whitefly), *G. somalense* and *G. thurberi* (bollworms) are known for possessing genes for resistance against cotton pests (Pandya and Patel, 1964; Sundramurthy, 1991). In addition to *G. somalense* and *G. thurberi, G. armourianum* and *G. rainmodii* are also known for resistance genes to bollworms. In general, cotton varieties belonging to *G. arboreum* and *G. barbadense* species have been reported as resistant sources for spotted bollworms (Butani, 1974). Similarly, varieties belonging to Cernnum and Bengalense races were found resistant to spotted bollworms (Duhoon and Singh, 1980). Screening of cotton genotypes for their reaction to key pests was an important programme. Several useful materials were identified and utilized in the breeding programmes. Many cultivars resistant/tolerant to cotton pests with economically viable productivity have been released for cultivation at farmer's fields In India (Table 17.3). Both morphological and biochemical mechanisms in *Gossypium* spp. have been found to mediate resistance against jassid, whiteflies and the bollworm. For instance, the pubescent varieties of cotton are highly resistant to cotton jassid but are susceptible to whitefly. Similarly, the glabrous leaved cottons greatly hamper oviposition by *H. armigera* and *B. tabaci*. Therefore, under such situations it is always desirable to lay more emphasis on exploiting other common characters which may be important for conferring resistance against majority of the insect species in a system. For example, in developing resistant varieties against cotton whitefly, major stress should be laid to select varieties having thin leaf lamina with high hair density. The end product of such studies will be helpful in reducing the damage due to both whitefly and jassid. These characters will be of vital importance for evolving transgenic cotton where these pests are likely to be of economic importance. Biochemical features such as high levels of gossypol, phenol, tannin and heliocides in squares and bolls have been found to impact host plant resistance to bollworm significantly (Table 17.4). While these features have been commercially exploited to a certain extent in the genotypes of the US (Jenkins, 1995), its utilization in the Indian context has been limited. Whitefly resistant varieties Kanchana, LK 861 and Supriya; the jassid resistant varieties, DHY286, Mahalakshmi, MCU15, Krishna, and Sujata;. Abhaditha – a bollworm resistant variety and Sahana, a bollworm tolerant cultivar were developed in India. In Pakistan, the main focus has been on developing early maturing insect resistant cultivars for pest resistance to escape late season whitefly and bollworm attack (ecological resistance). Insect pest resistant varieties can offer significant advantage in pest management programmes by reducing the need for pesticide applications but this has not been widely and effectively taken into account in cultivar breeding programmes in India (Dhawan 2004).

17.4.2 Natural enemies complex

Diversity of natural enemies is extremely important for stability of cotton ecosystem to have sustainable cotton production and to avoid outbreaks. Thus to understand the existing diversity among arthropods and pathogens population representing natural enemy complex in cotton is crucial to mitigate the ecological consequences of chemical based control and transgenic cotton as component of IPM. Vast difference exists in natural enemy complex in various cotton growing countries of world due to different insect pests and tools of integrated pest management. However, information on these is scanty. In general braconids, chrysopids, coreids, coccinellids, ichneumoids, syrphids, trichogrammatids and vespids are common entomophagous insect in cotton ecosystems. Cotton crop suffers from multiple pest attack from sowing to picking does not provide a sustainable niche for the long term establishment of natural enemies. Moreover, cotton has an indeterminate growth habit which provides a continuous source of food for wide range of insect pests throughout the season, necessitating insecticide use. Because of this reliance on chemicals to control a range of insect pests, cotton has not been a favourable environment for conservation of even naturally occurring beneficial organisms. However, biological control, being generally slow-acting in nature does not prevent larvae from damaging floral forms, resulting in losses. In the majority of the cotton cultivating countries the market availability of quality bio-control agents has been very poor and the production is very cumbersome and uneconomical. Over the past decade, results with commercial releases of bio-control agents in cotton have not been consistent and, despite the claims of being economic feasible, have not been popular with farmers. The efficacy that can be at best described as marginal and inconsistents, has not been enough to convince farmers to undertake their sustained use.

Parasitoids: Kranthi and Russell (2009) presented comprehensive review of the important naturally occurring parasitoids which includes:

***H. armigera**: Trichogramma chilonis* (Ishii), *Chelonus curvimaculatus* (Cameron)., *Campoletis chloridae* (Uchida), *Palexoristà laxa* (Curran), *Eucarceliaillota* (C.) and *Goniopthalmus halli* (Mesnil). Some major predators include *Geocoris ochropterus* (Fabricius).,*Coranus spiniscutis* (Reuter), *Chrysoperla carnea*(Stephens); *Orius* spp., *Polistes* spp., *Chilomenes sexmaculatus* (Fabricius) and spiders (*Oxyopes* spp., *Clubiona* spp. and *Thomisus* spp.).

Pink bollworm, *Pectinophora gossypiella* (Saunders):*Apanteles angaleti* (Mues.); *Chelonus* spp. and *Camptothlipsis* spp.

Spotted bollworm, *Earias vittella* (Fabricius) and the spiny bollworm, *Earias insulana* (Boisduval): *Trichogramma chilonis* (Ishii),*Apanteles angaleti* (Mues.) and *Rogas aligarhensis* (Q.).

Foliage eating lepidopterons pests: {The cotton leaf worm, *Spodoptera litura* (Fab.), cotton semilooper, *Anomis flava* (Fabricius) and leaf folder, *Syllepte derogata* (Fab.)}: *Trichogramma chilonis* (Ishii), *Glyptapanteles phytometrae* (Wilkinson), *Palexorista* spp., *Sysiropa Formosa* (Mensal) and *Carhops bicolour* (Czepligeti). Amongst the several parasitoids of the leaf folders, the most important ones are, *Apanteles significans* (Walker); *Phanerotoma syleptae* (Zettel), *Elasmus* spp., *Eurytoma syleptae* (Ferriere) and *Xanthopimpla punctata* (Fabricius).

Bemisia tabaci: *Encarsia* spp. and *Eretmocerus* spp. (Dhawan *et al.*, 2007)

Phenacoccus solenopsis Tinsley:*Aenasius bambawalei* Hayat (Rishi Kumar *et al.,* 2009).

In India, thirty four different species of egg (9), larval (19), egg-larval (2) and pupal parasitoids (4) are recorded from bollworms (pink, spotted and American bollworms) which cause parasitization of host to variable degree (Table17.5). The indigenous parasitoids are *Trichogramma achaeae* Nagaraja and Nagarkatti,

T.chilonis Ishii, *Bracon greeni* Ashmead, *B.hebator* Say, *B.brevicornis* Wesm., *Apanteles angaleti* Mues, *Chelonis* sp, *C.narayani* Subha Rao, Campoletis *chloride* Uchida and *Rogas aligarhensis* Quadri. *T.brasiliensis* Ashmead, *Bracon kirkpatricki* (Walk.) and *Chelonis blackburni* Camron are exotic parasitoids. The use of insecticides has reduced the population of parasitoids in cotton crop. Dhawan (1999a) observed only 0.78 percent parasitization of bollworms against 4.60 percent in 1975. Now with introduction of *Bt* cotton the total loss is expected due to non availability of host of these parasitiods.

Predators: Many general predators like coccinellids, chrysoipds, predatory bugs and spiders are recorded from cotton crop (Table 17.6). Amongst the many sucking insect pests that attack the cotton crop in its initial stages, jassids, *Amrasca devastans* (Distant), aphids, *Aphis gossypii* (Glover), whiteflies, *Bemisia tabaci* (Gennadius) and thrips (*Thrips tabaci* (Lindeman) are economically the most important in early vegetative phase and predators such as *Chrysoperla carnea* (Stephens) and *Geocoris ochropterus* (Fieber), *Chilomenes sexmaculatus* (Fabricius), *Scolothrips indicus* (Priesner) and *Scymnus* sps. can keep sucking pest populations under economic threshold levels if not disrupted with broad-spectrum insecticides. Among these chrypoids are most commonly exploited for management of cotton pests. In Punjab (India) the population of predators declined by 68.36 percent during 1975 to 1995. After introduction of *Bt* cotton there is slight increase in predator population (Table 17.7). In Uzbekistan, *Chrysopid* (Table 17.7) *carnea* and *Coccinella septempunctata* contribute to effective control of early season sucking pests (Jones *et al.,* 2000). First instar larvae hatching from egg masses of the cotton leaf worm, *Spodoptera littorals* in Egypt, are heavily_preyed upon by *Vesper/ Vistula*_wasps provisioning nests close to the cotton fields. Almost in all cotton growing countries, *Chrysoperla* and *Coccinella* and *Orius* species, nabbed bugs and spiders consistently predate on the eggs of number of pests.

Since a level of natural control exists in the ecosystem without the_need for human intervention, it is important, where practical and feasible, to design strategies aimed at conserving their populations with judicious and sensible use of selective insecticides. Van den Berg *et al.* (1993) showed that *Rheidol* ants frequently caused high mortality in *H. armigera* in Kenya. However, in spite of the evidence for the role of all_these natural enemies in reducing pest populations, the action of *Solenopsis*_fire ants_on boll weevils in Texas is almost the sole example of demonstrated irreplaceable_mortality caused by a key predator (Fillman and Sterling, 1983; Sterling, 1984). Kranthi *et al.* (2005) observed that avoidance of organophosphate insecticides for the first three months helps in build-up of entomophages populations such as *Chrysoperla, Campoletis chloridae, Microchilonis curvimaculatus* and Tachinids, which contribute to the management of *H. armigera*. First instar larvae hatching from egg masses of the cotton leaf worm, *Spodoptera littorals* in Egypt, are heavily preyed upon *Vespa/Vespula* wasps provisioning nests close to the cotton fields.

Releases of Parasites and Predators

A number of countries have attempted augmentative biological releases for bollworm control mostly with *Trichogramma*, with variable success. The species depends on the region of the world *e.g. T. pretiosum* Riley in the New World, *T.pinto* Vogel in Uzbekistan, *T. chilonis* Ishii in India. In 1992, in Uzbekistan, *T. pinto* Vogel and *Bracon hebetor* Say were applied to cotton fields for control of *H. armigera*, contributing to the decline in the insecticide use from 60,000 to 2000 tonnes during from 1975 to 1992 (Matthews 1993). In the USA, a number of on station trials had been conducted but the technology was not adopted in IPM programmes as it was not cost effective cate, 1985. Inadequate control with *Trichogramma* was reported from Queensland in Australia (Twine and Loyd, 1984) and hence its use was not encouraged in cotton IPM programmes. Some of the major reasons for the low control efficacy is poor searching ability, negligible recovery, and weak adaptability to temperature. The detailed examinations have been done on the actual mortality of pests resulting from *Trichogramma* releases, the results have generally been disappointing, to the extent that even major programmes like that in the southern USA have been abandoned (King *et al.,* 1985). Jones *et al.* (2000) found searching ability and heat tolerance to be major problems with factory reared *T. pintoi* in Uzbekistan and similar problems are evident in India. Rameis and Shanower (1996) reviewed the status of parasites and predators of *H. armigera* in India and concluded that the impact of classical or augmentative releases on pest numbers was very modest. *Chrysopa/Chrysoperla* lacewings, especially *C. carnea* (Stephens) and *Chrysoperla externa* (Hagen) but also *Chrysoperla chaquensis* (Navas) in Argentine are important in cotton (Polak *et al.* 2000).

Different experiments have been conducted on the use of natural enemies for the management of cotton in India. In India a trials on research stations were reported promising (Dhandapani *et al.* 1992) but records of field successes are rare. In Maharashtra, 7 releases of *T. brasiliensis, A.angaleti,* and *C. blackburni* at 10-days interval from 45 days age of crop reduced the infestation of bollworms by 55.5 per cent (Raodeo *et al.,* 1983). Mass releases of *T. chilonis*and *T. achaeae* @ 40,000 to 50,000 /ha and larval parasitoids, *C. blackburni* and *B. kirkpatricki*@ 15,000 to 20,000/ha during initial stage of formation of fruiting bodies, reduced the incidence of all three species of bollworms (Agarwal and Gupta, 1986). Inundative releases of *T. brasiliensis, B. kirkpatricki* and *C. blackburni* at 7-10 days interval parasitized 37-89 and 64.7-72.0 per cent population of bollworms (pink and spotted bollworms) in

Haryana and Punjab, respectively (Singh 1994a). *R. aligarhensis a* larval parasitoid of *E. vittella* parasitized 9.3-35.5 per cent larvae in Gujarat with peak activity in Ocober-November. *T. chilonis* has been reported to be effective against bollworms in the early phase of crop growth, *i.e.*, at low height of crop in Gujarat. The cultivation of semi-dwarf varieties like AKH 81, Anjali (LRK 516- 70 E) and dwarf hybrids (DHB 1 and CD HB 2) may, therefore, enhance the effectiveness of the parasitoids (Sundramurthy and Chitra 1992). The inundative releases *T. chilonis* @ 2 lakh/ha or *Chrysopa scelestes* Banks at 1 lakh/ha at weekly interval during peak oviposition period of pest have been suggested for control of *H. armigera* (Singh 1994b). In another studies, 11 inundative releases of *T. chilonis* @ 1 lakh/ha/week proved effective in early part of season but were not effective late in the season. In Punjab, eight releases of *T. chilonis* @ 1.5 lakh /ha/ release at 10-day interval during August-October coupled with 3 sprays was found effective in bio-intensive IPM of cotton (Verma and Gill, 1992). In large field trials, releases of *T.achaeae* and*T.chilonis* at 1-2 lakh/ha/release at 7-10 day interval during effective boll formation period on *hirsutum* cotton for control of pink and spotted bollworms were significantly better than untreated plots for the management of bollworms and in increasing the seed cotton yield. The parasitoids alone were, however, less effective than insecticides (Gill *et al.,* 1992). At Anand Gujarat, India, 8 induative releases of *T.chilonis* @ 1,00,000 /release/ha and 3 of *C.carnea* @ 1,00.000/ha or need based application of NPV of *H.armigera* and *Spodoptera litura* gave encouraging results (Thakur *et al.,* 1991).

For biological control interventions to be effective, continuous releases are required, which is neither possible nor economically feasible. In practice there has been almost no establishment of introduced natural enemies into cotton ecosystems worldwide (Russell, 2004).

17.4.3 Insecticide in pest management

Over the past five decades cotton cultivators had to rely on the conventional groups of insecticides such as organ chlorines (DDT, BHC), cyclodienes (aldrin, dieldrin, endosulfan), organophosphates (monocrotophos, quinalphos, chlorpyriphos, profenophos, dimethoate, phosalone, oxydemeton methyl, acephate, phorate, methyl parathion *etc.*), carbamates (carbosulfan, carbaryl, thiodicarb, methomyl) pyrethroids (cypermethrin, deltamethrin, fenvalerate, λ-cyhalothrin, *etc.*) and formamidines (chlordimeform and amitraz). Many insecticides have been used for *H. armigera* control over the years. Not all were equally effective, many have impacts on others insects pest and beneficial complex and many are to a greater or lesser extent harmful to human health. In the late 1990s four chemical classes dominated cotton crop protection in Asia, the synthetic pyrethroids, the organophosphates, the cyclodiene and the carbamates. Mixtures of insecticide with two or more active ingredients are widely used in Asian pest control programmes where different insect pests are being targeted simultaneously (*e.g.* sucking pests and lepidopterons) this may sometimes be justified. However, the use of mixtures for the control of caterpillars alone, especially *H. armigera* was more problematic.

Era of chlorinated hydrocarbons : The management of cotton pests with advent of DDT depend on chlorinated hydrocarbons till mid 1970s in different cotton growing area of the world. Organochlorines and cyclodienes (endrin, aldrin, dieldrin, endosulfan DDT, BHC, toxaphene, endosulfan) were used extensively to manage the sucking and bollworms mainly pink and spotted bollworm. Normally 5-7 sprays were used to manage the pests but the losses were 20-30 percent_in spite of the best management practices. The dominant pests were pink bollworm and jassid.

Era Organo phosphate : The most commonly used organophosphate were dimethoate, formothion, phophamidon, cabofuran, phorate for control of sucking pests. Monocrotophos, quinalphos, phenthoate, dicrotophos, chlorpyriphos, fenitrothion, profenophos, phosalone, acephate, and methyl parathion were used for management of bollworm complex on cotton. Monocrotophos and acephate were most widely used insecticides. The dominant pests were pink and spotted bollworms, cotton jassid. The minor pests were whitefly, leaf folder, semilooper *etc.*

Era of Carbamates : Carbaryl was introduced during 1970 and was followed by carbosulfan, thiodicarb, methomyl. Among these carbaryl was dominant insecticide and was in use till mid 1990s for control of bollworm complex.

Era of Pyrethroids : Fenvalerate, cypermethrin, permethrin and deltamethrin were first group of synthetic pyrethroids introduced for management of bollworms. Followed by fenpropathrin, fluvalinate, bifenthrin, lambda-cyhalothrin, beta cyflutrin etc. Among these fenvalerate, cypermethrin, and deltamethrin were widely used for management of bollworms and replaced the organophosphates and chlorinated hydrocarbons. The synthetic pyrethroids were introduced into India and several other countries in 1980-85 to control the bollworms and *Spodoptera* spp. and were highly effective at low dosages. They provide control of bollworms for longer periods (10-21 days) as compared to 7-10 days with organophosphates. The management of bollworms, pink and spotted bollworms was spectacular and population bollworms declined to all time low. The productivity increased tremendously by 25-40 per cent. However, the misuse resulted in development of resistance to American bollworm which was emerging as new pest from mid 1985s and resurgence of whitefly. The synthetic pyrethroids provided effective control of bollworm complex including

American bollworm which was in low intensity till mid 1990. Thereafter American bollworm emerged as major pests and the high development of resistance to synthelic pyrethroid was observed. The American bollworm *Helicoverpa armigera* was found to survive and cause extensive damage to cotton crop despite repeated applications of insecticides of even up to 30 sprays.

Combination of insecticides : Failure of synthetic pyrethroids to control the American bollworm which was dominant pests of cotton resulted in use of combinations of synthetic pyrethroids with chlorpyriphos, endosulphan, triazophos, ethion, amitraz *etc.* to manage the American bollworm (Table 17.8). But these did not provided desired control of bollworms and failure of cotton crop was observed on different cotton growing countries. Wu and Yang, writing in Russell and Kranthi (2006) concluded that few mixtures produce more mortality than the most effective component of the mixture on its own in Asia. Where resistance to pyrethroids is metabolically mediated some mixtures (especially those containing certain phosphorothionate organophosphates) can undermine the resistance and restore the efficacy of the pyrethroids but this effect is short lived. The employment of mixtures in controlling *H. armigera* can result in the simultaneous enhancement of multiple resistance mechanisms and significant cross resistance to other compounds. Mixtures may still be cost-effective for controlling insect pest complex in cotton, however rational use of mixture as insecticide resistance management strategy should be treated cautiously. Development of an anti-resistance mixture should base on full understanding of the genetic basis and mechanisms of resistance to each component in the mixture. In West Africa, use of pyrethroid/OP mixtures from the beginning resulted in suppressing the development of esteratic resistance to some extent, though oxidative resistance gradually developed in *H. armigera*(Martin *et al.* 2003). But in most cotton areas of China, India and Pakistan, both pyrethroids and OPs have been widely used and already have resistance problems. Employment of Pyrethroid/OP mixtures for resistance management in *H. armigera* is unlikely to be helpful in the long term. All these insecticides disrupt naturally occurring beneficial insect populations to variable extents.

Era of New Chemistry : Nicotinyl were introduced for management of sucking pests by seed treatment and foliar application. The provided effective control of jassid, aphids, whitefly, thrips etc. For bollworms indoxacarb, spinosad, pyridalyl, novaluron, were introduced and provided effective control but due to high cost and higher incidence of bollworms, the economics of cotton cultivation was not in favour of cotton grower (Table 17.9).

A number of newer insecticides with unique mode of action have been registered during the late 1990s and early 2000s for insect pest control in agriculture. These insecticides play an important role in Integrated Pest

Management (IPM) of many insects pests with good bio-efficacy, high selectivity and low mammalian toxicity, which make them attractive replacements for organochlorines, organophosphates, carbamates and pyrethroid insecticides (Cardwell *et al.*, 2005).

1. Neonicotinoids : Neonicotinoids are the most effective insecticides for the control of sucking insect pests cotton jassid, aphids, whitefly, thrips, Their broad spectrum of efficacy, together with systemic and translaminar action, pronounced residual activity and a unique mode of action, make the neonicotinoids the most rapidly expanding insecticidal class for control of jassid and replaced the dimethoate, oxy demeton methyl, phosphamidon *etc.* The first compound, was imidacloprid (Bayer Crop Science) followed by acetamiprid (Nippon Soda, 1995), thiamethoxam (Novartis, 1998), thiacloprid (Bayer Crop Science, 2000), clothianidin (Sumitomo Chemical Takeda Agro Company, Bayer Crop Science, 2000) and dinotefuran by Mitsui Chemicals, 2002 (Elbert *et al.*, 2008).

2. Indoxacarb : Indoxacarb is the first commercialized insecticide of the oxadizine group (Horowitz and Ishaaya, 2004). Indoxacarb has been recommended to control lepidopteron pests at a relatively low use rate. The indoxacarb was effective in controlling the *Heliothis* complex (*Helicoverpa zea* and *Heliothis virescens*), fall armyworm (*Spodoptera frugiperda*), beet armyworm (*S. exigua*), cabbage looper (*Trichoplusia nil*), and tarnished plant bug (*Lygus lineolaris*) (Sanchis *et al.* 1999). It inhibits the flow of sodium ions into nerve cell in insects that cause paralysis and death.

3. Diamides : (a) Flubendiamide: Flubendiamide (diamides group) is mainly effective for controlling *Helicoverpa* spp, *Spodoptera* spp.

(b) Chlorantraniliprole: Chlorantraniliprole, effective against lepidopteran bollworm, *Helicoverpa zea* (Boddie), fall armyworm, *Spodoptera frugiperda* and tobacco budworm, *Heliothis virescens* (F.). Chlorantraniliprole was harmless to the parasitoid wasp species and may be a useful tool in IPM programmes (Brugger *et al.* 2010).

(4) Flonicamid: Flonicamid (pyridine carboxamid) a novel compound with a unique mode of action, has contact activity and is upwardly systemic. Flonicamid effectively manages aphids, thrips, and leafhoppers.

(5) Spiromesifen: Spiromesifen tetronic acids derivative is especially active against whitefly (*Bemisia* spp.)

(6) Spinosad: Spinosad-bacterial fermentation product is active on various insect pests, especially lepidopteran. Avermectins emamectin benzoate; exhibits excellent activity against lepidopteran pests and a wide range of insect species *e.g.* American bolloworm (Helicoverpa armigera), tobacco budworm, *(H.virescens)*, and southern armyworm, (*Spodoptera* spp.).

(7) Pyridalyl Dichloropropenyl ethers: Pyridalyl provides effective control of *Helicoverpa armigera* and the tobacco cutworm, as well as against thysanopteran insects, including *Thrips palmi.* Pyridalyl is also less harmful than existing insecticides to various beneficial arthropods, so it should provide an important tool in IPM and insecticidal management programmes for the control of lepidopterous and thysanopterous pests.

(8) Novaluron: Novaluron is effective against lepidopteran larvae such as *Spodoptera littoralis* and *Helicoverpa armigera* (Hübner) and cotton whitefly nymphs of *B. tabaci*.

(9) Buprofezin: Buprofezin is effective for control of whitefly, *Bemisia tabaci* and mealy bug *Phenococcus_solenopsis*. Buprofezin suppresses embryogenesis and progeny formation of *B. tabaci* (Ishaaya *et al.*, 1988).

Economic Threshold Levels

The judicious use of insecticides is the most important component of pest management to increase the productivity with least possible disturbance to cotton ecosystem and to enhance the life of chemical insecticides. This concept is being advocated for management of insect pest in cotton for last five decades but adoption is low. In recent past due to failure of cotton it is essential to decide the right insecticide, dosages, time and method of application. Monitoring of pest population is very essential at micro level to avoid such failures in future. This help in reducing the number of sprays and hence the cost of plant protection, besides minimising the adverse effects of insecticide. IPM interventions are optimized only when the need for the intervention is justified based on the economic threshold levels (ETL) of the pest. Identifying these threshold levels requires a background understanding of the relationships between the pests, their natural enemies and the crop damage which may result from particular pest populations at particular stages of crop growth. The complete understanding of tritrophic interaction is required to have suitable ETL. However, this aspect is lacking in developing and under developing cotton growing countries. Moreover, counting insects remains a problem, particularly in small-scale farming systems. Samplings for majority of pest are not standardized even till today and counts are only relative and not absolute. The multiple pest attack may result in the adoption of ETL based sprays more difficult. Several studies have been conducted on sampling methods to define the optimum number of plants required to unambiguously determine ETLs for *H. armigera*_and whitefly (Silvie *et al.*, 2000, Traore *et al.*, 2000, Goze *et al.*, 2000 (Mazza *et al.*, 2000, Sekamatte *et al.*, 2003, Goze and Deguine, 2000). The population based ETL are not favoured by farmers in developing and under develop countries. The need is to make the concept of ETLs more farmer friendly for judicious use of insecticides.

Insecticide Resistance

***Sucking Pests*:** Resistance to organophosphates (monocrotophos and dimethoate) carbamates, endosulfan, synthetic pyrethroids (cypermethrin, deltamethrin and fenvalerate) in aphids, *Aphis gossypii*, have been reported by many workers (Kung *et al.* 1961, Furk *et al.,* 1980, Bobert *et al.*, 1994, Deguine, 1996, Delorme *et al.,* 1997, Ahmad *et al.*, 1999, Villatte *et al.,* 1999, Herron *et al.,* 2001, Gill and Dhawan, 2006d). High level of resistance was detected in cotton aphids from Xinjang and Shandong during 1995–1996 (Cheng *et al.,* 1997). The aphid population of western Australia displayed extreme resistance leading to control failure (Herron *et al.*, 2001). Dittrich and Ernast (1983) showed that Sudanese field strains of *B. tabaci* were highly resistant to dimethoate and monocrotophos. Cahill *et al.* (1996) reported resistance to monocrotophos and other organophosphate insecticides in *B. tabaci* strains from USA, Central America, Europe, Pakistan, Sudan and Israel. However, high resistance levels to monocrotophos during, 1992–1996 were lowered considerably by 2000 when the use of the product for whitefly control in Pakistan was reduced (Ahmad *et al.* 2002). Field strains of *B. tabaci* collected from 22 cotton_growing district across India exhibited high level of resistance to methomyl and monocrotophos and moderate resistance to cypermethrin (Kranthi *et al.*, 2002a).The cotton leafhopper, *Amrasca devastans* have developed resistance to endosulfan, monocrotophos, cypermethrin, phosphamidon, dimethoate (Santhini and Uthamasamy, 1997, Challam and Subbaratnam, 1999 and Jeyapradeepa, 2000).

Lepidoptera: A global history of insecticide resistance in *H. armigera* is given in Kranthi *et al.* (2005) and Kranthi and Russeell (2009). This was key factor for failure of cotton before introduction of transgenic cotton. The first reports of resistance to pyrethroids in *H. armigera* in India were in the late 1980s and resistance to organophosphates was almost negligible with highest resistance factors of 9-fold to quinalphos, and 3-fold to monocrotophos in *H. armigera* in India (McCaffery *et al.*, 1989; Armes *et al.,* 1992). Later, Armes *et al.* (1996) reported no resistance to monocrotophos, but resistance levels was 59-fold to quinalphos and >30-fold to methomyl. Kranthi *et al.* (2001a) reported high levels of *H. armigera* resistance to monocrotophos (65-fold); chlorpyrifos (82-fold), quinalphos (15-fold) and methomyl (22-fold). In China, *H. armigera* strains whichwere susceptible to monocrotophos until 1993 exhibited higher levels of resistance in 1995 (Wu *et al.,* 1995, 1996). High levels of >300-fold resistance to methomyl and >200-fold to monocrotophos were reported from China (Cheng and Lieu, 1996, Ren *et al.*, 2002) and 720-fold resistance to monocrotophos in Pakistan (Ahmad *et al.*, 1995). Resistance levels to endosulfan have generally varied at moderate in India (Armes *et al.* 1996, McCaffery *et al.*, 1989,Kranthi *et al.*, 2001a). Subsequent reports include, resistance levels of 25-205-fold to five pyrethroids in Pakistan (Ahmad *et al.,* 1997), 17-fold to cypermethrin in Turkey (Ernst and Dittrich, 1992), 1,361-fold to fenvalerate and 56,911 fold to deltamethrin in China (Shen *et al.,* 1993, Cheng and Lieu 1996), 6-fold to fenvalerate in Australia (Gunning, 1993) and 189-fold to deltamethrin in South Africa (Martin *et al.*, 2003). In India, reports on *H. armigera* development of resistance to pyrethroids, attributed field control failures (Dhingra *et al.*, 1988; McCaffery *et al.* ,1989). Subsequently, high levels of pyrethroid resistance were reported in several cotton growing regions of the country (Mehrotra and Phokela, 1992; Armes *et al.*, 1992, 1996; Sekhar *et al.,* 1996). In general, reports of *P. gossypiella* resistance to insecticides have been rare. For example, Tang *et al.* (1988) could not find any evidence of insecticide resistance in *P. gossypiella* in China. However, resistance to azinpho_methyl and permethrin_was reported from strains collected in Arizona and California (Osman *et al.,* 1991).Resistance in *E. vittella* was>70-fold to monocrotophos in Sriganganagar and Sirsa_strains of north India (Kranthi *et al.,* 2002a). Resistance in *Spodoptera litura* to endosulfan, carbaryl and malathion was reported in field strains from Andhra Pradesh (Ramakrishnan *et al.,* 1984).In Punjab (India) comprehensive studies on resistance of *H. armigera*_confirmed the high level of resistance to pyrethroids (Gill and Dhawan, 2006a, 2006b, 2006c).

Insecticide Induce Resurgence

The repeated use of synthetic pyrethroids, cypermethrin and deltamethrin against bollworms on cotton in Thailand caused the resurgence of whitefly, *Bemisia tabaci* (Gennadius) (Wanghoonkong 1981). Similarly, the continuous use of pyrethroids for the control of tobacco budworm and pink bollworm in south California resulted in increase in population of whitefly (Johnson *et al.* 1982). The increase in population of whitefly in Egypt, Sudan, Syria, Thailand, Israel, Zimbabwe, China, Pakistan and India is due to the increased use of synthetic pyrethroids. In India, the over dependence on synthetic pyrethroids for the control of bollworms increased the problem of whitefly resurgence in cotton. Jayaraj *et al.* (1987) observed outbreak of whitefly with twelve rounds of application of organophosphate and synthetic pyrethroids. The resurgence of whitefly with repeated application of fenopropathrin, fenvalerate, cypermethrin, deltamethrin and fluvalinate was observed by many workers (Reddy *et al.* 1986b, Natarajan *et al.*, 1987, Shelka *et al.*, 1987). Regu *et al.* (1990) concluded from that tank mixing of chlorpyriphos with cypermethrin and fenvalerate resulted in heavy build up of whitefly population and the build up increased with the increase in number of sprays. In Punjab, also heavy build of whitefly was observed with repeated use of fenvalerate, cypermethrin and deltamethrin (Buttar *et al.,* 1994; Dhawan and Simwat, 1997). However, alpha- cypermethrin did not cause whitefly resurgence (Dhawan, 1999a; Dhawan and Saini, 1998). The resurgence ratio was maximum after sixth spray and was much higher in cypermethrin @ 100 g a.i./ha than in other treatments.

Dhawan and Simwat (1997) observed the resurgence of cotton jassid with repeated use of chlorpyriphos and quinalphos reported it to be major cause for increase in population of jassid during the flowering phase in recent years due to the extensive use of these molecules for the management of *Helicoverpa armigera* (Hubner) on cotton. The repeated application of deltamethrin led to the build up of jassid while single application controlled this pest (Singh *et al.*, 1987). Resurgence of mealy bug, *Ferrisia virgata* (Cockerell) in the cotton took place with the two applications of fenvalerate and permethrin (Uthamasamy *et al.*, 1987). The increase in population of aphid, *Aphis gossypii* Glover on cotton with repeated use of synthetic pyrethroids has been observed by many workers (Dhawan *et. al.*, 2001, Sellammal Murugesan *et. al.* 1979, Balasubramanian *et al.*, 1980, Natarajan *et al.*, 1987, Patel *et al.*, 1987). Kidd *et al.*, (1996) observed the resurgence of cotton aphid with the use of lambda-cyhalothrin in Texas, USA. The aerial application of malathion resulted in aphid resurgence (Richter *et al.*, 1998). Rote *et al.* (1980) observed high infestation of mealy bugs in pyrethroid than in monocrotophos. Many workers observed the outbreak of mites with repeated application of pyrethroids (Pasupathy and Venugopal 1987, Reddy *et al.*, 1986a) observed the resurgence of aphid and *Tetranychus cinnabarinus* (Boisduval) with the application fenvalerate in China.

17.4.4 Safety of insecticides to natural enemies

Many workers observed insecticides safe to different natural enemies and are documented in Table 17.12.

Botanical Biopesticides

As many as 2,121 plant species have been reported to possess pest control properties. Of these, 1,005 plants have insecticidal and 384 antifeedant properties, 297 antifedant, 27 attractant and 297 repellent properties (Purohit and Vyas, 2004). Neem as tool of insect pest management in various crops including cotton has been subject of discussion at the nine global conferences on neem. India has an estimated 18 million neem trees with a potential of production of 0.7 million tonnes of fruit (Narwal *et al.*, 1997). Potential of neem in managemet of insect is reviewed by (Dhawan and Dhaliwal, 1996, Dhawan, 1999b, Dhawan *et al.*, 2013). The status in India is summarised in Table 17.12. Phadke *et al.*, (1988) observed the efficacy of neem products against *Bemisia tabaci* and *Heliothis* on cotton and observed that Neemark (0.4 %) was as effective as endosulfan in controlling both pests, while Neemark (0.5%) was more effective than fenvalerate in controlling *B. tabaci.* Flint and Parks (1989) tested Margosan-O for control of *B.tabaci*, *Frankliiella* spp., *Bucculatrix thurberiella* and *Tetranychus* spp. in small field plots of cotton (*Gossypium hirsutum*) at Phoenix, Arizona. There was significant reduction (60%) in population of immature stages of whiteflies on cotton leaves after applications of aqueous sprays containing azadirachtin (138.7 ml a.i./ha) but lower concentration were ineffective. Nimbalkar *et al.* (1990) observed that neem oil at 0.5 per cent and neem seed extract at 5 per cent reduced whitefly population to the extent of 56.36 and 43.70 per cent, respectively. According to Roychoudhry and Jain (1994) neem oil (2%) has good effect on the eggs and nymphs of whitefly after 24 hrs with 74.4 and 100 per cent mortality, respectively. Navneem and Neemark significantly reduced jassid (*Amrasca biguttulla*) populations during the vegetative phase. Repelin and Neemrich were more effective than Neemark in the field experiments against young larvae of *H. armigera* on *Gossypium hirsutum* (Dhawan and Simwat, 1996). Gupta *et al.* (1999) conducted field experiments to evaluate the bioefficacy of neem products against cotton bollworms viz. *Earias* spp. *Pectinophora gossypiella* and *H. armigera* and their impact on whitefly, *Bemisia tabaci* (Genn.). The application of neem alone or in combination with *Bt* formulations or board spectrum conventional insecticides failed to check the incidence of bollworms. However, neem in combination with one spray of synthetic pyrethroids gave significant control of bollworms. Most of the commercial neem formulations have strength of 0.03–0.3% of Azadirachtin. Neem formulations at the recommended rates were significantly inferior in efficacy when compared to 5% neem seed kernel suspensions. Combination of nimbecidine along with other control measures was recommended for the management of *Helicoverpa armigera* (Hubner) on cotton (Reddy and Manjunath, 2000) and even alternate application of neem based insecticides and synthetic insecticides led to the effective management of whitefly and bollworms in cotton (Mann *et al.*, 2001). Application of neem oil for the management of initial infestation of *Spodoptera* and whitefly has been advocated in Insecticide Resistance Management strategies (Kranthi and Russell, 2009). Neem oil @ 2.5 litre/ ha during the early part of the season is suggested for sucking pests under IRM strategies for *Bt* cotton and validated at farmers field under the Technology Mission during 2008 and 2009. One of the biggest constraints in adoption of botanicals is non acceptance of product by farmers as neem formulations do not give knockdown effect. Field trials with neem formulations in cotton have not shown satisfactory control and it has been seen that for a reasonable field efficacy even with seed kernel extracts, it is necessary to use repeated applications due to the low toxicity and rapid biodegradability. The second constraint is the availability of standard and economically viable formulation of neem and is rapidly inactivated by UV light, temperature, leaf surface pH conditions.

Microbial Control

Various microbials perform significant role in regulation of insect population. Among different micro agents, Nuclear Polyhedrosis Virus (NPV) and *Bacillus thuringiensis* (*Bt*), have been tested against bollworm complex of cotton. Baculoviruses have been recognized for the great potential for biological control and safety to natural enemies. The first commercial use of *H.zea* NPV (HzNPV), registered for commercial use by Sandoz in USA under name Elcar, was in 1975. The success was limited due the introduction of synthetic pyrethroids which provided improved kill, broad spectrum control and less cost (Smit, 1997). *H. Armigera* NPV was recommended in IPM programmes in Thailand (Ketunuti and Prathomrut 1989) and Indonesia (Ruchijat and Sukmaraganda 1992). The slow killing of host and high cost limited the use. In tropical countries, susceptibility to ultraviolent rays also affected the bioefficacy. When exposed in the open to direct solar radiation, half life of the NPVs of *Heliothis* spp. and *Spodoptera littoralis* (Boisduval) were less than one day or even only a few hours (in the USA) and less than one hour (in Egypt), respectively (Ignoffo and Couch, 1981; Jones, 1988). The average half-life of *Heliothis zea* (Boddie) NPV on the upper surface of *Gossypium hirsutum* was 13 hours and three quarter-life was 41 hours (Entwistle and Evans, 1985). Inactivation of the NPV also takes place on the alkaline leaf surface of cotton (Jones, 1988) resulting in reduced efficacy. Use of HaNPV for control of *H armigera* is part of specific window based IRM strategy in India. Many *Bt* formulations were effective against bollworm complex in north irrigated cotton belt (Dhawan and Simwat,1998). *Verticillium lecani, Metarhizium anisopliae*, and *Beauvaria bassiana*, in cotton IPM has been restricted by their relatively short field life. Commercial baculovirus strains have been developed against *Spodoptera exigua*, (Central America, South-East Asia), *S. littoralis* (Africa), *S. litura* F. (China, India), *S. frugiperda* (S America), *Helicoverpa armigera* (China, India Pakistan and Australasia), and *Heliothis* sp (North America) (Entwhistle 1998). *Spodoptera exigua* NPV was successful in Thailand, while *S. littoralis* NPV was successful in India (Jayraj *et al.,* 1981) and Egypt (Jones, 1994) but its use has not spread significantly. Yearian and Phillips (1983) reviewed the field efficacy of sprayed *B.t.* and the *Heliothis* baculovirus in Arkansas for several years and concluded that neither microbial product provided adequate control of *Heliothis* spp. in cotton. Sprayed *Bacillus thuringiensis* Berliner (commercial strain *B.t.* var. Kurstaki- HD-1) has been an important component of IPM programmes world over for the past 20 years. However, its field performance in cotton on *Heliothis* spp. had been far from satisfactory right from its introduction. Repeated field trials in the Delta and RioGrande Valley of Texas in the 1970s showed that *Bt.* failed to provide satisfactory commercial control (Allen, 1983) as *Heliothis* spp. feed only sparingly on the foliage and move rapidly on to squares and bolls where they feed internally thus escaping the toxins. Microbial control is slow in action and allows the larval damage to continue for the full period of infectivity with the exception of *Bacillus thuringiensis*, which is comparatively less slow in effect. Farmers find it unrealistic to tolerate the continuous damage to squares, which results in shedding and reduction in yields. Due to the unreliable performance, the commercial products have not been popular with the cotton farmers.

Cultural Control

The pest population can be managed effectively by adopting recommended agronomic practices during the crop season as well as in off-season. The adoption of such operations at appropriate time helps to reduce the damage (Table 17.13). However, the deviations from such package help the pest to multiply and adversely affect the yield.

Planting time : Planting of crop at optimum time is important to avoid condition that synchronises crop-pest association. Careful selection of planting date enables the plants to escape damage during susceptible growth stage, advance into tolerant stage before the pest attack occurs and reduces the number of generations of pests. In north India due to staggered sowing of cotton, pests are able to complete 1-2 generations more in the season. It is estimated that in case of pink bollworm the population increase by 20 fold from generation to generation and this indicates heavy build up of population during last 4-5 generations that could be avoided by timely sowing of cotton. Similarly, American bollworm and whitefly are late season pests and by sowing till mid May, crop can escape damage of these pests. High incidence of pink bollworm was observed in early and late sown cotton crop in South zone (Dhawan and Sidhu, 1985, Sundramurty and Chitra, 1992).

Seed purity : Due to the non-availability of certified seed, farmers mainly depend upon traders for seed requirement and usually get mixture of varieties with different maturity period that reduce the yield and extend growing period that favours the multiplication of pest (Dhawan, 1990, 1999a).

Plant spacing : The plant spacing has tremendous impact on host plant as it influences its health, growth and development as well as microclimate, which affect the build up of pest species. The plant to plant and row to row spacing varies greatly with cultivar and method of sowing. The normal spacing provides optimum space for proper plant growth. The close spacing result in bushy growth of crop that affect the penetration of light, results in vertical growth of plant, hinders the spraying operation and also results in higher relative humidity that favours the higher incidence of sucking pests and bollworms (Dhawan,1999a)

Fertilizers : The dose of fertilizers varies with the cotton cultivar, soil health, and rainfed or irrigated conditions. Higher yield of seed cotton usually require fertilizer application but excessive use of nitrogen, produce rank growth and increase the severity of infestation of sucking pests and bollworms. The optimum level of potassium and phosphorus is also essential for better yield. The higher application of NPK enhanced the incidence of jassids and bollworms (Jayaswal and Sundaramurthy 1992, Purohit and Deshpande, 1991, Dhawn *et al.*, 2012). In Central India the higher use of NPK increased the incidence of whitefly (Rote and Puri, 1992).

Intercropping : The American bollworm and whitefly which are serious pests of cotton are polyphagous with wide host range which enables them to find suitable host throughout the year. In North India, the cultivation of okra, mungi and pigeon pea in and around cotton crop is not recommended as it increases the build up of jassid, whitefly, spotted bollworms and American bollworm (Dhawan 1999a, 2000). The intercropping of cotton with black gram, green gram, onion and cowpeas is recommended in Southern and Central India to divert population of sucking pests and American bollworms from cotton (Venkatesan *et al.,* 1987). The effect of intercrop on *H. armigera* showed significantly higher incidence on cotton in cotton+clusterbean than cotton+maize (Jayaswal and Sundaramurthy, 1992). However, this intercropping reduced the leafhopper incidence by 50% (Rajendran *et al.*, 1996). Increase of *H. armiger* was low in cotton with intercrops like groundnut, soybean, mungbean and *Setaria* than to sole cotton situation (Rao *et al.*, 1995).

Trap Crops : The basic strategy involved in the cultivation of trap crops, which is the preferred host of pest for feeding, shelter and breeding, over a small area (not more than one-tenth of crop area) to attract and kill the pest. Marigold and *Nicotiana rustica* L. grown as trap crops are also preferred hosts of *H. armigera.* Planting of castor as trap crop diverted the population of *Spodoptera litura* from cotton (Reddy and Rosaiah, 1987). Besides these, trap crops like cowpeas, okra, corn, marigold, *Rustica* and castor can be used to increase the efficiency of natural enemies (Table 17.14) (Singh, 1994a). Improper management of pest on trap crop many a time results in migration of pest to cotton.

Cropping Pattern : Cotton is mostly grown in rotation with wheat that is not a host of major insect pest of cotton in North India. In Punjab, the cultivation of sunflower and tomato instead of wheat, in recent years in cotton belt is also responsible for increase carryover of jassid, whitefly and American bollworm from one season to another. The other crop rotations, *i.e.*, cotton-rice, cotton-groundnut, cotton-cowpeas or cotton-soybean are also less conducive for pest development on cotton (Sundramurty and Chitra, 1992).

Destruction of Alternate Weed Host Plant: Several weeds are preferred alternate hosts of American bollworm, spotted bollworm and whitefly. The removal and destruction of these hosts in the vicinity of cotton fields, on sides of roads, *etc.*, can help to reduce the multiplication and carryover of these pests (Jayaswal and Sundaramurthy, 1992, Dhawan, 1999b, 2000). Similarly, the destruction of weeds is suggested as effective tool in management of mealy bug, *P.solenopsis* (Dhawan *et al.,* 2011).

Termination of Crop: Cotton crop should be terminated as early as economically feasible to reduce the late season pest attack and carryover to next season. This can be achieved by (I) sowing at appropriate time (II) cultivation of short duration/early maturing cultivars (III) use of growth regulator (Dhawan, 2000).

17.5 Bt Cotton and Pest Management

Bollworm Management

Bollworms complex, pink bollworm [*Pectinophora gossypiella* (Saunders)], spotted bollworms [*Earias vittella* (Fabricius) and *E. insulana* (Boisdual)], American bollworm [*Helicoverpa armigera* (Hubner)] are serious pest of cotton. After the failure of insecticide dominated IPM due to development of resistance in *Helicoverpa* to commonly used insecticides, Helicoverpa emerged as most serious pest of cotton. Estimated losses due to bollworms were 50 % and later on losses resulted in failure of cotton crop in cotton growing countries of the world (Dhaliwal *et al.,* 2010). Main reason for adopting *Bt* cotton is the management of bollworm complex particularly *Helicoverpa armigera* and reduction in pesticide use. There is no doubt that farmers are satisfied with Bt cotton, as it has effectively brought the bollworm under control.. This is documented from all Bt cotton growing area across the world. No report of any additional sprays for management of bollworms, indicating bollworms have not developed resistance to level that additional measures are required. In future, in case 1-2 sprays may be required, but is not serious concern as new chemistry coming up can manage both sucking pests and bollworm to level where it will not cause economic damage. Bollgard II with *cryIAc +cry2Ab* also provided effective protection against *Spodoptera* spp. Fusion Bt also provided good control of young larvae of *S.litura* (Dhawan *et al.*, 2011). Econometric modelling of pesticide use in all fields surveyed also supports this hypothesis of declining bollworm infestation over time in different cotton growing areas of world. In Punjab with introduction of *Bt* cotton bollworms are not pests on cotton.

Secondary Pests

The secondary pests are likely to increase over time because of two factors: (1) the general ineffectiveness of *Bt* cotton against pests other than the bollworms and (2) a lowered usage of pesticides in *Bt* cotton. As a result of this, secondary pests that would otherwise not have survived have a chance to emerge and, without additional pest control, could potentially evolve into primary pests (Xu *et al.* 2008; Men *et al*., 2004). Further, in China econometric analyses show that fluctuation in mirid infestation is largely related to local temperature and rainfall (Huang *et al.,* 2010) and also decreased pesticide use due to *Bt* cotton has led to an increase in mirid bug (Lu *et al.,* 2010). The improved pest management strategies-integrating Bt with different methods of pest management are needed to control secondary pests. However, at present scenario are dependent on use of pesticides which increased the cost of production and excessive use may result in emergence/resurgence of secondary pests. In present scenario additional sprays are required for management of regular sucking pests. In India and Pakistan, sprays for bollworms during flowering phase used to manage sucking pests. Due to withdrawal of 7-12 sprays against bollworms, sucking pests particularly jassid attained the status of sucking pest during flowering phase. However, cost for 2-3 sprays against sucking pest is much lower than that of bollworms. Due to the photonic effect farmers are also using neonicotinoids as growth promoter. The cost of such sprays is compensated by higher yield even in the absence of pests. But additional benefits are that with introduction of Bt cotton the resurgence of whitefly is not of major concern. However, the plant type/ agronomic practices may favour the build up of whitefly. This will be difficult to manage due the lack of proper spray technology for bushy *Bt* hybrids late in season. With introduction of *Bt* cotton in Punjab, incidence of jassid and whitefly is on increase.

In the case of direct effects of *cry IAc* and *cry 2Ab* proteins on non-target species, the well established specificity of these proteins provides a clear safeguard that greatly reduces risks of direct effects on non-lepidopteran species. Bt cotton is not effective against secondary pests, which in non-transgenic cotton cultivation are usually killed due to heavier insecticide sprayings. However, as fewer pesticides are used, secondary pests might increase, thereby counteracting the effects of Bt cotton and gradually even evolving into primary pests themselves (Turnipseed *et al*., 1995). Although several studies have predicted the eventuality of secondary pests (Cannon 2000; Lang 2006; Wang *et al.,* 2006, 2008;Xu *et al*., 2008). Many of sucking pests (mirids, aphids, stink bugs, cotton stainers *etc.*) are the group most likely to show such effects in different cotton growing areas of world. The sucking pests have become a more significant part of the pest complex in *Bt* crops in some countries (Wilson *et al*., 2004; Wu *et al*., 2002) where the *Bt* cotton required additional spraying. In Australia, an average 60% reduction in sprays applied for *Helicoverpa* was accompanied by no change in sprays for mirids, aphids, mites and thrips (Fitt, 2004). Likewise Wu and Guo (2003) observed that *Bt*cotton in China help to prevent resurgence of aphid populations. In the south-eastern USA, stink bugs have assumed significant pest status in *Bt* cotton crops (Greene *et al*., 2001). The removal of broad spectrum sprays directed against *Helicoverpa* spp. has resulted in higher densities of plant bugs (*Creontiades* spp.) and stink bugs (*N. viridula*), and to a lesser extent leafhoppers (*Amrasca terraereginae* (Paoli)), present in Bt cotton. During the 2004/05 season, Bollgard®II crops were sprayed 1 to 3 times to control bug pests. Chemical options are available for these pests, but not IPM-friendly. In order to address this situation, increased emphasis on research into pest status, thresholds and management of these pests is required.

Natural Enemies

No significant effect of these *Bt*cottons on natural enemies was observed. Non-targets such as predators which do not feed on the plant are not directly exposed and no evidence exists for a secondary impact through consumption of intoxicated prey. Commercial experience with *Bt* cotton crops has now shown a 3-4 fold increases in beneficial insect abundance compared to conventional crops. Enhanced levels of beneficial species help to partially suppress secondary pests. Maintaining an appropriate balance and retaining the benefits of *Bt* cottons requires sustained efforts to enhance beneficial species and use of non-disruptive, short residual pesticides for key sucking pests. Overall the stability of these systems will require mobilization of the whole farm environment and greater understanding of the flows of impacts and services between intensive cropping systems and the surrounding landscape to conserve natural enemies.

Indirect effects on natural enemies' species may be mediated through changes in abundance and diversity of prey. The significance of a reduced density of noctuid larvae and pupae as food sources for predators or as hosts for parasitoids depends on the importance of *Helicoverpa* life stages in cotton in maintaining local populations of these beneficial. Clearly within transgenic cotton fields, the abundance of some predators and parasitoids may be reduced, particularly those whose survival is closely tied to the abundance of *Helicoverpa,* but this is unlikely to threaten their regional persistence since in the cropping systems where cotton is usually grown a significant proportion of the *Helicoverpa* population is also present on other crops and uncultivated hosts (Fitt 1989; Hearn and Fitt, 1992) where parasitoids are also active. None of the known predators that attack lepidoptera in cotton are specialists. *Helicoverpa* may be only incidental prey for some key predators whose within-field abundance is maintained by other prey. Other studies have sought further indirect effects of transgenic on parasitoids and non-target herbivores but have generally found no effect of the *Bt*plant itself (Schuler *et al.*, 2001). Within-field impacts on non-target insects, even if they do occur, are unlikely to be significant compared to the undoubted impacts from broad-spectrum pesticides (Fitt and Wilson, 2002). Observations over the past six years confirm that the abundance of beneficial invertebrates in commercial *Bt* cotton fields is markedly greater than in conventional fields.

There are no confirmed monitoring results of the negative effects of *Bt* cotton on insect pest predators (Men *et al.*, 2004; Zhang *et al.*, 2008). The population densities of major predator species, such as predatory spiders, coccinellids, chrysopids, and small flower bugs, in transgenic *Bt* cotton fields were not different from those in conventional *Bt* cotton fields, but were significantly greater than those in conventional cotton fields treated with insecticides (Sisterson *et al.*,2007; Sharma *et al.*, 2007; Dhillon and Sharma, 2009). In India, predator populations (*Chrysoperla* spp., *Orius* spp., *Coccinella* spp., *Brumus* spp., *Vespa* spp., *Lycosa* spp., and *Aranews* spp.) were similar on Bollgard I, Bollgard II, and conventional cotton (Mann *et al.*, 2010). As expected, the population densities of parasitic wasps (*Trichogramma confusum*, *Microplitis* spp., *Campoletis chlorideae*, and *Meteorus pulchricornis*) decreased significantly due to lower density of *Helicoverpa armigera* in *Bt* cotton fields (Wu and Guo, 2005; Yang *et al.*, 2005: Xia *et al.*, 2007). However, there is evidence of a negative influence of Bt cotton on soil biology and nutrient availability (Viktorov, 2008; Sarkar *et al.*, 2008).

Pesticide use

Bt cotton reduces the number and quantity of pesticide applications and was one of the aims. On average, the pesticide sprayings had decreased five to six times. *Bt* cotton in developing countries indicated that the benefits to smallholders because it decreases the number of pesticide sprayings and increases yields. For instance, in South Africa there had been a reduction in the average number of pesticide sprays per season for farmers who adopted Bt cotton. As a result, there were cost savings in the form of lower inputs for pesticide and labor (Bennett *et al.*, 2003). Early studies in northern China indicated that Bt cotton reduced pesticide use by 35.7 kg per hectare, or a reduction of 55% of pesticide between 1999 and 2001 (Huang *et al.*, 2002a). Studies in India, Burkina Faso, and China have reached similar conclusions (Qaim 2003; Pray *et al.*, 2001; Vitale *et al.*, 2008). Bt cotton is only effective against the bollworms, while secondary pests have increased is well documented. Wang *et al.*(2009) concludes that the increase in insecticide use for the control of secondary insects is far smaller than the reduction in total insecticide use due to *Bt* cotton adoption. Another study contradicts these findings, based on a 3-year experimental field study in one province in China (Henan), Men *et al.* (2004) concluded that there is no difference in the total pesticide application between *Bt* cotton and conventional varieties because of additional sprayings to control secondary pests. The impact of *Bt* on reducing pesticide use indicates that adoption of *Bt* cotton can substantially reduce risk and incidence of pesticide poisoning. Similarly, in Argentina farm survey reveals that the technology leads to a considerable decline in pesticide application rates. On average, adopting farmers use 50 per cent less insecticides on their Bt cotton than conventional cotton. The insecticide reductions occur in highly toxic chemicals. Moreover, *Bt* adopters benefit from significantly higher yields, which is due to insufficient pest control in conventional cotton (Table 17.14).

In Australia, *Bt*cottons expressing the *CryIAc* endotoxin (Ingard®) were commercialised in 1996/97 and gradually increased in 30% of the cotton area. Two gene (*Cry IAc/Cry 2Ab*) varieties (Bollgard II) have been commercialised from 2004/05 and have now completely replaced Ingard varieties. Reduced reliance on conventional insecticides has been the most outstanding contribution of Bt cotton. In its first six years of commercial use, Ingard® reduced total insecticide use by between 45 and 60% compared to conventional cotton (Pyke, 2003). In 2003/04, when Bollgard®II, Ingard® and conventional cotton were grown, Bollgard®II delivered a 90% reduction in chemical use compared with conventional cotton, while Ingard® reduced chemical use by 56% (Anon., 2004).

Resistance Management

Bt resistance development in pest populations, would influence the technology's sustainability. Although significant resistance build-up has not been observed so far in commercial cotton production, entomological studies indicate a high risk of rapid insect adaptation to the *Bt* toxin. Resistance development is also one of the main concerns of environmentalists with respect to *Bt* crops. It would not only render the transgenic technology useless, but would also imply loss of *Bt* as an ecologically friendly microbial insecticide which is widely used in organic agriculture. To avoid the development of resistance against *Bt* cotton, it is advised to plant 20% of *Bt* cotton acreage with non-transgenic varieties as a refuge area to avoid resistance development or 5 % non Bt as unsprayed crop. This is part of strategy in India, S. Africa, China, Pakistan and Argentina. In Argentina even 10 % area is found effective. Australia is the country which strictly follow refugia requirement and has given various options like 10 % unsprayed conventional cotton, 5 % irrigated unsprayed pigeon pea, 15 % irrigated unsprayed sorghum, and 20 % irrigated unsprayed irrigated maize. However, particularly in developing countries, the enforcement of such refuge areas has been problematic due to weak regulatory structures and small, fragmented agricultural plots on which cotton cultivation generally takes place. In India, each pack contains 450g of *Bt* cotton contains 120 g non *Bt* as refugia for the farmers to grow. Unfortunately, seed industry is not serious in implementing the refugia. Seed of non *Bt* does not match the agronomy of *Bt* cotton. Moreover in most of the case the non *Bt* seed is of hybrid which does not bear any fruiting and no action is taken by implementing agencies for this serious lapse. Most of the farmers do not use any non *Bt* cotton seed, which is resulting huge national in loss. No non-*Bt* is sown as refugia in cotton belt of the country. In USA, refugia concept was followed with Bollgard I , but with introduction of Bollgard II and research by various institutes indicate that many alternate hosts are available for bollworms and there is no need of refugia.

In addition to this, the development of resistance in jassid (*Amrasca biguttulla*) to insecticides is serious issue in cotton growing belt of India. This is due the treatment of *Bt* cotton seed which is marketed with chloronicotinyl compound (CNC). This has resulted in development of resistance to jassid and field dose of CNC is 2-3 times more which was initially recommended. This will further increase cost of cultivation. Moreover, in north India, the seed treatment has no additional benefit as spray against this pest starts in July irrespective of seed treatment or not. However, farmer prefer as it act as growth promoter which help in good early vegetative growth.

Ecological Concerns

There is scientific uncertainty concerning the long-term sustainability of Bt technology with regard to its ecological effects. The uncertainty of the environmental impact of *Bt* cotton lies in the fact that ecological change may take many years to manifest. Ecological effects of *Bt* cotton can appear in several ways: (1) the development of the bollworm's resistance against *Bt* cotton, (2) a negative impact of *Bt* cotton on the bollworm's natural predators and non-target species, (3) adverse effects on soil nutrients and biota, (4) the increase of secondary pests, such as lygus bugs, aphids, and mites. Controlled field conditions have provided evidence for the bollworm's potential to build up resistance against Bt cotton (Downes *et al.,* 2007; Gujar *et al.*, 2007; Wu, 2007). Various researchers have pointed to the potential environmental risks of *Bt* cotton (Wang *et al.* 2008; Qaim, 2003). Due to the lack of scientific understanding of *Bt* cotton's ecological impact and the fact that ecological changes can only be monitored and evaluated on a long term, it is vital to adhere to the precautionary principle when bio safety issues are at stake. *Bt* cotton has a good potential in improving worldwide cotton production due to its effective resistance against the bollworm. At the same time, because of an evident rise of secondary pests, it is critically important to closely follow and assess its commercial production in the field. Scenario results demonstrate that rapid resistance build-up and associated pest outbreaks are unlikely if minimum non-*Bt* refuge areas are preserved. These results suggest that the economic and ecological advantages of *Bt* cotton could be maintained if in the medium to long run concept of refugia is strictly implemented. Under current circumstances where implementation is weak high risk will continue to prevail. The effectiveness of pest control due to lowered and varying *Bt* gene expression is also a great concern (Jost *et al.,* 2008; Xu *et al.*, 2008).

Lower Productivity of Bt cotton

There are reports of lower productivity of *Bt* from various cotton growing countries. *Bt* cotton is to provide protection against bollworms and productivity will be function of genetic characters, seed quality and agronomic practices. This is evident from the decline in productivity with time. *Bt* cotton is providing effective control of bollworm, therefore technological gaps in adoption seem to be one of the reasons including large number of hybrids with low productivity level. The low productivity may be due to (1) decline in *Bt* efficacy as a result of being backcrossed into more varieties by numerous public- and private-sector plant breeders or as a result of the development of bollworm resistance to *Bt* and (2) growth of secondary pests into major pest problems. Using historical data beginning in the 1950s, yield growth and decline in yield volatility after 1995 correspond with the introduction and spread of *Bt* cotton. Lack of good agronomic practices for large number of hybrids released, low yield potential of released *Bt* hybrids, lack of precision cultivation technology, seed vigour and imperfect input deliver system may result in low yield and even decline in area under cotton. In Punjab, India, yield has declined with adoption of *Bt* cotton and need to be addressed.

Economics

Bt cotton seed prices are higher than conventional cultivars across the cotton growing areas , but differences in seed costs were offset by reductions in expenditures on pesticides and labor (ASSOCHAM Survey, 2007). This resulted in overall decreases in production costs for *Bt* cotton, as compared to non-Bt cotton, and increased net revenue (Huang *et al.* 2002). In India, the price of hybrid *Bt* cotton seed at US$18 for a 450 gram packet, including technology fee and is less than of US $29 per 450 gram packet in USA. Revenue earned by seed firms was especially affected (falling from 27 % to 2 %), perhaps due to two reasons a reduction in seed prices by nearly 50 to 60 % combined with increased cost of seed production by 35 to 40 %. Royalties paid before price controls were as high as US$ 40 per packet of Bt cotton seeds in 2002-2003. This was reduced to US$ 9 per packet with the onset of price control and in 2009-2010 royalties went down to as low as US$1 per packet for some *Bt* cotton seeds. In India, price controls, introduced mid-2006, are partially responsible for increased sales of *Bt* cotton seeds Introduction of BG II and new *Bt* events from JK Agri Genetics and Nath seeds also increased farmer choice. By 2006-07, the average Bt seed price declined by 68 % (Sadashivappa and Qaim, 2009). This resulted in higher seed demand, seed sales and expansion of area under *Bt* cotton. *Bt* cotton in India is commercialized in hybrids, and farmer has to buy seed every season. Thus, the private sector profits from selling the GM cotton hybrids are higher than in China. *Bt* technology target increased yield due low pest damages. Qaim and Zilberman (2003) observed that Mahyco Bt hybrids with the same hybrid without *Bt* in three major cotton-growing states of India (Tamil Nadu, Maharashtra and Madya Pradesh) were sprayed three times less (70%) and yield increased by 80 to 87%. Bennett *et al.* (2006) found a significant reduction in pesticide expenditure, 72-83 % in 2002 and 2003. Seed costs were higher, however, increased yields of *Bt* cotton of 45-63 % compensated for higher seed costs. In Maharashtra also reported a 79.2% higher profit from *Bt* cotton cultivation, compared to non *Bt* cultivation under irrigated conditions (Vaidya, 2005). In study of cotton-growing states of Gujarat, Maharashtra, Andhra Pradesh and Tamil Nadu, Gandhi and Namboodiri (2006) found significantly higher yields of *Bt* cotton with reduced pesticide costs under both irrigated and rain-fed conditions. The average yield of *Bt* cotton increased to 560 kg/ha in 2007-2008 as compared to 300 kg/ha in 2001-2002 after introduction of *Bt* cotton. Frisvold and Reeves (2007) concluded that Bt adoption in India led to a more than US$ 200 billion gain in India. The increased worldwide production led to a 3 % decline in world cotton prices. The widespread adoption of *Bt* cotton in India and other South Asian countries will result in additional regional welfare gains on the order of $1 billion per year. Traditionally, cotton production has required significant insecticide use. In the USA and China, *Bt* cotton was commercialized in the mid 1990s, and today the technology covers about 30–40 per cent of the cotton area in both countries. Recent studies show that USA and Chinese Bt adopters realize significant pesticide and cost savings in most cotton-producing regions (Pray *et al.*, 2002; Huang *et al.*, 2002a). Preliminary benefits of Bt cotton have also been reported for South Africa (Thirtle *et al.*, 2003; Isma¨el *et al.*, 2002) and Mexico (Traxler *et al.*, 2001). Nonetheless, relatively little is known about Bt–insecticide interactions and productivity effects under different agroecological conditions (GRAIN, 2001). The broader impacts of GM crops in general, and Bt cotton in particular, are still a matter of controversy, especially with respect to long-term environmental implications and sustainability (Batie and Ervin, 2001; Benbrook, 2001; UK Soil Association, 2002). This holds true both in developed and developing countries. Data presented in Table 17.14 and 17.15 indicates higher profits from *Bt* cotton adoption.

17.6 Dissemination of IPM Technology

17.6.1 IPM Programs in Latin America

In Latin American countries there are many successful examples of IPM. Cotton pest management in Peru and Nicaragua in the mid 1950s and early 1970s amply proved that sustainable pest management is possible by adopting a combination of pest management tactics. In Brazil, there are numerous pests that attack the cotton plant. These can be divided in two groups: a) pests which attack the plant during the establishment phase, (root borers, thrips, stem borers, stink bugs, aphids, and leaf sucking insects), and b) pests that occur at flowering and boll forming. (Alabama, whitefly, various types of caterpillar, mites, and boll weevils). The most logical method to control these pests is through IPM practices, such as destroying crop residues, concentrated planting windows; use of varieties resistant to attacks; crop rotation and complete control of crop border areas where most infestation begins; as well as monitoring, using pheromones, *etc.*; and localized controls. Chemical control is used in accordance with the monitoring of the types and specific species of pests. In Brazil, the use of agricultural chemicals requires that prescription and use be controlled by an agronomist, who will recommend using a systemic, biological or contact product. During late 1970s, the resistance of cotton boll worm, *Heliothis virescens* (Fabricius) to organophosphates was a major problem. At that time cotton received 15-20 insecticide applications per season. The agricultural research and extension services in Brazil initiated an intensive program to transfer IPM technology to cotton farmers (Bleicher *et al.,* 1979). The launching of the IPM program, helped significantly to reduce insecticide applications on average to six sprays per season helping to optimize profits through lower production costs for the same level of yield (Pimentel and Bandeira, 1981; Seganmllar and Hewson, 2000). But in 1983, with the introduction of cotton boll weevil, *Anthonomus grandis*, Boheman to Brazil, the number of insecticide_applications again rose to 10-12 applications per season. However, after local behaviour_patterns were established for the new pest, and IPM adapted accordingly, the number of applications decreased again to an average of 8 per season (Seganmullar and Hewson,2000). IPM has produced excellent economic, social and ecological results in Brazil. Later, the chemical pesticide industry began to collaborate in an IPM program_which did not show desired results (Ramalho, 1994). The Latin American Association for Cotton Research and Development should establish working groups on research and extension in all member countries for exchange of information on IPM in cotton. In Peru, the Peruvian Action Network of Alternatives to Agrochemicals (Spanishacronym RAAA) in 1992 started a training programming Can˜ete Valley to strengthen farmers' interest in IPM, under the theme "Ecological Pest Management for Cotton Growers". In 1997, a small project on organic cotton production was also set up (CABI, 2000). The National Agricultural Research Institute's (INRA) cotton IPM program in Argentina is researching, implementing, validating_and monitoring_the mass production of the predator_*Chrysopa* spp. for aphid and cotton pest management. The IPM concept adopted based on biological control by cotton growers resulted in reducing insecticide applications from 11-12 per season to <4 (Williamson, 1999). The first attempt to organize farmer training along discovery – learning methods was in Peru. The IPM intervention in cotton resulted in reduction in pesticide use by 50-70 percent (Castro *et al.*, 1997; Van Elzakker, 1999). Most training in IPM programs in Latin America had been based on result demonstration methods with little active farmer participation.

Africa

IPM strategies were refined in African countries, to ensure that IRM principles were incorporated (Ochou and Martin, 2003). In West Africa, Government or semi-government cotton companies provide technical expertise, arrange for reliable inputs such as seeds, pesticides, fertilizers etc., and supply them to cultivators from time to time depending on the need. In east and central African countries such as Zambia and Uganda, ginneries and Government cotton companies emerged as the main source of expert technical advice and input providers to farmers co-operative organisations (Burgess 2001, Jarvis 2001). In the African countries, there is a move towards 'liberalisation' of the production channels under pressure from the IMF/World Bank structural Adjustment Programmes, as a result of which West Africa for example, is moving towards multiple private company (or grower co-operative) cotton companies. Staff of the Ministry of Agricultural Manages IPM in Egypt. Single varieties are grown in particular governorates and' engineers' of the Ministry of Agriculture supervise pest management programmes. Thus the extension systems organize area-wide IPM implementation to ensure long term impact. West Africa appeared to avoid pyrethroid resistance for many years, perhaps because the pyrethroids/ organophosphate mixtures used prevented the selection of esterase-based metabolic detoxification resistance through the impact on esterases of the OP component of the mixtures (Martin *et al.,* 2002). Problems were first identified in 1996 (Vassal *et al.,* 1997) with a 20–100 fold resistance identified across the region by 1998. The West African pyrethroid resistance action network (PR-PRAO) quickly suggested the replacement of the first two calendar sprays of the OP/pyrethroid mixtures by endosulfan. This was widely implemented across West Africa thereby leading to problems of insecticide resistance, pest resurgence, accumulation of harmful residues and toxicity to non-target organisms. *Helicoverpa armigera* declined rapidly in importance in cotton across the region the following year and remained low thereafter (Ochou and Martin, 2003). *Helicoverpa armigera* is notoriously variable in pest pressure, but this is probably a rare example of successful regional co-operation, combined with the benefits of a bulk pesticide purchase and provision system by a well organised cotton company. A three window strategy is now being proposed with spinosad for the first two sprays as it has a better profile against *Earias* and phytophagous mites and the beneficial coccinellids. This followed by two sprays of the tradition OP/pyrethroid mixtures for control of mites and the endocarpic bollworms. A final two applications of indoxacarb made in the late season when cotton stainer are of greater importance.

USA and Australia

In the developed countries such as the USA and Australia, where the average land holding per farmer is more than 800 ha, private consultants provide pest management expertise. The IPM strategies are devised by technical experts from scientific organizations and are implemented by the private consultants. In South America, large cotton companies oversee the cotton pest management programmes. The first successful example of formulating and implementing IRM guidelines came from Australia in 1983. The strategies were carefully incorporated into IPM methods so as to ensure that unnecessary selection pressure by chemical groups was avoided. The strategies have been refined continuously each year with scientific expertise from the CSIRO and other institutions. The recent recommendations incorporate soft chemistries to minimize adverse effects on beneficial fauna and set up a three window rotational approach for insecticide use on *H. armigera*. Soft chemistries such as endosulfan, methoxyfenozide, Bt, NPV and Amitraz are permissible through the first two windows, with Bt, NPV and Amitraz extending until the end of third window. Spinosad is used from 10th December onwards and Avermectins, Emamectin and Abamectin are used from 15th November until the end of the 2nd window. Indoxacarb is permissible from 20th December until the end of February (mid point of 3rd window). Other insecticides such as Chlofenapyr, pyrethroids (with or without BO), Organophosphates (Chlorpyriphos and Profenophos) extend through the $2^{nd}$and 3rd windows. Carbamates (Thiodicarb and Methomyl) are used exclusively in the 3rd Window.

Asia

The IPM program_was implemented in, China, India, and Pakistan during mid 1970 by different agencies and emphasis was on judicious use of insecticides for management of insect pests. The calendar based spray schedules were replaced by timing of sprays on basis of economic threshold level. Due to various operational difficulty in adoption of ETL farmers preferred fixed spay schedule. Due to the lack of resistance varieties, constraints in adoption of cultural practices and non availability of effective bioagents, dependence was on use of insecticides. This farmer approach provided better return to farmers and continued till 1990s. IPM programmes in India and Pakistan are implemented mainly by the Agricultural extension system. In India, state agricultural departments were expected to play major role in popularizing the IPM concepts and implementing them. However, instead of reducing dependence subsidies were given on insecticide and aerial sprays in Punjab, India. IPM was not adopted on a large scale as non chemical approach was not popular with cotton growers and were able to mange with insecticide with high return. Farmers generally obtain guidance and pest management inputs from an extremely extensive, unlicensed, pesticide dealer network, moderated by technical advice from the Agricultural Universities and the Government Cotton Research Institutions. In China, the extension system is active with state extension employees and licensed input dealers delivering technical advice.

China

China has put forward the concept "integrated control" as early as the early 1950s (Jing, 1997). Chinese plant protection scientists formulated the principle of plant protection "Focus on Prevention and Implement Integrated Control", namely the IPM framework in 1975 and coordinated as well as arranged a number of research and promotion programs on IPM and has made great success (Zhang *et al.*, 2001). According to Wang and Lu (1999) implementing the development process of IPM framework in China can be generally divided into three stages :(i) Pest-centered IPM, *i.e.*, the first-generation IPM. For example, during the period of "The Sixth Five Year Plan" (1981-1985), each of the main pests on a certain crop was controlled below the economic threshold using physical, chemical and biological control methods. (ii) Crop-centered IPM, i.e., the second-generation IPM. For example, during "The Seventh Five Year Plan" (1986-1990), with crop as the center, a variety of major pests on the crop were controlled. At this stage, IPM gave full play to the natural control of ecosystem and IPM systems began to be established. During "The Eighth Five Year Plan" (1991-1995), a large number of IPM systems have been developed, assembled, improved and applied in China. (iii) Ecosystem-centered IPM, i.e., the third-generation IPM. The entire field or regional ecosystem was the focus of IPM; a large quantity of advanced scientific information and data were collected and used, and advanced technologies were developed in IPM practices. Overall and global benefit was expected to be increased with the natural control of ecosystems as the main force. At present, China is in the transition phase of the second and third-generation IPM. During 1993-1996, the ADB Cotton IPM Program was implemented in Tianmen, Hubei Province, under the auspices of the National Agricultural Technology Promotion Center of China (Zhang *et al.*, 2002). Cultivation benefits of cotton increased by implementing IPM. For example, implementing IPM cotton can increase income by US 206.4/ha and reduce the use of chemical pesticides. Pesticide applications cotton were reduced 12.2 times. According to the survey, there were totally 2,325 predatory natural enemies on IPM cotton but only 1,168 predatory natural enemies on non-IPM cotton (Zhang *et al.*, 2002). On the whole the applications of IPM technologies in China are still highly localized. Pesticide misuses are still common. Excessive use of pesticides in cotton production reached 50% (Chen and Han, 2005). The lack of applications of IPM in China is attributed to: (i) under the household contract system, agricultural intensification and on-scale operation could not be realized easily, the farmers have less demand on IPM technologies. (ii) IPM technical extension services systems are insufficient. (iii) Pesticides markets are not ordered, the social environment for IPM application has not yet been established. (iv) Short of theoretical researches and application technologies of IPM. At present, IPM technologies are not perfect, and monitoring effectiveness and forecasting accuracy are at a lower level (Chen and Han, 2005).

India

During the green revolution in mid 1960s, use of pesticides, high yielding varieties and fertilizers were on agenda to increase productivity. The traditional methods of pest management were replaced with insecticide based insect pest management as the sole pest control strategy. Most of the pesticide was consumed in the green revolution was in Punjab, Haryana, Andhra Pradesh and 50 per cent in cotton crop which is cultivated on a mere 5 percent of the total cultivable land of 176 million hectares. Soon the policy planners, entomologists and extension agencies realised the negative impact of pesticide use brought fore by Rachel Carson in her book 'Silent Spring "in 1962 and evolved pest management option to use all options from sowing to picking to keep pest below economic threshold an use insecticide as last option. Many projects were initiated to transfer the IPM technology.

Operational Research Projects

Punjab

Operational Research Project (ORP) on Integrated Pest Management in 15 villages Punjab in 1975 to 1990 under World Bank funded project by Indian Council of Agricultural Research. Main objectives ORP project were to test and demonstrate the efficacy, practicability and economics of IPM strategy. Farmers were educated by ORP staff about different insect pest problems, side effects of excessive use of insecticides and cultivation of susceptible long duration varieties and concept of ETL/IPM strategy. The first main success was shifting from long duration varieties, mainly LSS (*G. hirsutum*) to medium duration and insect-tolerant variety *Bikaneri Narma* early maturing varieties. Prior to start of project, are under long duration and short duration varieties 88.8 and 9.9 per cent. By 1978, LSS variety was completely replaced by *Bikaneri Narma* / F 414. Induction of early maturing variety improved the economics cotton-wheat rotation. The adoption of IPM technology resulted in 73.7 and 12.4 per cent reduction in the number of insecticide sprays for control of sucking pests and bollworms respectively (Sidhu *et al.*, 1984). The reduction of bollworm incidence in ORP villages was 38.55 per cent higher than non-ORP villages leading to 23.1 per cent higher yield and 31.1 per cent higher net income. The yield potential of seed cotton in demonstration fields was 20.45 q/ha as against 10.17 q/ha in other fields (Sidhu *et.al.,* 1990).

Tamil Nadu

A similar ORP project as in Punjab on cotton was initiated at Coimbatore, Tamil Nadu in 1980. Various IPM option from sowing to picking were demonstrated and adopted by farmers which resulted in reducing the use of insecticides without affecting the yield of seed cotton, at the same time encouraging the natural enemies and reducing the environmental pollution. The mean quantity of insecticides used in the project villages was 3.82 kg a.i/ha in 6 applications at spray interval of 15-18 days, which was 58.6 per cent less than the amount (9.23 kg a.i./ha) used in non-project villages with 11 sprays at an interval of 7-8 days . The abundance of native natural enemies by three fold reduced the cost of insecticide and environmental pollution by 50.3 and 53.4 per cent respectively. IPM technology in 48 villages was demonstrated by Department of Agriculture, Tamil Nadu in 2650 ha cotton area during 1984. The farmers in IPM villages harvested on an average yield of 26.70 q/ha as against 22.25 q/ha in control villages (Sundaramurthy and Chitra, 1992). These studies have demonstrated the efficacy and economics of IPM on cotton.

Insecticide Resistance Management based IPM programme

Central Institute of Cotton Research, Nagpur, Asian Development Bank – Commonwealth Agricultural Bureau International (ADB-CABI) and Directorate of Plant Protection, Quarantine and Storage (DPPQS) conducted season long training of the IPM extension worker since 1994 to promote IPM in the entire country. DPPQS through its Central IPM Centre (CIPMCs) promoted the concept of IPM and developed IPM packages for cotton. Agroecosystem analysis (AESA), Farmer Field Schools (FFSs), scouting and monitoring *etc.* were popularized. Various State Agriculture Departments implemented IPM from mid-1990s on the ICDP grant from the centre. Incentives were provided to the farmers to adopt the IPM practices and critical inputs arranged. From 2000-2001 Government of India launched Technology Mission on Cotton (TMC) with the main objective of improving yield and quality of cotton, reduce the cost of cultivation and post harvest aspects (Barik *et al.*, 2002). The CICR, Nagpur launched the IRM programme in six cotton-growing states of India, focusing on 26 districts, which among them, were consuming 80 per cent of insecticides for cotton crop. IRM programmes were implemented by state agricultural universities and some other All India Coordinated Cotton Improvement Project (AICCIP) centers of different states. The impact of adoption of IRM strategies resulted in the reduction of insecticidal spray and increased in seed cotton yield (Table 17.15 to 17.17).

PAU, State Department of Agriculture and Sir Rattan Tata Trust- led project

PAU Ludhiana in collaboration with State Department of Agriculture adopted entire state of Punjab during 2008-2010 to disseminate IPM technology. The impact of adoption of IPM technology is summarized in Table 17.8. Department of Entomology, PAU Ludhiana adopted 24 villages in four districts of Punjab to disseminate the IPM technology in *Bt* cotton in SRTT funded project. Department organized different training programmes on IPM of cotton during 2008-2011 in which 1500 scouts were trained employed by Sir Rattan Tata Trust and State department of Agriculture to implement IPM at village level. The impact of dissemination of IPM technology resulted in the reduction in the spray cost and cost of cultivation in the IPM field over non-IPM fields.

Andhra Pradesh

The results of several studies have revealed the utility of on-farm testing and demonstration of IPM technology. Divakar *et al*. (1992) have demonstrated IPM on hybrid cotton against *H. armigera* in Andhra Pradesh. Pest population was monitored at weekly interval by scouting. Control measures included parasitoids, manual picking of eggs and larvae and need based use of insecticides. The first IPM plot received 10 sprays of conventional insecticides, second one 9 sprays including 3 sprays of synthetic pyrethroids as against 18 sprays including 10 sprays of synthetic pyrethroids in case of control plot with farmer's practices. The cost benefit ratio was higher in IPM plots (1: 4.5) than non-IPM plot (1: 1.15).

17.7 Conclusion

The cultivation of high yielding varieties, increase in area under irrigated cotton and better pest management due to availability of toxic and effective insecticides particularly, synthetic pyrethroids resulted in higher productivity before introduction of *Bt* cotton. The excessive exploitation of energy subsidies is responsible for volatile economy of cotton crop. This resulted in wiping out of natural biocontrol mechanism and development of insecticide resistance and resurgence of pests with use of insecticides, making the pest management very difficult. With the result, that in spite of increase in cost of pest management the yields are declining. The continuous failure of cotton crop has shattered the economy of cotton belt of world. *Bt* cotton was introduced to manage the bollworms particularly American boll worm [*Helicoverpa_armigera* (Hubner)] as this pest developed resistance to insecticides and threatened the cultivation

Table 17.1 : Arthropod fauna in irrigated cotton agro-ecosystem (India)

S. No. (1)	Common/Scientific name (2)	Family (3)	Status of the arthropod species# (4)	Plant part harboured (5)
Order: Coleoptera				
1.	Lady bird beetle, *Coccinella septempunctata* (Linnaeus)	Coccinellidae	B	Aphids*
2.	*Menochilus sexmaculatus* (Fabricius)	Coccinellidae	B	Aphids*
3.	*Brumoides suturalis* (Fabricius)	Coccinellidae	B	Aphids*
4.	*Rodolia cardinalis* (Mulsant)	Coccinellidae	C	Leaves
5.	*Rodolia fumida* (Mulsant)	Coccinellidae	C	Leaves
6.	*Harmonia dimidata* (Fabricius)	Coccinellidae	C	Leaves
7.	*Zygogramma bicolorata* (Pallister)	Chrysomelidae	C	Leaves
8.	Blister beetle, *Mylabris phalerata* (Pallas)	Meloidae	P	Squares and flowers
9.	Click beetle, *Dicronychus* spp.	Elateridae	C	Leaves
10.	*Altica cyanea* (Weber)	Chrysomelidae	C	Leaves
11.	Red pumpkin beetle, *Aulacophora foveicollis* (Lucas)	Chrysomelidae	C	Leaves
12.	*Myllocerus discolor* var. variegate (Boheman)	Curculionidae	P	Leaves
13.	*Myllocerus sabulosus* (Mshll.)	Curculionidae	P	Leaves
14.	*Myllocerus undecimpustulatus* var. maculosus (*Desbrocher*)	Curculionidae	P	Leaves
15.	*Neocleonus sannio* (Herbst)	Curculionidae	P	Leaves
16.	*Oxycetonia versicolor* (Fabricius)	Cetonidae	C	Squares and flowers
17.	Rove beetle, *Paederus fuscipes* (Curtis)	Staphilinidae	B	Small insects*
18.	*Chiloba acuta* (Widemann)	Scarabaeidae	C	Leaves
19.	*Monolepta signata* (Olivier)	Galeuricidae	P	Green bolls
20.	*Palorus* spp.	Tenebrionidae	P	Flowers
21.	Callosobruchus spp.	Bruchidae	P	Leaves, bolls
22.	*Formicomus rufipennis* (Gyllenhal)	Anthicidae	B	Dead insects*
Order: Dictyoptera				
23.	European mantid, *Mantis religiosa* (Linnaeus)	Mantidae	B	Caterpillars, other insects*
24.	*Creobrator pictipennis*	Mantidae	B	Caterpillars, other insects*

Contd...

Order: Diptera				
25.	Syrphid fly, *Eristalis quinquestriatus* (Fabricius)	Syrphidae	B	Aphid*
26.	Syrphid fly, *Ischiodon scutellaris* (Fabricius)	Syrphidae	B	Aphid*
27.	Robber fly, *Philodicus femoralis* (Ricardo)	Asilidae	B	Cutworms and grubs*
28.	*Synthesiomyio nudisteta* (Wulp)	Muscidae	C	Squares and flowers
29.	Cotton leaf miner, *Liriomyza trifolli* (Burgess)	Agromyzidae	P	Leaves
30.	*Agromyza obtuse* (Mall.)	Agromyzidae	P	Leaves
Order: Hemiptera				
31.	Cotton whitefly, *Bemisia tabaci* (Gennadius)	Aleyrodidae	P	Leaves
32.	Cotton aphid, *Aphis gossypii* (Glover)	Aphididae	P	Leaves
33.	Cotton jassid, *Amrasca biguttula biguttula* (Ishida)	Cicadellidae	P	Leaves
34.	Dusky cotton bug, *Oxycarenus hyalipennis* (Costa)	Coreidae	P	Seeds and open bolls
35.	*Riptortus* spp.	Coreidae	C	Leaves
36.	Tur pod bug, *Clavigralla gibbosa* (Spinola)	Coreidae	C	Leaves
37.	*Corizus bengalensis* (Dall.)	Coreidae	P	Leaves
38.	*Agraphopus orientalis* (Dist.)	Coreidae	P	Leaves
39.	Anoplocnemis phasiana (Fabricius)	Coreidae	C	Leaves
40.	Big eyed bug, *Geocoris* spp.	Geocoridae	B	Whitefly, bugs bollworms, mites*
41.	Red cotton bug, *Dysdercus koenigii* (Fabricius)	Pyrrhocoridae	P	Leaves, Green and open bolls
42.	*Antilochus coqneberti* (Fabricius)	Pyrrhocoridae	P	Green and open bolls
43.	Assassin bug, *Zelus* spp.	Reduviidae	B	Whitefly, bugs bollworms, mites*
44.	Minute pirate bug, *Orius* spp.	Anthocoridae	B	Whitefly*
45.	Green potato bug, *Nezara viridula* (Linnaeus)	Pentatomidae	P	Leaves
46.	*Dolicoris indicus* (Stal.)	Pentatomidae	C	Leaves
47.	*Halys* spp.	Pentatomidae	C	Leaves
48.	Painted bug, *Bagrada hilaris* (Burmeister)	Pentatomidae	P	Leaves
49.	*Albizzia* spp.	Lygaeidae	C	Leaves
50.	*Graptostethus servus* (Fabricius)	Lygaeidae	P	Green and open bolls
51.	*Creontiades pallidus* (Rambir)	Miridae	C	Squares and

Contd...

				flowers
52.	*Malacocoris* spp.	Miridae	C	Leaves
53.	*Camplobrochis signatus*	Miridae	C	Leaves
54.	*Campylomma* spp.	Miridae	C	Leaves
55.	*Oxyrhachis subjecta* (Walker)	Membracidae	P	Leaves and stem
56.	Cotton mealy bug, *Phenococcus solenopsis* Stanley	Coccidae	P	Leaves, stem and bolls
Order: Hymenoptera				
57.	Giant/rock bee, *Apis dorsata* (Fabricius)	Apidae	O	Flowers, floral buds
58.	Little honey bee, *Apis florea* (Fabricius)	Apidae	O	Flowers, floral buds
59.	Italian honey bee, *Apis mellifera* (Linnaeus)	Apidae	O	Flowers, floral buds
60.	Yellow wasp, *Polistes olivaceous* (DeGeer)	Vespidae	B	Insects*
61.	Red wasp, *Vespa orientalis* (Linnaeus)	Vespidae	B	Insects*
62.	*Xylocopa pubescens* (Spinola)	Xylocopidae	O	Flowers, floral buds
63.	*Xylocopa fenestrata* (Fabricius)	Xylocopidae	O	Flowers, floral buds
64.	Carpenter bee, *Pithitis smaragdula* (Fabricius)	Anthophoridae	O	Flowers, floral buds
65.	Black ant, *Polyrhachis simplex* (Mayr)	Formicidae	B	Small insects*
66.	*Monomorium indicum* (Forel)	Formicidae	P	Leaves, stem
67.	Small red house ants, *Dorylus labiatus* (Schukard)	Formicidae	P	Leaves, stem
68.	*Oecophylla smaragdina* (Fabricius)	Formicidae	B	Small insects*
69.	Leaf ant, Solenopsis geminata (Fabricius)	Formicidae	P	Leaves
70.	*Camponotus compressus* (Fabricius)	Formicidae	B	Jassid*
71.	*Camponotus sericeus* (Fabricius)	Formicidae	B	Jassid*
72.	*Megachile anthracina*	Megachilidae	O	Flowers, floral buds
73.	*Megachile lanata* (Fabricius)	Megachilidae	O	Flowers, floral buds
74.	*Scolia* spp.	Scoliidae	B	Scarabaeid larvae*
75.	*Chalybion bengalense* (Dahlborn)	Sphecidae	B	Grass-hoppers, beetles
76.	Sceliphron coromendalicum	Sphecidae	B	Grass-

Contd...

				hoppers, bugs, plant hoppers, beetles*
77.	Digger wasp, Sphex subtruneatus (Dahlbom)	Sphecidae	B	Grass hoppers, crickets, plant hoppers, beetles*
78.	*Chlorion lobatum* (Fabricius)	Sphecidae	B	Grass hoppers, bugs, plant hoppers, beetles*
79.	*Cerceris instabilis* (Smith)	Sphecidae	B	Bugs, plant hoppers, Grasshoppers, beetles*
80.	*Larra nigriventris* (Menon)	Sphecidae	B	Mole cricket and small insects*
81.	*Eretmocerus* spp.	Aphelinidae	B	Whitefly*
82.	*Encarsia formosa* (Gahan)	Aphelinidae	B	Whitefly*
83.	*Apanteles angaleti* (Muesbeck)	Braconidae	B	Bollworms*
84.	*Chelonus blackburni* (Cameron)	Braconidae	B	Bollworms*
85.	*Campoletis chloridae* (Uchida)	Ichneumonidae	B	American bollworm*
86.	*Xanthopimpla punctuator* (Linnaeus)	Ichneumonidae	B	Cotton leaf roller*
87.	*Sphecodes* spp.	Halicitidae	O	Flowers and floral buds
88.	*Halictus matheranensis*	Halicitidae	O	Flowers and floral buds
Order: Isoptera				
89.	*Microtermes obesi* (Holmgren)	Termitidae	P	Squares, stem, flowers, roots
90.	*Odontotermes obesus* (Rambur)	Termitidae	P	Squares, stem, flowers, roots
Order: Lepidoptera				
91.	Spiny bollworm, *Earias insulana* (Boisduval)	Noctuidae	P	Floral buds, flowers, green bolls

Contd...

92.	Spotted bollworm, *Earias vittella* (Fabricius)	Noctuidae	P	Floral buds, flowers, green bolls
93.	American bollworm, *Helicoverpa armigera* (Hubner)	Noctuidae	P	Leaves, green bolls, squares, flowers
94.	Tobacco caterpillar, *Spodoptera litura* (Fabricius)	Noctuidae	P	Leaves, flowers, bolls
95.	Tobacco caterpillar, *Spodoptera exigua* (Hubner)	Noctuidae	P	Leaves, flowers, bolls
96.	Green semilooper, *Anomis flava* (Fabricius)	Noctuidae	P	Leaves
97.	Cotton semilooper, *Tarache notabilis* (Walker)	Noctuidae	P	Leaves
98.	Black cutworm, *Agrotis ipsilon* (Hufnagel)	Noctuidae	P	Seedlings, leaves
99.	Tobacco budworm, *Heliothis virescens* (Fabricius)	Noctuidae	P	Green bolls, squares, flowers
100.	Red gram leaf webber, *Maruca testulalis* (Geyer)	Pyraustidae	C	Leaves
101.	Red hairy caterpillar, *Amsacta moorei* (Butler)	Arctiidae	P	Seedlings, leaves
102.	Bihar hairy caterpillar, *Spilosoma obliqua* (Walker)	Arctiidae	P	Leaves
103.	Akk butterfly, *Danaus plexippus* (Linnaeus)	Nymphalidae	C	Flowers, floral buds
104.	Rice skipper, Pelopidas mathias (Fabricius)	Hesperidae	C	Leaves
105.	Cotton leaf roller, *Sylepte derogata* (Fabricius)	Pyralidae	P	Flowers
106.	Cabbage butterfly, *Pieris brassicae* (Linnaeus)	Pieridae	C	Flowers
107.	Grass yellow, *Eurema hecabe* (Linnaeus)	Pieridae	C	Flowers, floral buds
108.	Castor hairy caterpillar, *Euproctis lunata* (Walker)	Lymantridae	P	*Leaves*
109.	Pea blue butterfly, *Lampides boeticus* (Linnaeus)	Lycaenidae	C	Flowers

Order: Neuroptera

110.	Green lace wing, *Chrysoperla carnea* (Stephens)	Chrysopidae	B	Aphid, jassid, mites whitefly, small insects*

Order : Odonata

111.	*Pantala flavescens* (Fabricius)	Libellulidae	B	Small insects*
112.	Acisonia panorpoides	Libellulidae	B	Small insects*

Contd...

113.	*Coenagrion puella* (Linnaeus)	Coenagriidae	B	Small insects*
114.	Trithemis pallidinervis (Kirby)	Libellulidae	B	Small insects*
Order: Orthoptera				
115.	Surface grasshopper, *Chrotogonus trachepaterus* (Blanchard)	Acrididae	P	Seedlings, leaves
116.	*Acrida exaltata* (Walker)	Acrididae	P	Leaves
117.	Ak grasshopper, *Poecilocerus pictus* (Fabricius)	Acrididae	P	Leaves
118.	*Oxya hylahyla* (Serville)	Acrididae	P	Leaves
119.	*Atractomorpha crenulata* (Fabricius)	Acrididae	P	Leaves
120.	*Atractomorpha* spp.	Acrididae	P	Leaves
121.	*Apilogous thalassimus* (Fabricius)	Acrididae	P	Leaves
122.	Desert locust, *Schistocerca gregaria* (Forskall)	Acrididae	P	Leaves
123.	Field cricket, *Gymnogryllus* spp.	Gryllidae	P	Seedlings, leaves
124.	*Trigonidium humbertianum* (Sauss)	Gryllaeridae	P	Leaves
125.	*Elimaea nigropunctata* (Br.W.)	Tettigoniidae	P	Leaves
Order: Thysanoptera				
126.	Cotton thrips, *Thrips tabaci* (Lindeman)	Thripidae	P	Leaves
Order: Araneae (Spiders)				
127.	Orb spider, *Araneus inustus* (Koch)	Araneiidae	B	Bollworms, bollweevil*
128.	Jumping spider, *Phidippus* spp.	Salticidae	B	Bollworms, bollweevil*
129.	Lynx spider, *Oxyopes javanus* (Throell)	Oxyopidae	B	Bollworms, bollweevil*
130.	Lynx spider, *Oxyopes lineatipes* (Koch)	Oxyopidae	B	Bollworms, bollweevil*
131.	Wolf spider, *Lycosa pseudoannulata* (Boesenberg and Strand)	Lycosidae	B	Bollworms, bollweevil*
132.	Long jawed spider, *Tetragnatha maxillosa* (Thorell)	Tetragnathidae	B	Bollworms, bollweevil*
Mites				
133.	Red spider mite, *Tetranychus urticae* (Koch)	Tetranychidae	P	Leaves
134.	Predatory mites, *Ambyseius* spp.	Phytoseeidae	B	Thrips, spider mites*

P- pests of cotton (major, minor and sporadic herbivorous arthropods), B- beneficial arthropods/ natural, enemies, C- casual visitors, O-pollinators)

* Host of arthropod species/prey of the beneficial arthropod species

(*Source* : Bal and Dhawan 2007)

Table 17.2 : Wild gene sources and nature of insect resistance and contributing characters

Wild/racial gene source	**Contributing characters**	**Insect pests**
G.annomalum	Raised gossypol glands, hairiness, narrow bract	Jassid, mite, *Earias* spp.
G.sturtiannum	Waxy epidermis, glanded	Bollworms
G.australe and G.bickii	Pubescent bolls, delayed morphogenesis of glands, high gossypol	Bollworms
G.armourianum	Double palisade, smooth, caducous bract	Jassid, *Helicoverpa*, whitefly
G.harknessii	Double palisade, smooth, caducous bract	Bollworms
G.davidsonii	Thick leaf epidermis	Aphids
G.raimondii	More glanded, pubescent	Jassid, thrips, leaf roller, *Heliothis*, bollworms
G.gossypioides	Foliar nectariless	Sucking pests, *Helicoverpa*
G.somalense	Gossypol high	Bollworms
G.tomentosum	Tomentose hairs , nectariless	Jassids,thrips,bollworms
Cernuum	Gossypol high, long bolls, hard rind	Pink bollworm
Bengalense	Red plant, gossypol	Pink bollworm, *Helicoverpa*
Rozi	Red plant, gossypol	Bollworms
Nadam	Gummy exudation	Stem weevil
Mariegalante and Brasiliense	Gummy exudation	Stem weevil
Palmeri	Palmate leaf	Sucking pests,bollworms
Punctatum	Gossypol	Bollworm, jassid
G.longicalyx	Nectariless	Whitefly
G.labatum	Small bracts	Jassids
Texas NPP races	Gossypol and other unknown factors	*Heliothis,* pink bollworm
G.thurberi	Deeply dissected leaf, more glanded	Bollworms

Source : Narayanan *et al.*(1992)

Table 17.3 : Resistant/ tolerant cotton cultivar recommended for cultivation since 1985 in different agroclimatic conditions.

Gossypium spp	Year	Cultivar	Pest species	Zone/State
G.arboreum	1985	LD327	Jassid, Whitefly PBW, SBW	Punjab
	1995	LD 491	Jassid, Whitefly Bollworms	Punjab
G.hirsutum	1987	Kanchan	Whitefly,ABW	South Zone
		LK 861	Whitefly,ABW	South Zone
		AKH 081	PBW,SBW	Maharashtra
	1988	Abadhita	PBW,SBW	Tamil Nadu
		JK 119	PBW,SBW	Karnataka
	1992	LRK 516	PBW,SBW	Gujarat Maharashtra

Source : Dhawan, 1999

Table 17.4 : Resistance potential of various plant traits against different insect pests of cotton

Plant traits	American bollworm	Spotted bollworm	Pink Bollworm	Jassids	Thrips	Aphids	White-fly	Spider mite
A. Morphological characters								
Okra	-	R	R	-	-	-	R	R
Rapid fruiting	-	E	E	S	-		R	-
Frego bract	-	R	-	-	-	-	-	-
Nectariless	R	R	R	-	-	-	-	-
Glandless	S	S	S	-	-	-	-	-
Hirsute	S	-	-	R	-	-	S	-
Pilose	S	-	S	R	R	S	S	-
Glabrous leaf	R	R	R	S	S	S	R	-
Hairy leaves	-	S	-	R	-	-	S	-
Low bract teeth	R	R	R	-	-	-	-	-
Boll rind thickness	-	-	R	-	-	-	-	R
Red body	-	R	R	-	-	-	R	-
Glanded plant-cum-glandless seed	R	R	R	-	-	-	-	-
Naked seed	-	-	R	-	-	-	-	-

B.Biochemical characters								
High gossypol	R	R	R	R	-	-	-	-
Heliocides (H1 to H 4)	R	R	-	-	-	-	R	-
Silicated leaf	-	-	-	R	-	-	-	-
High tannins	R	R	R	-	-	-	R	R
High bud sucrose	S	-	S	-	-	-	-	-

R = Relative resistance; S= Relative susceptibility; E=Escape

Source : Butter *et al.* 1992, Ilango and Uthamaswamy 1989, Narayanan *et al.*1988,1990,1992, Rao *et al.* 1990, Uthamasamy1996 185, 241.

Table 17.5 : Parasitoids of major insect pests of cotton

Parasitoids	Host	Parasitism (%)	Period of activity
		A.Egg parasitoids	
Aphelinus sp	E	1-10	September-October
Erythmelus empoascae	E	5	September-October
Gonatocerus sp	E	2	November
Trichogramma achaeae	PE	5.9-27.24.0-60.8	September-October
T.brasiliensis	E	5	October-November
T.chilonis	EA	30-3535-83	September-November
T.chilotraeae	PE	30-352.7-15.0	September-November
Telenomus remus	S	-	
*Trichogrammatoidae*sp near *guamensis*	PE	21.4-25.0 4.9-41.9	October-November
		B. Larval parasitoids	
Agathis fabiae	E,P	2	September-October
Apanteles angaleti	P	25	January-December
Bracon chinensis	P	5	September-November
Bracon greeni	PE	7.0-10.02.0-6.3	October -January
Bracon kirkpatricki	E	-	-
Bracon brevicornis	E	-	-
Bracon hebetor	E	-	-
Camptolithlipsis gossypiella	P	5	September-November
Rogas aligarhensis	P ,EPE	3359.070.4-1.0	August-December
Scambus liniepes	PA	Negligible0.3-16.7	December

Contd...

Goniozus sp	P	0.3-16.7	
Carcelia illota	A	6.1-12.0	
Campoletis chloridae	A		
Elasmus johnstoni	P	10-12	August-November
E.philippinensis	S	1.7	October
Cremastus	S	1.0	September-October
Afrovoria indica	P,H		
Sisropa formosa	H		
Eucorcelia sp	H		
Drino sp	H		
Strobliomyia	P		
Pyemotes ventricosus (mite)	P	26.0-39.0	August-November
C.Egg-larval parasitoids			
Chelonus sp	PEH	0.5-3.63.07.6	September-November
C.blackburni	P	5	September-November
D.Pupal parasitoids			
Xanthopimpla punctata	S	5.5	October
Brachymeria sp.n. euploeae	S	16.5	October
B.apantelesi	E	Negligible	October -November
B.lasus	E	40	October
B.nephantidis	E	15	October -November
B.nosatoi	E	5	October

Source : Dhawan, 1990, 1999b, 2011; Natrajan and Seshadri, 1989; Natrajan and 'Sundaramurthy, 1990, Rao *et al*., 1989; Singh, 1985,1994a, 1994b; Sundaramurthy and Chitra, 1992; Jayaswal and Sundaramurthy, 1992.

Table 17.6 : Predators of cotton major sucking pests of India

Insect Pests	Predator	Stage attacked	Period of activity
Amrasca biguttula	*Chrysoperla carnea*	Nymph	Throughout the season
(Ishida)	*C.cymbela*	Egg,Nymph	-
	Coccinella septumpunctata	Egg,Nymph	July-October
	Cheilomenes sexmaculatus	Egg,Nymph	July-October
	Geocoris sp	Egg,Nymph	June-October
	Zelus sp	Egg,Nymph	June-October
	Camponotus sp	Egg,Nymph	-
	Spiders	Egg,Nymph	Throughout the season
Bemisia tabaci	*Chrysoperla carnea*	Egg,Nymph	Throughout the season
(Gennadius)	*C.cymbela*	Egg,Nymph	-
	Geocoris bicolor	Egg,Nymph	Throughout the season
	Orius sp	Egg,Nymph	Throughout the season
	Encarsia sp	Nymph	Throughout the season
	Mallada boninensis	Egg,Nymph	Throughout the season
	Geocoris sp	Egg,Nymph	Throughout the season
	Brumus saturalis	Nymph	Throughout the season
	Cheilomenes sexmaculatus	Nymph	Throughout the season
	Spiders	Nymph,Adult	Throughout the season
Aphis gossypii	*Mallada astur*	Nymph,Adult	Throughout the season
(Glover)	*Cheilomenes sexmaculatus*	Nymph,Adult	Throughout the season
	Syrphus confracter	Nymph,Adult	Throughout the season
	S.baleatus	Nymph,Adult	September-November
	S.searius		
	Spiders	Nymph,Adult	Throughout the season
	Coccinella septumpunctata	Nymph	Throughout the season
	Scymnus nubilis	Nymph	Throughout the season
	Ceolophora bissellata	Nymph	-
	Pullus xeramphelinus	Nymph	-
	Sphaerophoria javana	Nymph	-
	Ischiodon scutellaris	Nymph	-

Source : Dhawan, 1990, 1999b, 2011; Singh, 1985,1994a, 1994b; Sundaramurthy and Chitra, 1992; Jayaswal and Sundaramurthy, 1992.

Table 17.7 : Population of different predators in cotton agroecosystem under Punjab conditions in different years

Parasitoids	Population /5 plants (Mean ± S. D.)			
	1975	1985	1995	2005
Predatory bugs	1.07 ± 0.68	0.81 ± 0.74 (24.29)	0.22 ± 0.42 (79.43)	0.25± 0.32 (76.63)
Chrysoperla sp.	1.19 ± 1.08	0.96 ± 0.98 (19.32)	0.56 ± 0.64 (52.94)	0.60± 0.46 (49.57)
Coccinellids	1.04 ± 0.96	0.81 ± 0.74 (22.11)	0.33 ± 0.48 (68.26)	0.44± 0.34 (57.69)
Spiders	1.78 ± 1.58	1.41 ± 0.89 (20.78)	0.52 ± 0.58 (70.78)	0.62 ± 0.39 (65.16)
Misc.	0.41 ± 0.64	0.37 ± 0.56 (9.75)	0.26 ± 0.53 (36.58)	0.32± 0.23 (21.95)
Total	**5.46 ± 2.80**	**4.41 ± 1.93 (19.23)**	**1.73 ± 1.03 (68.36)**	**2.03± 0.98 (62.82)**

Figures in parentheses are reduction in population over 1975 as base year.

Source : Modified after Dhawan 1999.

Table 17.8 : Potentiating effect of insecticide mixtures against bollworms on cotton

Mixtures	Contents	Promising dosage		Recommended dosage
		Cf (l/ha)	G ai/ ha	(g/ha)
Polytrin CCy44 EC	permethrin 40 + profenophos 400	1.25	50+500	Cypermethrin 50 Profenophos 600
Spark 36 EC	Deltamethrin 10+ Triazophos 350	1.00	10+350	Deltamethrin 10 Triazophos 600
Nurella D 505	Cypermethrin 50+ Chlorpyriphos	500 1.00	50 +500	Cypermethrin 50 Chlorpyriphos 500
Chlorguard 50.4 EC	Alphamethrin 24+ Chlorpyriphos 480	0.80	24 +384	Alphamethrin 25 Chlorpyriphos 1000
Decidan 32.8 EC	Deltamethrin 8+ Endosulfan 320	1.00	8 +320	Deltamethrin 10 Endosulfan 1000
Cyphos 202 EC	Cypermethrin 50+ Chlorpyriphos 500	2.50	50+500	Cypermethrin 50 Chlorpyriphos 1000
Duet 17 EC	Alphamethrin 10+ Chlorpyriphos 160	2.50	25 +400	Alphamethrin 25 Chlorpyriphos 1000

Cf= Commercial formulation

Source : Dhawan,1999, .2004

Table 17.9 : Newer insecticides for management of cotton pests

No. of class	Class of the Insecticides	Name of the insecticides
1.	Neonicotinoids	Imidacloprid, Acetamiprid, Thiacloprid, Thiamethoxam, Clothianidin, Dinotefuran, Nitenpyram
3.	Oxadiazines	Indoxacarb
6.	Diamide	Flubendiamide, Chlorantriniliprole
7.	Pyridine Carboxamid	Flonicamid
9.	Tetronic Acid Derivatives	Spiromesifen
10.	Tetramic Acid Derivative	Spirotetramat
11.	Bacterial Fermentation Products	Spinosyns,
12.	Dichloropropenyl Ethers	Pyridalyl
13.	Benzoylphenyl Urea	Novaluron, Buprofezin

Table 17.10 : Insecticides safe to natural enemies of cotton pests

Safe Insecticide	Natural enemy	Stage tested
Endosulfan	*Aleiodes algharensis*	All stages
	Camopoletus cloridae	Adult,cocoon
	Mallada boninsis	Larvae
	Chrysoperla carnea	Immature stages
Carbaryl	*Aleiodes algharensis*	Stage tested
	Chrysoperla carnea	Immature stages
	Camopoletus cloridae	Adult,cocoon
Phosphamidon	*Apanteles angaleti*	All stages
	Chrysoperla carnea	Immature stages
Phosalone	*Apanteles angaleti*	All stages
	Camopoletus cloridae	Adult,cocoon
	Trichogramma achaeae	Adult
	T. chilonis	Adult
	T.brasiliensis	All stages
	Bracon kirkpatricki	All stages
	Chrysoperla carnea	Immature stages
Acephate	*Camopoletus cloridae*	Adult, cocoon
Monocrotophos	*Camopoletus cloridae*	Adult, cocoon
	Chrysoperla carnea	Immature stages
	Trichogramma achaeae	Adult
	T.brasiliensis	All stages

Oxy demeton methyl	*Camopoletus cloridae*	Adult, cocoon
	Chrysoperla carnea	Immature stages
	Mallada boninsis	Larvae
	Trichogramma achaeae	Adult
	T. chilonis	Adult
Dimethoate	*Chrysoperla carnea*	Immature stages
	Trichogramma chilonis	Adult
Fenthion	*Camopoletus cloridae*	Adult, cocoon
	T.chilonis	Adult
Permethrin	*Bracon kirkpatricki*	All stages
	Chelonus blackburni	Adult
	Trichogramma achaeae	Adult
Cypermethrin	*Bracon kirkpatricki*	Adult
	Chrysoperla carnea	Immature stages
	Mallada boninsis	Latvae
Deltamethrin	*Bracon kirkpatricki*	Cocoon
	Chelonis blackburni	Adult
	Trichogramma chilonis	Adult
Fenvalerate	*Chelonis blackburni*	Adult
	Mallada boninsis	Larvae
	Trichogramma achaeae	Adult
	T.chilonis	Adult
	T.brasiliensis	All stages
	Chrysoperla carnea	Immature stages
Fluvalinate	*Trichogramma armigera*	All stages
	Mallada boninsis	Larvae

Source : Modified after Singh 1994 a,b

Table 17.11 : Insecticides causing resurgence of insect and mite pests of crops

Scientific name	Insecticide(s)	Mode of application	Reference(s)*
Amrasaca buguttula (Ishida)	Chlorpyriphos, quinalphos	FFoliar	Dhawan and Simwat (1997b)
Aphis gossypii Glover	Carbaryl	DDust	Thimmaiah and Kadapa (1987)
	DDT	Foliar	Natarajan *et al.* (1987a)
	Monocrotophos	Foliar	Balasubramanian *et al.* (1980), Rote *et al.* (1980), David (1981)
	Phorate, disulfoton, dimethoate	Soil	Sithanantham (1968), Navaneethan (1970), Regupathy and Jayaraj (1973), Gajendran (1985), Jayaraj and Regupathy (1987)
	Cypermethrin, deltamethrin, fenvalerate, flucythrinate, permethrin, S-524	Foliar	Balasubramanian *et al.* (1980), Rangarajan *et al.* (1987), Surulivelu and Sundaramurthy (1987), Natarajan *et al.* (1987b), Sellammal Murugesan *et al.* (1979), Patel *et al.* (1987), Jayaraj *et al.* (1987), Singh *et al.* (1987), Nandihalli *et al.* (1992), Pasupathy and Regupathy (1992), Ravindhran and Xavier (1997)
Bemisia tabaci (Genndius).	DDT, dimethoate	Foliar	Joyce (1955), van den Laan (1961), Dittrich *et al.* (1985)
	Monocrotophos	Foliar	Balasubramanian *et al.* (1980), Natarajan *et al.* (1987a), Jayaraj *et al.* (1987)
	Cypermethrin, deltamethrin, fenvalerate	Foliar	Wangboonkong (1981), Johnson *et al.* (1982), Hafex *et al.* (1983), Gerling and Horowitz (1984), Dittrich *et al.*(1985), Thakare *et al.*(1986), Patil (1987); Jayaraj *et al.*(1987), Katiyar (1997), Natarajan *et al.*(1987a), Jayaswal and Singh (1987); Ajri *et al.*(1986), Gour (1987), Dhawan and Simwat (1997b), Dhawan and Saini (1998)
	PP-321	Foliar	Reddy *etal.* (1987), Ajri *et al.* (1986), Jayaswal and Singh (1987)
	Toxaphene	Foliar	Watve and Clower (1976), Penman and Chapman (1980)
Tetranychus cinnabarinus (Boisduval).	DDT	Foliar	Huffaker and Spitzer (1950), Cone (1963), Duncomb (1972), Ongoren *et al.* (1975)

	Carbaryl	Foliar	Hoyt (1962)
	Cypermethrin, fenvalerate, fluvalinate	Foliar	Ramesh Babu and Azam (1983), Lhostele and Piedalu (1977), Hoyt *et al.* (1978), Honsy nd Abbassy (1980), Reddy (1981), Ajri *et al.* (1986), Reddy *et al.* (1987), Pasupathy and Venugopal (1987), Patil (1987), Gu-Zy *et al.* (1993)
Ferrisia virgata (Cockerell)	Deltamethrin, cypermethrin, fenvalerate, permethrin	Foliar	Patel *et al.* (1987), Uthamasamy *et al.* (1987)

*For detail references see Dhawan *et al.*, 2004.

Table 17.12 : Botanical insecticides evaluated against different insect pests of cotton

Insect pest	Formulation	Current status
Amrasca biguttula	Nimbecidine, IndNe, NO, NSKE, Navneem, Karanj oil, Repelin	Low –Moderate control
Bemisia tabaci	Castor seed oil, Karanj oil, Nimbecidine, IndNe, NO, NSKE, Neem rich. Navneem, Karanj oil, Repelin, Margocide CK, Neemgold, Achook, Rakshak gold*, Neemazal T/S*, Econeem*, Aza Fortune*	Low –moderate control Aza enrich formulations (3000-10,000 ppm) effective control
Aphis gossypii	NO, NSE, NO+Karanj oil	Moderate control
Bollworm complex	Nimbecidine, IndNe, NO, NSKE, Neem rich, Navneem, Karanj oil, Repelin, Margocide CK, Neemgold, Achook, Aza Fortune*, Rakshak gold*, Neemazal T/S*	Aza enrich formulations (3000-10,000 ppm) gave low- moderate controlS Early interventions in IPM gave effective control of insect pest complex

NO - Neem oil; NSE-Neem seed extract; NSKE- Neem seed kernel extract;

* - Aza enrich formulations

Source: Modified after Dhawan, 1999a, b; Puri *et al.,* 1999; Dhawan and Dhaliwal, 1996, Gupta, 2001; Dhawan *et al.*, 2013.

Table 17.13 : Recommended/ approved cultural and mechanical control strategies for cotton insect -pests

S. No.	Physical/cultural/mechanical practices	Target pests	Impact
1.	Uprooting and destruction of weeds during off season	Mealy bug, whitefly and CLCuV	Reduces carry over infestation
2.	Deep ploughing for soil exposure before sowing and after harvest	American boll worm, tobacco caterpillar	Pupal mortality through birds and other factors
3.	Early sowing (North India)	American boll worm	Escape mechanism
4.	Proper plant spacing and maintenance of plant population	Sucking pests, bollworms and foliage feeders	Reduces incidence, proper application of insecticides through foliar application
5.	Growing bajra/ sorghum/ maize as barrier crop	Mealy bug	Act as a physical barrier and check spread
6.	Growing of (refugia) non Bt around *Bt*-cotton field	Boll worm complex	Delay of resistance to Bt mechanism
7.	Use of recommended dose of nitrogenous fertilizer	Jassids, whitefly and bollworms	Congenial over dose for population build up
8.	Non judicious use of irrigation water	Entire pest complex	Good for population build up of pests
9.	Cultivation of undescript material	Entire pest complex	More damage
10.	Installation of yellow sticky trap	Whitefly	Monitoring and managing whitefly population
11.	Collection and destruction of bad opened bolls at the end of season.	Pink Bollworm	Reduces carry over Cultivation of adjoining crops in and around cotton crop
12.	Okra, moong and Pigeon Pea	Jassid, whitefly, spotted and american bollworm	Most preferred host, help in increase in population build up and incidence on cotton crop
13.	Okra, moong, castor and dhaincha	Tobacco caterpillar	up and incidence on cotton crop
14.	Okra, moong and cluster bean	Mealy bug	Uprooting and destruction of weeds during the season
15.	*Sida* sp. and *Abutilon* sp	Whitefly, mealybug *Trianthema* sp. *Congress grass*	Reduces carry over and population build up Tobacco caterpillar, mealybug

16.	Do not throw uprooted infested plant/ weeds in cotton plant/ water channel/ common place	Mealy bug	Check further spread
17.	Collection and destruction of egg masses and early instar feeding gregariously along with leaves	Tobacco caterpillar	Reduces population build up and incidence
18.	Removal of infested terminal shoots	Spotted bollworm	Reduces incidence
19.	Early termination of crop	Bollworms	Reduces carry over
20.	Grazing of cattle, sheep, goat after last pick to feed on plant debris and unopened bolls	Pink bollworm	Reduces carry over
21.	Before stacking dislodge and burn the burs and unopened bolls	Pink bollworm, mealybug	Reduces carry over
22.	Stacking of cotton stalks in open in villages	Pink bollworm & mealy bug	Reduces carry over
23.	Do not allow the movement of farm animals	Mealy bug	Reduces carry over
24.	Restrict the movement of farm workers in infested fields	Mealy bug	Reduces carry over
25.	Prevent the movement of sticks from the infested areas to the new area	Mealy bug	Reduces carry over
26.	Do not cultivate cotton as a ratoon crop.	Pink Bollworm	Reduces carry over

Modified after Dhawan,1990,1999, Dhawan *et al.*, 2011, 2012

Table 17.14 : Benefits of *Bt* cotton in India for the years (2006-09)

Publica-tion	**ICAR FLD 2006**	**Andhra University 2006**	**CESS 2007**	**Subra-manian & Qaim 2009**	**Sada-shivappa& Qaim 2009**	**Qaim *et al.*, 2009**
Period studied	2005	2006	2004-2005	2004-2005	2006-2007	1998-2006
Yield increase	30.9%	46%	32%	30-40%	43%	37%
Reduc-tion in no. of spray	–	55%	25%	50%	21%	41%
Increased profit	–	110%	83%	-	70%	89%
Average increase in profit/hectare	–	$223	$225	$156	$148	$131

Sources:

1. Front line demonstrations on cotton 2005-06. Mini Mission II, Technology Mission on Cotton, Indian Council of Agricultural Research (ICAR), New Delhi, India.
2. Ramgopal N. 2006. Economics of Bt cotton vis-à-vis Traditional cotton varieties (Study in Andhra Pradesh)," Agro-Economic Research Center, Andhra University, A.P.
3. Dev SM and NC Rao. 2007. "Socio-economic impact of *Bt* cotton", CESS Monographs, Centre for Economic and Social Studies (CESS), Hyderabad, A.P.
4. Subramanian A and M Qaim. 2009. Village-wide Effects of Agricultural Biotechnology: The Case of *Bt* Cotton in India, *World Development.* 37 (1): 256–267.
5. Sadashivappa P and M Qaim. 2009. Bt Cotton in India: Development of Benefits and the Role of Government Seed Price Interventions, *AgBioForum*. 12(2): 1-12.6.1.
6. Qaim M, A Subramanian and P Sadashivappa. 2009. Commercialized GM crops and yield, Correspondence, *Nature Biotechnology.* 27 (9) (Sept 2009).

Cited from : Choudhary, B. and Gaur, K. (2011).

Table 17.15: Impact of IPM in reduction in number of sprays and additional profit over non-IPM villages in Punjab

Parameter	**2002**	**2003**	**2004**	**2005**	**2006**	**2007**	**2008**	**2009**
No. of villages	32	30	45	121	213	230	225	225
No. of IPM farmers	300	567	900	1044	14804	27196	12919	12120
Area under IPM (ha)	1152	28282	19429	37032	44848	45262	31735	35182
Seed cotton yield (kg/ha)	1585	1969	2441	2452	2435	2260	2433	2586
Number of sprays*	10.6	9.5	5.4	3.4	3.1	4.7	4.2	3.61
Spray cost (Rs/ha)	5460	7937	4021	1793	2163	2714	1761	1955
Additional profit due to IPM (Rs/ha)	7670	7844	13111	5509	5371	9864	12456	15699

Mean of three years

* The average number of sprays in 2001 was 21.0

** *Bt* cotton was adopted as component of IPM

Source : Dhawan 2011

Table 17.16 : Benefits of the IRM program in India

Year	No of villages	No of farmers	Reduction in no. of insecti cides(%	Yield increase $ US/ha	Profit increase farmers	Total benefit to participating ($ US million)	Benefit to Cost ratio
2004-05	444	20,525	46%	11%	$193	$11.5 m	28 : 1
2005-06	565	46,400	48%	12%	$183	$24.6 m	32 : 1
2006-07	1,062	72,783	52%	10-15%	$174	$33 m	44 : 1

Table 17.17 : Impact adoption of IPM strategies in Punjab

Sr. No.	Parameter	IPM	Non-IPM	Per cent increase/ decrease over non-IPM
1.	Number of villages			800
2.	Number of farmers			48537
3.	Number of sprays	3.65	6.70	37.17
4.	Spray cost (Rs/ha)	1917	4282	62.16
5.	Seed cotton yield (q/ha)	2663	1922	38.55
6.	Net profit (Rs/ha)	44442	30753	39.94

of cotton crop. *Bt* cotton reduced the dependence on insecticide use and increased the yield and improved the economies of cotton growers. However, new problem emerged like increase in incidence of sucking pests and reduction in yield of seed cotton. Adoption of IPM with cotton growers is low due to lack of precise technology, non availability of solution to new problems and dissemination of technology which is acceptable and adoptable. The non adoption of IPM will again increase dependence on insecticides and may repeat the past history.

References

Agarwal, R.A. and Gupta, G.P. 1986. Recent advances in cotton pest management. *Pl. Prot. Bull.* 38(1-4): 375-378.

Ahmad, M., Arif, M. I. and Ahmad, Z. 1995. Monitoring insecticide resistance of *Helicoverpa armigera* (Lepidoptera: Noctuidae) in Pakistan. *J. Econ. Entomol.* 88: 771–776.

Ahmad, M., Arif, M. I. and Ahmad, Z. 1999. Insecticide resistance in *Helicoverpa armigera* and *Bemisia tabaci*, its mechanisms and management in Pakistan. In: *Pro. ICAC-CCRI Regional Consultation on Insecticide Resistance Management in Cotton*. Central Cotton Research Institute, Multan, Pakistan. June 28–July 1, 1998: 33–39.

Ahmad, M., Arif, M. I. and Attique, M. R. 1997. Pyrethroid resistance of *Helicoverpa armigera*(Lepidoptera: Noctuidae) in Pakistan. *Bull. Entomol. Res.* 87: 343–347.

Ahmad,M., Arif, I., Zahoor, A. and Denholm. I. 2002. Cotton whitefly (*Bemisia tabaci*)resistances to organophosphate and pyrethroids insecticides in Pakistan. *Pest Mgmt. Sci.* 58:203–208.

Allen, T.C. 1983. The role of *Bacillus thuringiensis* var *Kurstaki* in *Heliothis* spp. in Lousiana and the Lower Rio Grande Valley of Texas. In: *Proc.Beltwide Cotton Prod. Res. Conf.*, San antonio,Texas, 2–6 January 1983, pp. 214–216.

Anonymous. 2004. *Annual Report 2003/04.* Cotton Research and Development Corporation, Australia.

Armes, N. J., Jadhav, D. R. and de Souza, K. R. 1996. A survey of insecticide resistance in *Helicoverpa armigera* in the Indian sub-continent. *Bull. Entomol. Res.* 86: 499–514.

Armes, N. J., Jadhav, D. R. and Lonergan, P. A. 1995. Insecticide resistance in *Helicoverpa armigera* (Hubner): status and prospects for its management in India.. *In* Constable, GAL. and Forrester, N.W. (eds.*) Challenging the future. Proc. World Cotton Res. Confer.* I, 14–17 Feb. 1994. Brisbane, Australia. CSIRO, Melbourne, pp. 522–523

Armes, N. J., Jadhav, D. R., Bond, G.S. and King, A.B.S. 1992. Insecticide resistance in *Helicoverpa armigera* in south India. *Pestic. Sci.*34: 355–364.

Arora, R, Dhawan A K and Jindal, V. 2007. Ecological basis for the outbreak of tobacco caterpillar *Spodoptera litura* on cotton. *Indian J. Ecol.*34(1):88-89.

Arshad, M. and Suhail, A. 2010. Studying the sucking insect pests community in transgenic Bt cotton. *Int. J. Agric. Biol.* 12, 764–768.

ASSOCHAM - Association of Chambers of Commerce and Industry of India. 2007. Economic benefits of Bt cotton cultivation in India. In *Bt cotton farming in India*. New Delhi, India. Retrieved from *Monsanto website athttp://monsanto. Media room.com/index.php*

Bal, H. K. and Dhawan, A.K. 2008. Effect of intercropping on population of pests and natural enemies in Bt and non-Bt cotton under sprayed and unsprayed conditions. *Indian J. Ecol.* 35(1): 59-63.

Bal, H. K. and Dhawan, A.K. 2009a. Effect of transgenic cotton on arthropod diversity under sprayed and unsprayed conditions in cotton agro-ecosystem. *Pestic. Res. J.* 21 (2): 126-132.

Bal, H. K. and Dhawan, A.K. 2009b. Arthropod fauna in cotton agro-ecosystem in Punjab state of India. *Indian J. Entomol.*71 (1): 29-34.

Bal, H. K. and Dhawan, A.K. 2009c. Diversity of arthropod communities and insect sub communities in Bt and non-Bt cotton. *J. Insect Sci.* 22 (3): 238-247.

Bal, H. K. and Dhawan, A.K. 2009cd. Impact of transgenic cotton on diversity of non-target arthropod communities in cotton agro-ecosystem. *J. Insect Sci* 22 (2): 130-138.

Bal, H. K. and Dhawan, A.K. 2010. Impact of intercropping on arthropod diversity in Bt and non-Bt cotton. *J. Insect Sci.* 23 (2): 136-140.

Balasubramanian, M., Sellammal Murugesan and Parameswaran, S.1980. Synthetic pyrethroids in the control of cotton pests and seed cotton yield. *Proc. Symp. ETL Key Pests Use Synth. Pyreth. Cotton.* September 19-20, 1980. Central Institute of Cotton Research, Nagpur, p.186.

Barik, A., Singh, R. P. and Joshi, S. S. 2002. *Technology Mission on Cotton in Nutshell.* Directorate of Cotton Development, Mumbai.

Bartlett, A. 2005. Farmer Field Schools to promote Integrated Pest Management in Asia: The FAO Experience, *Case Study presented to the Workshop on Scaling Up Case Studies in Agriculture*, International Rice Research Institute, 16-18 August 2005, Bangkok.

Batie, S.S. and Ervin, D.E. 2001. Transgenic crops and the environment: missing markets and public roles. *Environ. Develop. Econ.*6: 435–457.

Belloti, A.C., Cardona, C. and Lapointe, S.L. 1990. Trends in pesticide use in Colombia . *J. Agric. Entomol.* 7: 191-201.

Benbrook, C. 2001. 'Do GM crops mean less pesticide use?' *Pesticide Outlook* 12:204–207.

Bennett, R., Buthelezi, T. J., Ismael,Y. and Morse, S. 2003. Bt cotton, pesticides, labor and health: A case study of smallholder farmers in the Makhathini Flats, Republic of South Africa. *Outlook on Agriculture* 32:123–128.

Bennett, R., Kambhampati, U., Morse, S. and Ismail, Y. 2006. Farm-level economic performance of genetically modified cotton in Maharashtra, India. *Rev. Agric. Econ.*,28:50-71.

Bleicher, E., Silva, A. L., Calcagndo, G., Nakano, O. and Freire, E. C. 1979. Sistema de controle das pragas do algodociro paras a regiao centro sul do Brazil *EMBRAPA/ CNDPCRC Tee* 2. pp 21. (Original not seen) Cited by Ramalho F S. 1994. *Annu. Rev. Entomol.*39:563.

Bobert, G. H., Tabashnic, B. E., Ulliman, D. E., Marshal, W. J. and Messing, R. 1994. Resistance of *Aphis gossypii* (Homoptera:Aphidae) to insecticides in Hawaii, spatial patterns and relation to insecticide use. *J. Econ. Entomol.*87: 293–300.

Brugger,K.E.,.Newman,I.C., Parker,N., Scholz,B., Suvagia, P., Walker, G. and Hammond, T.G. 2010. Selectivity of chlorantraniliprole to parasitoid wasps. Pest. 66: 1075-1081.

Burgess, M. W. 2001. Key success factors in integrated crop management systems in Africa. In:*Proc. Technical Seminar at the 60th plenary meeting of ICAC.* International Cotton Advisory Committee, Washington, USA, 1997: 3–5.

Butani, D.S. 1974. Insect pests of cotton XVII- Effects of cotton varieties, cultural practices and fertilizers, on infestations by pink bollworm. *Cotton Fibr. Trop.* 29: 237-240.

Butter, N.S., Vir, B.K., Gurdeep Kaur, Singh, T.H. and Raheja, R.K. 1992. Biochemical basis of resistance to whitefly, *Bemisia tabaci* Genn. (Aleyrodidae, Hemiptera) in cotton.*Trop. Agric.* 69(2): 119-122.

Buttar, N.S., Kular, J.S. and Singh, T.H. 1994. Management of cotton pests due to humic acid. In: *National Seminar on Cotton Production Challenges in 21st Century*, April 18-20,1994, Haryana Agricultural University, Hisar, pp 69-70.

CABI. 2000. Learning to Cut the Chemicals in Cotton: Case Studies and Exercises in Farmer Focused Cotton IPM from Around the World. Commonwealth Agricultural Bureau International

Cahill, M., Groman. K., Day. S., Denholm. I., Elbert, A. and Nauen, R. 1996. Baseline determination and detection of resistance to imidacloprid in *Bemisia tabaci* (Homoptera:Aleyrodidae).*Bull. Entomol. Res.* 86: 343–349.

Cannon, R. J. C. 2000. BT transgenic crops: Risks and benefits. *Integrated Pest Mgmt. Rev.*5:151–173.

Cardwell, G., Larry, D. and William, E. 2005. Various novel insecticides are less toxic to humans, more specific to key pests. *California Agric.* 59: 29-34

Castro, Z.J., Loayza, C.F., Castro, M.T., Meza, P.M., Pena, V.L. and Molinari, N.E. 1997. *Control Integrado de Plagasy Produccion de Controladores Biologicos en el Valle de Ica y el Callejon de Huaylas.* CEDEP/RAAA, Lima, Peru. 149pp.

Cate, J. R. 1985. Cotton: status and current limitations to biological control in Texas and Arkansas. In: Hoy,M.A. and Herzog, D.C. (eds) *Biological Control in Agriculture IPM systems*. Academic Press, New York, USA: 536–556.

Challam, M.S. and Subbaratnam, G.V. 1999. Insecticide resistance in cotton leaf hopper *Amrasca biguttula biguttula* (Ishida) in Andhra Pradesh. *Pest Mgmt. Econ. Zool.* 7: 105–110.

Chen, J.L., Han, Q.X. 2005. Influencing factors and countermeasures of implementing integrated pest management in China. *J. Zhongkai Univ. Agricu. Technol.* 18(2): 51-58.

Cheng, G. and Lieu, Y. 1996. Cotton bollworm resistance and its development in northern cotton region of China 1984–1985. *Resistant Pest Mgmt. Newsl.*8(1): 32–33.

Cheng, G.L., Liu, R. andMingjiang, H. 1997. Comparison of cotton aphid resistance level between Xiang and Shandong populations. *Resistant Pest Mgmt. Newsl.* 9: 10–12.

Deguine, J. P. 1996. The evolution of insecticide resistance in *Aphis gossypii* (Homoptera:Aphidae)in Cameroon, *Resistant Pest Mgmt. Newsl.* 8: 13–14.

Delorme, R. D., Auge, M. T., Bethenod, and Villatte. F., 1997. Insecticide resistance in a strain of *Aphis gossypii* from Southern France. *Pestic. Sci.*49: 90–96.

Dhaliwal, G. S. , Jindal, V and Dhawan, A. K. 2010. Insect pests problems and crop losses: Changing trends. *Indian J. Ecol.*37(1):1-7.

Dhandapani, N., Kalyanasundaram, M., Swamiappan, M., Sundara Babu, P.C. and Jayraj, S. 1992. Experiments on management of major cotton pests of cotton with biocontrol agents in India.*J. Appl. Entomol.* 114: 52–56.

Dhawan,A.K. 1990. Management of cotton bollworms. In: Proc. Summer Institute on *"Key Insect Pests of India, Their Bioecology with Special Reference to Integrated Pest Management* " June 6-15, 1990, PAU, Ludhiana, pp.153-171.

Dhawan, A.K. 1993**.** Impact of different insecticides on pest management and productivity of cotton in Punjab. *Pestology* 17(3): 7-15.

Dhawan. A. K. 1999a. Potential of Neem in Cotton Pest Management in India : An update, pp.63-76 In: Agnihotri N P, Walia S C and Gajbhiye V T (eds.) *Green Pesticides Crop Protection and Safety Evaluation, Society of Pesticide Science*, India Division of Agricultural Chemicals Indian Agricultural Research Institute New Delhi-110012.

Dhawan, A.K.1999b. Major insect pests of cotton and their integrated management. In: R.K. Updadhyay, K.G. Mukerji and R.L. Rajak (eds). *IPM System in Agriculture* Vol.6- *Cash Crops*. Aditya Books Pvt.Ltd., New Delhi, pp.165-225.

Dhawan, A.K. 2000a.Impact of some new insecticides on natural enemy complex of cotton ecosystem. *Pestology*24(5): 8-14.

Dhawan, A.K. 2000b. Cotton pest scenario in India: Current status of insecticides and future perspectives. *Agrolook*1(1): 9-26.

Dhawan,A.K. 2004. Insect Resistance in Cotton. In: Dhaliwal, G.S. and Ram Singh (eds). *Host Resistance to Insects-Concepts and Applications*, Panima Corporation, New Delhi, pp. 263-315.

Dhawan,A.K., Butter, N.S. and Narula, A.M. 2007. *The Cotton Whitefly Bemisia tabaci(Gennadius).* Punab Agricultural University, Ludhiana, 99 pp.

Dhawan, A K 2011. *Bt* cotton in Punjab-Economic Impact and Risk Analysis. Soc Sustainable Cotton Production, Ludhiana, Punjab, India.

Dhawan, A. K. and Bal, H. K. 2007. Effect of transgenic cotton on arthropod diversity in cotton agroecosystem *Indian J Ecol.*34(1):1-7.

Dhawan.A.K., Dhaliwal,G.S. and Chelliah,S. 2000. Insecticide-induced resurgence of insect pests in crop plants. In: G.S.Dhaliwal and B.Singh (eds). *Pesticides and Environment.* Commonwealth Publishers, New Delhi, pp.86-127.

Dhawan, A.K. and Dhaliwal, G.S.1996. Cotton Pest Management. In: S.S. Narwal, P.Tauro and S.S.Bisla(eds).*Neem in Sustainable Agriculture*, Scientific Publ. Jodhpur, pp 215-28.

Dhawan, A K, Kumar,V.,Grewal, G K and Singh, J. 2011. Effect of artificial diet of fusion Bt on larval mortality of tobacco caterpillar *Spodoptera litura* (Fab.). *Indian J Ecol.*38(2):239-41.

Dhawan, A.K., Kumar, V., Singh,K. and Saini, S. 2011. *Pest Management Strategies in Cotton.* Soc. Sustainable Cotton Production, Ludhiana, 127 pp.

Dhawan, A. K. , Kumar, V. and Grewal, G.K. 2013. Potential of neem (*Azadirachta indica* Juss) (Meliaceae:Rutales) in insect pest management. In: A. K. Dhawan, Singh, B., Bhullar, M.B. and Arora A. (eds). *Integrated Pest Management*, Scientific Publishers (India), Jodhpur, pp 370-402.

Dhawan, A.K. and Saini, H.K. 1998. Impact of synthetic pyrethroids on biology of whitefly, *Bemisia tabaci* (Genn.) *Proc. National Seminar* on *Entomology in 21st Century,* College of Agriculture, Udaipur, India.

Dhawan, A. K. and Saini, S. 2009. First record of *Phenacoccus solenopsis* Tinsley (Homoptera: Pseudococcidae) on cotton in Punjab. *J. Insect Sci.*22 (3): 309-310.

Dhawan, A..K, Kumar, V. and Shera, P.S 2012. Managment of Insect Pests of Cotton: Retrospect and Prospect. In: Arora, Singh, B. and Dhawan, A.K. (eds). *Throry and Practice of Integrated Pest Management,* Scientific Publishers (India) Jodhpur, pp 274-297.

Dhawan, A.K, Shera, P.S. and Kumar, V. 2011. *Bt* cotton in India: Adoption and Impact Analysis. In; In: A. K. Dhawan, Singh, B., Arora A and Bhullar, M.B. (eds). *Integrated Pest Management*, Indian Soc. Adv. Insect Sci., Ludhiana, pp17-33.

Dhawan, A.K. and Sidhu, A.S. 1984. Assessment of capture threshold of pink bollworm moths for timing insecticidal applications on *Gossypium hirsutum* Linn. *Indian J. Agric. Sci.* 54(5): 426-433.

Dhawan, A.K. and Sidhu, A.S.1985. Effect of time of sowing on incidence of pink bollworm (*Pectinophora gossypiella* Saunders) in *hirsutum* cotton. *J. Res. Punjab Agric. Univ.*22(1): 63-66.

Dhawan, A.K. and Simwat, G. S. 1996. Field evaluation of some botanical insecticides alone and in combination with other insecticides for management of bollworm complex on cotton. In *Neem and Environment*,(eds. R.P. Singh, M.S. Chari, A.K. Raheja and W. Kraus)Oxford and IBH Publ Co Pvt Ltd, New Delhi, pp. 485-491.

Dhawan, A.K. and Simwat, G.S. 1997. Insecticide induced resurgence of sucking pests in upland cotton in Punjab, India. In: *Abstr. Third International Congress of Entomological Sciences,* March, 18-20,1997, National Agricultural Center, Islamabad, p12.

Dhawan A K, Simwat G. S. and Dhaliwal, G. S. 1998. Evaluation of different biopesticides against cotton bollworm, *Helicoverpa armigera* (Hubner).pp.274-280. In: Dhaliwal G S, Arora R, Randhawa N S and Dhawan A K(eds.), Proceedings of International Conference on *Ecological Agriculture: Towards Sustainable Development*, Chandigarh

Dhawan, A K, Singh, K., Saini, S., Mohindru, B.Kaur, A., Singh, G and Singh, S. 2007. Incidence and damage potential of mealy bug , *Phenococccus solenopsis* Tinsley on cotton in Punjab. *Indian J. Ecol.*34(2):166-72.

Dhillon, M.K. and Sharma, H.C. (2009) Impact of Bt-engineered cotton on target and non-target arthropods, toxin flow through different trophic levels and seed cotton yield. *Karnataka J. Agric. Sci.* 22:462–466.

Dhingra, S., Phokela, A. and Mehrotra, K. N. 1988. Cypermethrin resistance in the populations *Heliothis armigera*. In: *National Academy of Sciences, India; Science letters.*(11): 123–125.

Dittrich. V. and Ernast. G.H., 1983. The resistance pattern in white flies of Sudanese cotton. *Mitteilumgen deer Deutschemark Gesellschatt Fuir Allgemeine and Augewandte and Entomologie* 4: 96–97.

Divakar, B.J.,Sarma, P.V., Ragunathan, V., David, B.V., Reddy, G.R.S and Swamy, S.V. 1994. Integrated pest management in cotton. *Indian J. Pl. Prot.* 22(1): 98-104.

Downes, S., Mahon, R. and Olsen, K. 2007. Monitoring and adaptive resistance management in Australia for Bt-cotton: Current status and future challenges. *J. Invertebrate Pathol.* 95:208–213.

Elbert, A,; Haas, M,; Springer, B.; Thielert, W. and Nauen, R. (2008) Applied aspects of neonicotinoid uses in crop protection. *J. Med. Chem* 64: 1099-1105

Entwhistle, P. E. 1998. A world survey of virus control of insect pests. In: Hunter-Fujita, F.R., Entwhistle, P,F, Evans, H.F. and, Crook ,N.E. (eds) *Insect viruses and pest management.* Wiley, UK; pp. 189–200.

Entwistle, P.F. and Evans, F. 1985. Viral control. In: Kerkut, G.A. and Gilbert, L.I. (eds) *Comprehensive Insect Physiology and Pharmacology*, Pergamon Press. Oxford 12: 347–412.

Ernst, G. H. and Dittrich, V. 1992. Comparative measurements of resistance to insecticides in three closely-related old and new bollworm species. *Pestic. Sci.* 34: 147–152.

Fillman, D. and Sterling, W. L. 1983. Killing power of the red imported fire and (Hym.: Formicidae): a key predator of the boll weevil (Col.: Curculionidae). *Entomophaga* 28:339–344.

Fitt, G. P. 1989. The ecology of *Heliothis* species in relation to agroecosystems. *Annu. Rev. Entomol.*34:17-52.

Fitt, G. P. 1994. Cotton Pest Management (Part 3) An Australian perspective. *Ann. Rev. Entomol.*39: 543–562.

Fitt, G. P. 2004. Implementation and Impact of Transgenic *Bt* cottons in Australia. *In* “Cotton Production for the New Millennium. Proceedings of the third *World Cotton Res. Conf.,* 9-13 March, 2003, Cape Town, South Africa”, pp. 371-381

Fitt, G. P., and Wilson, L. J. 2002. Non-Target Effects of *Bt*-cotton: A Case Study from Australia. *In*:R. J. Akhurst, C. E. Beard, and P. A. Hughes (eds). *Biotechnology of Bacillus thuringiensis and Its Environmental Impact*, pp. 175-182.

Fitt, G. P.; Mares, C. L. and Llewellyn, D. J.1994. Field evaluation and potential ecological impact of transgenic cotton (*Gossypium hirsutum*) in Australia . *Biocontrol Sci. Technol.* 4: 535-548.

Frisvold, G. B. and Reeves, J. M. 2007. Economy-wide impacts of *Bt* Cotton. Proc. Beltwide Cotton Conf., January 2007.

Furk, C., Powell, D. F. and Heyd, S.1980. Pirimicarb resistance in the melon and cotton aphid *Aphis gossypii* Glover. *Plant Pathol.* 29: 191–196.

Gandhi, V. P. and Namboodiri. 2006. The adoption and economics of *Bt* cotton in India. Preliminary results from a study. Research and Publications, Indian Institute of Management, Ahmedabad, India, pp. 2-25.

Gill, H. K., and A. K. Dhawan. 2006a. Insecticide resistance in *Helicoverpa armigera* (Hübner) populations from Punjab. *J. Insect Sci.*19 (Special):1-3.

Gill, H. K., and A. K. Dhawan. 2006b. Monitoring of insecticide resistance to *Helicoverpa armigera* (Hübner) in Cotton growing areas of Punjab. *Pesticide Res. J.*18(2):150-53.

Gill, H. K., and A. K. Dhawan. 2006c. Insecticide resistance profile of *Helicoverpa armigera* (Hübner) to different insecticides on cotton in Punjab. *J. Insect Sci.*19 (Special):109-116.

Gill, H. K., and A. K. Dhawan. 2006d. Global status of insecticide resistance in *Helicoverpa armigera* on cotton. *J. Cotton Res. Development* 20 (2): 226-231. (Review article).

Gill, J.S., Varma, G.C., Sekhon, B.S., Shenhmar, M. and Brar, K.S. 1992. Efficacy of *Trichogramma* species for the control of *Pectinophora gossypiella* Saund. and Earias species on cotton. In: Proc. National Symp. "*Recent Advances in Integrated Pest Management*." October 12-15, 1992, PAU, Ludhiana, pp.73-74.

Gouse, M., Pray, C. and Schimmelpfenning ,D. 2004. The distribution of benefits from Bt cotton adoption in South Africa. *AgBioForum* 7:187–194.

Goze, E. and Deguine, J. P. 2000. Spatial and probability distribution of *Aphis gossypii* infestation in West Africa: application to non-random field sampling. In: *Proc. World Cotton Conf. – 2*, Athens, Greece, Sept. 6–12, 1989: 885–886.

Goze, E., Nibouche, S. and Deguine, J. P. 2000. Bollworm sampling for action thresholds in sub-Saharan Africa: spatial and probability distribution. In: *Proc. World Cotton Conf. – 2*, Athens, Greece, Sept. 6–12, 1989: 883–884.

GRAIN. 2001. '*Bt* cotton through the back door', *Seedling* 18, Barcelona: Genetic Resources Action International.

Greene, J. K., Turnipseed, S. G., Sullivan, M. J., and May, O. L. 2001. Treatment thresholds for stink bugs in cotton. *J. Econ. Entomol.*94, 403-409.

Gujar, G. T., Kalia, V., Kumari, A., Singh, B. P., Mittal, A. and Nair, R., et al. 2007. H*elicoverpa armigera* baseline susceptibility to Bacillus thuringiensis Cry toxins and resistance management for Bt cotton in India. *J.Invertebrate Pathol.* 95:214–219.

Gunning, R. V. 1993. Comparison of two bioassay techniques for larvae of *Helicoverpa* spp. (Lepidoptera: Noctuidae). *J. Econ. Entomol.*86: 234–238.

Gupta, G.P. 1996. Use of neem in management of pest complex in cotton. In: R.P. Singh, M.S. Chari, A.K. Raheja and W. Kraus (eds). *Neem and Environment,* Oxford and IBH Publ. Co Pvt Ltd , New Delhi, pp. 469-474.

Gu-Zy, Hans, L.J., Wang, Q, Huang, X.L. and Xu-XL. 1993. Monitoring the rate of the cotton aphid and carmine spider mite on cotton and resurgence of cotton aphid under pyrethroid pressure. *Acta Phyto. Sincia*20 (4): 319-324.

Hargreaves, H. 1948. *List of Recorded Cotton Insects of the World*. Commonwealth Institute of Entomology, London, pp. 50.

Hearn, A. B. and Fitt, G. P. 1992. *Cotton Cropping Systems*. In :C. Pearson (ed). *Field Crop Ecosystems of the World*. Elsevier Press, Amsterdam, pp. 85-142.

Herron, G. A., Powis, K. and Rophial, J. 2001. Insecticide resistance in *Aphis gossypii* Glover (Homoptera: Aphidae), a serious threat to Australian cotton. *Australian J. Entomol.*40: 85-91.

Horowitz, A.R. and Ishaaya, I. 2004. Biorational insecticides – Mechanisms, selectivity and importance in pest management. In: Horowitz, A.R. and Ishaaya. (eds.) *Insect Pest Management*. Springer-Verlag, Berlin Heidelberg, pp. 1-28.

Huang, J., Hu, R., Rozelle, S., Qiao, F. and Pray, C.E. 2002a. Transgenic varieties and productivity of smallholder cotton farmers in China. *Australian J. Agric. Resour. Econ.*46(3) :367-387.

Huang, J., Rozelle, S., Pray, C and Wang, Q 2002b. 'Plant Biotechnology in China', *Science* 295: 674–677.

Huang,J.K., Mi, J.W., LIN, H., Wang, Z.J., Chen, R.J., Hu, R.F., Rozelle, S. and Pray, C. 2010. A decade of Bt cotton in Chinese fields: Assessing the direct effects and indirect externalities of Bt cotton adoption in China. *Sci. China Ser. C-Life Sci.*53:981–991.

Hussain, H S and Trehan, K.N. 1933. Observations life history, bionomics and control of whitefly of cotton *Bemisia gossypiperda* (M& L). *Indian J. Agric. Sci.*6: 701-753.

Ignoffo, C. M. and Couch, T. L. 1981. The nuclear polyhedrosis virus of *Heliothis* species as a microbial insecticide. In: Burges, H.D. (ed) *Microbial control of pest and plant diseases* 1970–1980. Academic Press., London: 329–362.

Ilango, K. and Uthamasamy, S. 1989. Biochemical and physical basis of resistance to bollworms complex in cotton varieties. *Madras Agric. J.* 76(9): 73-77.

Ishaaya, I.; Mendelson, Z. and Melamed-Majar, V. 1988. Effect of Buprofezin on embryogenesis and progeny formation of sweet potato whitefly (Homoptera: Aleyrodidae). *J. Econ. Entomol.*81: 781-784.

Jarvis, R. 2001. Impact of seed quality and crop management practices on fibre quality. In: *Proc.Technical Seminar at the 60th plenary meeting of ICAC*. International Cotton Advisory Committee, Washington, USA: 12–16.

Jayaraj, S., Santharam, G., Narayanan, K., Sundarajan, K. and Balagurunathan, R. 1981. Effectiveness of polyhedrosis virus against field population of tobacco caterpillar, *Spodoptera litura* on cotton. *Andhra agric. J.*27: 26-29.

Jayaraj, S.1987 (ed). *Resurgence of Sucking Pests*. Tamil Nadu Agricultural University, Coimbatore, 262 pp.

Jayaraj, S., Rangarajan, A.V., Sellammal Murugesan, Santharam, G., Vijayaraghavan, S. and Thangaraju, D.1987. Studies on the outbreak of whitefly, *Bemisia tabaci* (Gennadius) on cotton in Tamil Nadu. In: S.Jayaraj. (ed.). *Resurgence of Sucking Pests.* Tamil Nadu Agricultural University, Coimbatore, pp.103-112.

Jayaswal, A.P. and Sundaramurthy, V.T. 1992. Achievements in insect management in cotton. In: *All India Coordinated Cotton Improvement Project. Silver Jubilee (1967-1992),* September,17-19,1992, Central Institute of Cotton Research, Nagpur, pp.117-151.

Jenkins, J. N. 1995. Host plant resistance to insects in cotton. In: G.A. Constable and N.W. Forrester (eds.) *Challenging the future*, *Proc.First World Cotton Confer.*, CSIRO, Narrabri, Australia: 359–372.

Jeyapradeepa, S. 2000. Studies on insecticide resistance in cotton leaf hopper *Amrasca devastans*(Distant). M Sc(Agric) Thesis. Tamil Nadu Agricultural University, Madurai, 82 pp.

Jing D.C. 1997. Sustainable control of agricultural pests: situation and perspective of IPM. *Guizhou Agric. Sci.*25(Suppl.): 66-69.

Johnson, M.W., Tascano, N.C., Reynolds, H.T., Sylvester, E.C., Kenkido and Natwick, E.T.1982. Whitefly causes problems for southern California growers. *Calif. Agric*. 36:24-26.

Jones, K. A. 1994. Use of bacculovirus for cotton pest control. In: Matthews, G.A. and Tunstall, J.P. (eds.) *Insect Pests of Cotton*. CAB International, Wallingford, UK: 477–504.

Jones, K. A., Verkerk, R. H. J. and Asanov, K. 2000. Prospects for the integration of non-chemical and chemical pest management in cotton. In: *Proc. Brighton Crop Prot. Confer. 2000:Pests and Diseases*. British Crop Protection Council: 199–204.

Jones, K.A. 1988. Studies on the persistence of *Spodoptera littoralis* NPV on cotton in Egypt. Ph.D. thesis, University of Reading, UK: 390 pp

Jost, P., Shurley, D., Culpepper, S., Roberts, P., Nichols, R. and Reeves, J., *et al.* 2008. Economic comparison of transgenic and non transgenic cotton production systems in Georgia. *Agron. J.*100: 42–51.

Joyce, R. J. V. and Roberts. 1959. Recent progress in entomological research in the Sudan Gezira, Emp. *Cotton Gr. Rev* 36: 3.

Karihaloo, J.L. and Kumar. P.A. 2009. *Bt Cotton in India—A Status Report (2nd ed)*. Asia-Pacific Consortium on Agricultural Biotechnology (APCoAB), New Delhi , India . pp. 1–56.

Ketunuti, U. and Prathomrut, S. 1989. Cotton bollworm larva control by *Heliothis armigera* nuclear polyhedrosis virus. In: *Abstracts of the First Asia-Pacific Conference of Entomology*, Mai, Thailand, Bangkok. The Secretariat APCE, Nov. 8–13 1898: 16.

Khan, Q. and Rao, V. P. 1960. *Cotton in India-a monograph*. Indian Central Cotton Committee Publication. Vol 2, pp. 217-301.

Kidd, P.W., Rmmel, D.R. and Thorvilson. 1996. Effect of cyhalothrin on field population of aphid, *Aphis gossypii* Glover, in Texas High Plains. *Southwestern Ent.*21(3): 293-301.

King, A. B. S. 1994. *Heliothis/Helicoverpa* (Lepidoptera: Noctuidae). *In* Mathews G.A. and and Tunstall, J.P. (eds) *Insect Pests of Cotton*. CAB International Wallingford, Oxon, UK: 39–106.

King, E. G., Coleman, R. J., Phillips, J. R. and Dickerson, W. A. 1985. *Heliothis* spp. and selected natural enemy populations in cotton: a comparison of three insect control programs in Arkansa s(1981–82) and North Carolina (1983). *Southwest Entomol Suppl* 8: 71–98.

Kranthi, K R., Jadhav, D. R., Wanjari, R. R., Kranthi, S. and Russell, D. 2001a. Pyrethroid Resistance and Mechanisms in Field Strains of *Helicoverpa armigera* Hubner (Lepidoptera: Noctuidae).*J. Econ. Entomol.*94: 253–263.

Kranthi, K. R., Jadhav, D. R., Wanjari, R. R., Shakir Ali, S. and Russell, D. A. 2001b. Carbamate and organophosphate resistance in cotton pests in India, 1995–1999. *Bull. Entomol. Res.* 91: 37–46.

Kranthi, K. R., Jadhav, D. R., Kranthi, S.,Wanjari, R. R., Ali, S. and Russell, D. 2002a. Insecticide resistance in five major insect pests of cotton in India. *Crop Protec.* 21: 449–460.

Kranthi, K. R., Russell, D.,Wanjari, R., Kherde,M.,Munje, S., Lavhe, N. and Armes, N. 2002b. In season changes in resistance to insecticides in *Helicoverpa armigera* (Lepidoptera: Noctuidae) in India. *J. Econ. Entomol.*.95: 134–142.

Kranthi, K. R., Kranthi, S., Banerjee, S. K., Raj, S., Chaudhary, A., Narula, A.M., Barik, A., Khadi, B.M.,Monga, D., Singh, A. P., Dhawan, A. K., Rao, N. H. P., Surulivelu, T., Sharma, A., Suryavanshi, D. S., Vadodaria, M. P., Shrivastava, V. K. and Mayee, C. D. 2004. IRM – revolutionizing cotton pest management in India. *Resistant Pest Management*: 32:44.

Kranthi, K. R., Jadhav, D. R., Kranthi, S. and Russell, D. A. 2005. Insecticide resistance management strategies for *Helicoverpa*. In: Sharma H.C. (ed) *Heliothis/Helicoverpa management –Emerging trends and strategies for future research*. Oxford and IHB Publishing Co, New Delhi, India.: 405–430.

Kranthi, K. R. and Russell, D.A. 2009. *Changing Trends in Cotton Pest Management. In* R. Peshin, A.K. Dhawan (eds.). Integrated Pest Management: Innovation-Development Process, Springer pp 499-542.

Kung, K. Y., Chang, K. L. and Chai, K.Y. 1961. Detecting and measuring resistance of cotton aphids to systox. *Acta Entomologia Sinica* 13: 19–20.

Lang, S. 2006. Seven-year glitch: Cornell warns that Chinese GM cotton farmers are losing money due to 'secondary' pests. http://www.news.cornell. edu/ stories/ july06/bt.cotton. china. ssl.html.

Lei T, Khan M, Wilson L 2003. Boll damage by sucking pests: An emerging threat, but what do we know about it? In: Swanepoel, A.(ed). World Cotton Research Conference: Cotton for the New Millennium. Agricultural Research Council–Institute for Industrial Crops, Cape Town , South Africa . pp. 1337–1344.

Li G.P., Feng, H.Q., Chen, P.Y., Wu, S.Y., Liu ,B. and Feng, Q. 2010. Effects of transgenic Bt cotton on the population density, oviposition behavior, development, and reproduction of a non-target pest, *Adelphocoris suturalis* (Hemiptera: Miridae). *Environ. Entomol.* 39:1378–1387.

Lu ,Y.H., Qiu, F., Feng, H.Q., Li, H.B., Yang, Z.C., Wyckhuys, K.A.G and Wu, K.M. 2008. Species composition and seasonal abundance of pestiferous plant bugs (Hemiptera: Miridae) on Bt Cotton in China. *Crop Prot.*27:465–472.

Lu, Y., Wu, K., Jiang,Y., Xia,B., Li, P., Feng, H., Wyckhuys, K.A.G. and Guo, Y. 2010. Mirid Bug Outbreaks in Multiple Crops Correlated with Wide-Scale Adoption of *Bt* Cotton in China. *Science,* 328:1151-1153.

Mann, R.S., Gil,l R.S., Dhawan, A.K. and Shera, P.S. 2010. Relative abundance and damage by target and non-target insects on Bollgard and Bollgard II cotton cultivars. *Crop Prot.* 29:793–801.

Mann, G. S., Dhaliwal, G. S. and Dhawan, A. K. 2001. Effect of alternate application of neem products and its impact on bollworm damage in upland cotton. *Pl. Prot. Sci.* 9 (1): 22-25.

Martin T, Chandre F, Ochou OG, Vassayre M, Fournier D, 2002. Pyrethroid resistance mechanisms in the cotton bollworm, *Helicoverpa armigera* (Lepidoptera: Noctuidae) from West Africa. *Pestic. Bioch. Physiol.* 74:17–26.

Martin, T., Ochou, G.O.,Vaissayre, M. and Fournier, D. 2003. Organophosphorus Insecticides Synergize Pyrethroids in the Resistant Strain of Cotton Bollworm, *Helicoverpa armigera* (Hubner) (Lepidoptera: Noctuidae) from West Africa. *J. Econ. Entomol.* 96: 468–474.

Matthews, G.A. 1993. Biological control takes precedence in Uzbekistan cotton. *Pesticide Outlook*.4 (4): 36–38.

Mazza, S. M., Contreras, G. B., Simonella, M. A., Polak. G. M. A., Schroeder, J. A., Tannure, C. J. and Royo, O. M. 2000. Sampling techniques for the evaluation of *Aphis gossypii* Glover (Homoptera: Aphididae) Infestation in cotton *Gosypium hirsutum* (L.) In: *Proceedings of the World Cotton Conference-2*, Athens, Greece, Sept. 6–12, 1998: 909–913.

McCaffery, A. R., King, A. B. S., Walker, A. J. and El-Nayir, H. 1989. Resistance to synthetic pyrethroids in the bollworm, *Heliothis virescens* from Andhra Pradesh, India. *Pestic. Sci.* 27: 65-76.

McCaffery, A. R., Maruf, G. M., Walker, A. J. and Styles, K, 1988. Resistance to pyrethroids in *Heliothis* spp.: bioassay methods and incidence in populations from India and Asia. In: *Proc. Brighton Crop Protec. Confer.– Pests and Diseases* 1988:;433-438

Mehrotra, K. N. and Phokela, A. 1992. Pyrethroid resistance in *Helicoverpa armigera* (Hubner). V. Response of populations in Punjab cotton. *Pestic. Res. J.*4: 59–61.

Men, X., Feng, G., Edwards, C.A. and Yardim, E. N. 2004.Influence of pesticide applications on pest and predatory arthropods associated with transgenic Bt cotton and non transgenic cotton plants in China. *Phytoparasitica* 32:246–254.

Men, X., Ge, F., Liu, X. and Yardim, E.N. 2003. Diversity of arthropod communities in transgenic Bt cotton and non-transgenic cotton agroecosystems. *Environ. Entomol.* 32:270–275.

Nagrare, V.S., Kranthi, S., Biradar, V.K., Zade, N.N., Sangode, V., Kakde, G., Shukla, R.M., Shivare, D., Khadi, B.M. and Kranthi, K.R. 2009. Widespread infestation of the exotic mealybug species, *Phenacoccus solenopsis* (Tinsley) (Hemiptera: Pseudococcidae), on cotton in India. *Bull. Entomol. Res.*99:537–541.

Naranjo, S.E.2005. Long-term assessment of the effects of transgenic Bt cotton on the function of the natural enemy community. *Environ. Entomol.* 34:1211–1223.

Narayanan, S.S., Singh, V.V., and Kothandaraman, R. 1990. Cotton genetic resources in India. In: A.K.Basu, N.D.Manikar and S.S.Narayanan eds.), "*Cotton Scenario In India (Souvenir)* ICAR, New Delhi, pp 10-28.

Narayanan, S.S., Singh, V.V., Kothandaraman, R. and Singh, P. 1988. Role of genetic resources, earliness and plant type on bollworm resistance in cotton. Proc. "*Group Discussion on Bollworm resistance in cotton*". Central Institute of Cotton Research Nagpur, India, pp. 120-133.

Narayanan,S.S., Basu, A.K., Mehta, N.P. and Mandloi, K.C. 1992. Breeding achievements in central zone. In: *All India Coordinated Cotton Improvement Project. Silver Jubilee (1967-1992).* September,17-19,1992, Central Institute of Cotton Research, Nagpur, pp.19-38.

Narwal, S.S., Tauro, P. and Bisla, S.S. 1997. *Neem in sustainable Agriculture*. Scientific Publishers, Jodhpur

Natarajan, K. and Seshadri, V. 1989. Abundance of natural enemies of cotton insects under intercropping system. *J. Biol. Cont.* 2: 3-5.

Natarajan, K., Sundaramurthy, V.T. and Chidambaram, P. 1987. Whitefly and aphid resurgence in cotton as induced by certain insecticides. In: S.Jayaraj, (ed.). *Resurgence of Sucking Pests.* Tamil Nadu Agricultural University, Coimbatore, pp.137-143.

Nimbalkar, S.A., Khodke, S.M., Taley, Y.M. and Patil, K.J. 1990. Bioefficacy of some neem insecticides including neem seed extract and neem oil for control of whitefly, *Bemisia tabaci* Genn on cotton. In M.S. Chari and G. Ramaprasad (eds.) *Botanical Pesticide in Integrated Pest Management*,Indian Society of Tobacco Science, Central Tobacco Research Institute, Rajahmundry, p.256-260.

Ochou, O. G. and Martin, T. 2003. Activity spectrum of spinosad and indoxacarb: rationale for aninnovative pyrethroid resistance management strategy in West Africa. *Resistant Pest Mgmt.Newsl.*12: 75–81.

Pandya, P.S. and Patel, C.T. 1964. Possibilities of imparting resistance to pests in cotton by use of wild species of *Gossypium. Indian Cotton Grow. Rev*. 18: 175-176.

Pasupathy, S. and Venugopal, M.S.1987. Resurgence of spider mite, *Tetranychus cinnabarinus* on cotton, treated with cypermethrin electrodyn formulation. In: S. Jayaraj (ed.). *Resurgence of Sucking Pests.* Tamil Nadu Agricultural University, Coimbatore, pp.184-190.

Patel, B.K., Rote, N.B. and Mehta, N.P.1987. Resurgence of sucking pests by the use of synthetic pyrethroids on cotton. In: S. Jayaraj (ed.). *Resurgence of Sucking Pests.* Tamil Nadu Agricultural University, Coimbatore, pp.197-201.

Perkins, J.H. 2002. History. *In:* Pimentel, D. (ed.) Encyclopedia of Pest Management, Marcel and Dekker, New York, USA, 368-372pp.

Pimentel, C. R. M. and Bandeira, C. T. 1981. Beneficious obtidos pelo productor atraves do manejo de pragas na cultura algodoeira na regiao centro-sul. *EMBRAPA/CNPA Comun Tec* 18: 13.

Polak, M. G. A., Conteras, G. B., Maranich., M. J., Royo, O. M., Simonella, M. A. and Poisson, J. A. F., 2000.Mass rearing and use of a new species of *Chrysoperla* (Neuroptera: Chrysopidae) in cotton crops in Argentina. In: *Proc. World Cotton Res. Confer.-2*, Athens, Greece, Sept. 6-12, 1998: 672–674.

Prasad, V. D., Bharati,M. and Reddy, G. P. V. 1993. Relative resistance to conventional insecticides three populations of cotton white fly *Bemisia tabaci* (Genn.) in Andhra Pradesh. *Indian J. Pl. Protec.* 21:102–103.

Praveen, P.M. 2003. Studies on insecticide resistance in early season sucking pests of cotton in Tamil Nadu. Ph D Thesis.Tamil Nadu Agricultural University, Coimbatore, 116 pp.

Pray, C., Huang, J. K., Hu, R. F., and Rozelle, S. 2002. Five years of Bt cotton in China: The benefits continue. *The Plant J.*31:423–430.

Purohit, S.S. and Vyas,S.P. 2004. Medicinal Plant Cultivation-A Scientific Approach. Agrobios, Jodhpur, India pp 158-179.

Purohit, M.S. and Deshpande, A.D. 1991. Effect of inorganic fertilizers on population density of cotton whitefly (*Bemisia tabaci*). *Indian J. Agric. Sci.* 61(9): 696-698.

Pyke, BA. 2003. The performance of Bt transgenic (Ingard®) cotton in Australia over six seasons. Proceedings 3rd *World Cotton Res. Confer.* Capetown, South Africa. pp 1273-1280.

Qaim, M. 2003. Bt cotton in India: Field trial results and economic projections. *World Development*31:2115–2127.

Qaim, M. and Zilberman, D. 2003. Yield Effects of Genetically Modified Crops in Developing Countries. *Science* 299:900–902.

Rajendran, M., Nalni, R. and Muthusankar Narayanan, A. 1996. Evaluation of influence of intercropping and plant protection on the incidence of leafhopper on rainfed cotton. *J. Cotton Res. & Dev.* 10(1): 76-80.

Ramakrishnan, N., Saxena, V. S. and Dhingra. 1984. Insecticide resistance in the population of *Spodoptera litura* (F) in Andhra Pradesh. *Pesticides* 18: 23–27.

Ramalho, F. S. 1994. Cotton Pest Management: Part 4 - A Brazilian perspective. *Annu. Rev. Entomol.*39: 563-78.

Rameis, J. and Shanower, T.G. 1996. Arthropod natural enemies of *Helicoverpa armigera* (H¨ubner) (Lepidoptera: Noctuidae) in India. *Biocont. Sci. Tech.* 6: 481–508.

Rao, N.V, Raja Sekhar,P., Venkataiah, M. and Rajasri, M. 1995. Influence of habitat on *H. armigera* (Hubner) in cotton ecosystem. *Indian J. Plant Prot.*. 23: 122-125.

Rao, N.V., Reddy, A.S. and Rao, K.T. 1989. Natural enemies of cotton whitefly, *Bemisia tabaci* Genn. in relation to pest population and weather factors. *J. Biol. Cont.*3(1): 10-12.

Rao, N.V., Reddy, A.S., Ankaiah, R., Rao,V.N. and Khasim, S.M. 1990. Incidence of whitefly (*Bemisia tabaci*) in relation to leaf characters of upland cotton (*Gossypium hirsutum*). *Indian J. agric. Sci.* 60(9): 619-624.

Raodeo, A.K., Sarkate, M.B., Kaunsale, P.K., Bilapate, G.G. and Shinde, K.S. 1983. Possibilities of biological control of cotton bollworms. *Cotton Dev.* 13 (2): 31-34.

Reddy, A.S. and Rosaiah, B. 1987. Insect pest management. In: *"Plant Protection in Field Crops."* In: M. Veerabhadra Rao and S. Sithanantham(eds.).Publ. 4, Plant Protection Association of India, Hyderabad, pp.293-299.

Reddy, A.S., Rosaiah, B. and Bhaskara Rao, T.1986a. Seasonal occurrence of whitefly, *Bemisia tabaci* (G.) on cotton in Andhra Pradesh. *Proc. Group Discussion on Cotton Whitefly.* April 29-30,1986.pp.2-3.

Reddy, A.S., Rosaiah, B., Bhaskara Rao, T., Rama Rao, B. and Venugopal Rao, N.1986b. The Problem of whitefly on cotton.*Proc. Seminar on Problem of Whitefly on Cotton.*, March 14,1986, Pune.

Reddy, G. V. P. and Manjunatha, M. 2000. Laboratory and field studies on the integrated pest management of *Helicoverpa armigera* (Hubner) in cotton, based on pheromone traps catch threshold. *J. Appl. Entomol.*124 (5-6): 213-221.

Regu, K., Tamilselvan, C., Sundaraja, R. and David, B.V. 1990. Influence of certain insecticides on population build up of whitefly, *Bemisia tabaci* (Genn.) on cotton. *Pestology* 14(4): 8-10.

Ren, X. X., Han, Z. J. and Wang, Y. C. 2002. Mechanisms of monocrotophos resistance in cotton. bollworm, *Helicoverpa armigera* (Hubner). *Archives Insect Biochem. Physiol.* 51: 103–110.

Richter, D. 1998. Aerial application of fipronil (Regent R) vs. malathion in replicated field tests for boll weevil. In: D.K.Reed, M.Christian, R.G.Jones, and P. Duggar. (eds.) *Proc. Beltwide Cotton Conf.* Vol 2. San Diego, California. pp 1262-1264.

Rishi Kumar, Kranthi,K.R., Monga, D. and Jat, S. L. 2009. Natural parasitization of *Phenacoccus solenopsis* Tinsley (Hemiptera: Pseudococcidae) on cotton by *Aenasius bambawalei* Hayat (Hymenoptera: Encyrtidae). *J. Biol. Control.* 23 (4): 457-460.

Rote, N.B. and Puri, S.N. 1992. Effect of fertilizer application on incidence of whitefly on different cotton cultivars. *J. Maharastra Agric. Univ.*17: 45-48.

Rote, N.B., Mehta, N.P., Patel, B.K. and Shah, A.H.1980. Comparative efficacy of different synthetic pyrethroid insecticides against bollworms on irrigated hybrid-4 cotton. *Proc. Symp.ETL Key Pests Use Synth. Pyreth. Cotton.* September 19-20, 1980. Central Institute for Cotton Research, Nagpur, p.47.

Ruchijat, E. and Sukmaraganda, T. 1992. National integrated pest management in Indonesia; Ist *International pest management in the Asia-Pacific region.* CAB International, Wallingford, UK: pp 329–347.

Russell, D. A. 2004. Integrated pest management for insect pests of cotton in less developed countries. In ARE Horowitz and I. Ishaaya *Insect Pest Management – Field and Protected Crops.* Springer Verlag, Berlin, Heidelberg, New York, 141–180.

Russell, D.A. and Kranthi, K.R. 2006. Handbook of sustainable control of cotton bollworm *Helicoverpa armigera*. Technical Bulletin 45 Common Fund for Commodities, Amsterdam. 160 pp

Sadashivappa, P. and Qaim, M. 2009. Bt cotton in India: Development of benefits and the role of government seed price interventions. *AgBioForum*12(2):172-183.

Santhini, S. and Uthamasamy, S. 1997. Susceptibility of cotton leaf hopper (*Amrasca devastans*) to insecticides in Tamil Nadu. *Indian J.Agric. Sci.* 67: 330–331.

Sarkar, B., Patra, A. K., and Purakayastha, T. J. 2008. Transgenic Bt-cotton affects enzyme activity and nutrient availability in a sub-tropical Inceptisol. *J.Agron. Crop Sci.*194(4):289–296

Sawicki, R. M. and Denholm, I. 1987. Management of resistance to pesticides in cotton pests. *Trop. Pest Mgmt.* 33: 262–272.

Schuler, T. H., Denholm, I., Jouanin, L., Clark, S. J., Clark, A. J. and Poppy, G. M. 2001. Population-scale laboratory studies of the effect of transgenic plants on non-target insects.

Seganmullar, A. and Hewson, R.T. 2000. Global implementation of ICM in cotton. *Proc the Brighton Crop Conference 2000: Pest and Disease.* pp 193-198. British Crop Protection Council, Farnham, Surrey, UK.

Sekamatte, M. B., Russell, D. A. and Luseesa, D. 2003. Extending IPM practices into Ugandan cotton management. In: *Proc. World Cotton Confer. – 3*, Cape Town, S Africa, March 9–13 2003: 1560–1567.

Sekhar, P. R., Venkataiah, M., Rao, N. V., Rao, B. R. and Rao, V. S. P. 1996. Monitoring of insecticide resistance in *Helicoverpa armigera* (Hubner) from areas receiving heavy insecticidal applications in Andhra Pradesh (India). *J. Entomol. Res.* 20: 93–102.

Sellammal Murugesan, Parameswaran, S. and Balasubramanian, M.1979. Efficacy of five different synthetic pyrethroids in the control of bollworms and aphids in cotton. *Pesticides* 13:15-17.

Sharma, H. C. 2001 Cotton bollworm/Legume Pod Borer, *Helicoverpa armigera* (Hubner) (Noctuidae Lepidoptera): Biology and Management. *Crop Protection Compendum.* Wallingford, UK: CAB International. 72 pp.

Sharma, H.C., Arora, R., and Pampapathy, G. 2007. Influence of transgenic cottons with *Bacillus thuringiensis* Cry1Ac gene on the natural enemies of *Helicoverpa armigera. BioControl* 52:469–489.

Sharma, O.P., Bambawale, O.M., Dhamdapani, A., Tanwar, R.K., Bhosle, B.B., Lavekar, R.C.and Rathod, K.S. 2005. Assessing of severity of important diseases of rainfed Bt transgenic cotton in southern Maharashtra. *Indian Phytopath.* 58:483–485.

Shelka, S.S., Mali, A.R. and Ajri, D.S. 1987. Effect of different schedules of insecticidal sprays on pest incidence, yield of seed cotton and quality of seed in Laxmi cotton. *Curr. Res. Rep. Mahatma Phule Agric. Univ.* 3 (2): 39-45.

Shen, J., Wu, Y., Tan, J., Zhou, B., Chen, J. and Tan, F. 1993. Comparison of two monitoring methods for pyrethroid resistance in cotton bollworm (Lepidoptera: Noctuidae). *Resist. Pest Manag.* 5: 5–7.

Sidhu, A.S., Sandhu, M.S., Arora, R., Brar, D.S., Dhaliwal, C.S. and Bal, R.S. 1990. Operational research project on the integrated control of cotton pests in the Punjab. Paper presented at " *ICAR-Transfer of Technology Projects Workshops*." May 27-28, 1990, PAU, Ludhiana.

Silvie, P., Deguine, J. P., Nibouche, S., Michel, B. and Vaissayre, M. 2000. Procedures advantages and constraints of staggered targeted control programmes on cotton in West Africa. In: *Proc. World Cotton Confer.-2*, Athens, Greece, Sept. 6–12, 1998: 829–832.

Singh, J.P., Banerjee, S.K. and Lather, B.P.S.1987. Effect of single and repeated application of contact insecticides on the incidence of cotton leafhopper. In: S.Jayaraj (ed.). *Resurgence of Sucking Pests*. Tamil Nadu Agricultural University, Coimbatore, pp.205-208.

Singh, S.P. 1994a. *"Fifty years of AICRP on biological control"*, Project Directorate of Biological Control, Bangalore.

Singh, S.P.1994b. Biointensive integrated pest management in cotton. In: *"Cotton Production Challenges in 21st Century."* CCS HAU, Hisar April,18-20, 1994, pp. 126-139.

Sisterson, M.S., Biggs, R.W., Manhardt N.M., Carrière ,Y., Dennehy, T.J. and Tabashnik, B.E. 2007. Effects of transgenic Bt cotton on insecticide use and abundance of two generalist predators. *Entomol. Exp. Appl.*124:305–311.

Sivakumar, S.D., Srinivasan, N.and Gailce Leo Justin, C. 1993. Pesticide use pattern in cotton- An economic analyses. *Pestology* 17(6): 12-15.

Sivasubramanian, P. and Uthamasamy, S.1992. Impact of cultural practices on pink bollworm infesting cotton. In: Proc. National Seminar on *"Changing Scenario in pest and pest management in India"* January 31 -February 1, 1992, Hyderabad, p. 41.

Smith, R. F. and Allen, W. W. 1954. Insect control and the balance of nature. *Scientific American* 190: 38-92.

Smits. H. 1997. Insect pathogens: their suitability as biopesticide. In: Microbial Insecticides: Novelty or Necessity? 1997,*BCPC Sym. Proc. no.* 68, pp 21-28.

Stadler, T. 2001. Integrated pest management of the cotton boll weevil in Argentina, Brazil and Paraguay. *ICAC Recorder* 19(4):14–19.

Sterling, W. L. 1984. Action and inaction levels in pest management. *Texas Agric Sta Bull* B- 1480, 20 pp.

Stern, V.M., Smith, R.F., van den Bosch, R. and Hagen. K.S. 1959. The integration of chemical and biological control of the spotted alfalfa aphid. The integrated control concept. *Hilgardia* 29:81–101.

Sundramurthy, V.T. 1991. Host plant resistance in cotton insect pest management. Paper presented in the *Summer Institute on Host Plant Resistance to Insect Pests and its Application in Insect Pest Management*, June 6-15, 1991, Tamil Nadu Agricultural University, Coimbatore.

Sundramurthy, V.T. and Chitra, K. 1992. Integrated pest management in the cotton. *Indian J. Plant Prot.* 20: 1-17.

Tang, Z. H., Gong, K. Y., You, Z. P., 1988. Present status and countermeasures of insecticide resistance in agricultural pests in China. *Pestic. Sci.*23: 189–198.

Thakur, B.S., Bhanorkar, B.K., Puri, S.N. 1991. Bioefficacy of some insecticides against whitefly infesting cotton. *J. Maharashtra Agric. Univ*. 16(3): 432-433.

Traxler, G., Godoy-Avila, S, J., Falck-Zepeda, and Espinoza-Arellan, J. 2001. *Transgenic Cotton in Mexico: Economic and Environmental Impacts*, Auburn: Auburn University.

Turnipseed, S. G., Sullivan, M. J., Mann, J. E., and Roof, M. E. 1995. Secondary pests in transgenic Bt cotton in South Carolina. *Beltwide Cotton Conferences (USA)*, January 4–7, 1995. Texas: San Antonio.

Twine, P.H. and Loyd, R.J. 1984. Observations on the effect of regular releases of *Trichogramma* spp. in controlling *Heliothis* spp. and other insects in cotton. *Queensland J. Agri. Animal Sci.*39: 159–167.

UK Soil Association. 2002. Seeds of Doubt: North American Farmers Experience of GM.

Uthamasamy, S. 1995. Host resistance to the leaf hopper, *Amrasca devastans* in cotton, Gossypium spp. In: *Challenging the future Proceedings of the First World Cotton Conference*, Ed G.A. Constable and N.W. Forrester, CSIRO, Narrabri, Australia: 494–498.

Uthamasamy, S., Bharathi, M. and Abdul Kreem, A.1987. Insecticides induced occurrence of mealy bug *Ferisia virgata* (Cockrell) on cotton. . In: S.Jayaraj (ed.). *Resurgence of Sucking Pests*. Tamil Nadu Agricultural University, Coimbatore, pp.214-216..

Uthamasamy, S.1996. Biochemical basis of resistance to insects in cotton, *Gossypium* spp. In: Proc. Symp. " *Biochemical basis of host plant resistance in insects* " March 15-16, Pune, pp. 2-3.

Vaidya, A. 2005. Monsanto's cotton has deficiencies: Study. Retrieved from *The Times of India* at:http://timesofindia.indiatimes.com/city/pune/Monsantos-cotton-has-deficiencies -study/ articleshow.

Van den Berg, H., Cock, M. J., Odour, G. I. and Onsongo, E. K. 1993. Incidence of *Helicoverpa armigera* (Lepidoptera: Noctuidae) and its natural enemies in small-holder crops in Kenya. *Bull Entomol Res* 83:321–328.

van Elzakker, B. 1999. Comparing the costs of organic and conventional cotton. In: Mayers, D. and Stolton, S. (Eds.) *Organic cotton 'From Field to Final Product'*. Intermediate Technology Publication, London, UK, pp86-110.

Vassal J-M, Vaissayre and M, Martin T 1997. Decrease in the susceptibility of *Helicoverpa armiger*(H¨ubner) (Lepidoptera: Noctuidae) to pyrethroid insecticides in Cote d'Ivoire. *Resist Pest Manag* 9:14–15.

Verma, G.C. and Gill, J.S. 1992. Manipulation of natural enemies in management of insect pests of sugarcane and cotton. In *Proc. National Symp. "Recent Advances in Integrated Pest Management"* , October 12-15,1992, PAU, Ludhiana, pp 161-162.

Viktorov, A. G. 2008. Influence of *Bt*-plants on soil Biota and Pleiotropic effect of delta-endotoxin-encoding genes. *Russ. J. Plant Physiol.*55:738–747.

Villatte, F., Auge, D., Touton, P., Delorme, R. and Fournier, D. 1999. Negative cross-insensitivity in insecticide-resistant cotton aphid *Aphis gossypii* Glover. *Pesticide Biochem. Physiol.*65: 55–61.

Vitale, J., Glick, H., Greenplate, J., Abdeennadher, M. and Traoré, O. 2008. Second-generation Bt cotton field trials in Burkina Faso: Analyzing the potential benefits to West African farmers. *Crop Science*48:1958–1966.

Wang Q.D. and Lu G.Q. 1999. A brief analysis on IPM in China. *Hubei Plant Protection* 6: 30-32.

Wang, S., Just, D. R. and Pinstrup-Andersen, P. 2006. Damage from secondary pests and the need for refuge in China. *Natural Resource Management and Policy*30:625–637.

Wang, S., Just, D. R. and Pinstrup-Andersen, P. 2008. *Bt* cotton and secondary pests. *Inter. J. Biotech.*10:113–121.

Wang, Z.J., Lin, H., Huang, J.K., Hu, R.F., Rozelle, S and Pray, C. 2009. *Bt* cotton in China: Are secondary insect infestations offsetting the benefits in farmer field?*Agr. Sci. China* 8:101–105.

Wanghoonkong, S.1981. Chemical control of cotton pests in Thailand. *Trop. Pest Mgt*.27: 495-500.

Williams, M.R. 2006. Cotton insect losses 2005. In: *Proc. Beltwide Cotton Conference*, National Cotton Council, Memphis , TN , USA , pp. 1151–1204.

Williamson, S. 1999. Report on Organic Cotton and Farmer participatory Training in Latin America. CABI Bioscience, UK.

Wilson, L. J., Mensah, R. K. and Fitt, G. P. 2004. Implementing Integrated Pest Management in Australian Cotton. *In:*A. R. Horowitz, and I. Ishaaya, (eds). "*Insect Pest Management: Field and Protected Crops*,. Springer, Berlin, Heidelberg, New York., pp. 97-118.

Wilson, L., Hickman, M. and Deutscher, S. 2006. Research update on IPM and secondary pests. In: "*Proceedings of the 13th Australian Cotton Res. Confe.*", Broad beach, Queensland, Australia, pp. 249–258.

Wolfenbarger, L.L., Naranjo, S.E., Lundgren, J.G., Bitzer, R.J and Watrud, L.S. 2008. *Bt* crop effects on functional guilds of non-target arthropods: A meta-analysis. *PLoS ONE***3**:1–11.

Wu, K. M. 2007. Monitoring and management strategy for *Helicoverpa armiger* resistance to *Bt* cotton in China. *J. Invertebrate Pathol.*95:220–223.

Wu, K., and Guo, Y. 2003. Influences of *Bacillus thuringiensis* cotton planting on population dynamics of the cotton aphid, *Aphis gossypii* Glover, in northern China. *Environ Entomol.*32: 312-318.

Wu, K., Li, W., Feng, H. and Guo, Y. 2002. Seasonal abundance of the mirids, *Lybux lucorum* and *Adelphocoris* spp. (Hemiptera: Miridae) on Bt cotton in northern China. *Crop Protec.*21: 997-1002.

Wu, K.M and Guo, Y.Y. 2005. The evolution of cotton pest management practices in China. *Annu. Rev. Entomol.*50, 31–52.

Wu, Y., Shen, J., Chen, J., Lin, X. and Li, A. 1996. Evaluation of two resistance monitoring methods in *Helicoverpa armigera*: topical application and leaf dipping method. *J. Pl.Protec.*5: 3–6.

Wu, Y., Shen, J., Tan, F. and You, Z. 1995. Mechanism of fenvalerate resistance in *Helicoverpa armigera* (Hubner). *J. Nanjing Agricola Univ.*18: 63–68.

Xu, N., Fok, M., Bai, L., and Zhou, Z. 2008. Effectiveness and chemical pest control of Bt-cotton in the Yangtze River, Valley, China. *Crop Protec.*27:1269–1276.

Yearian, W. C. and Phillips, J.R. 1983. Effectiveness of microbials in insect control programs in Arkansas. In: *Proc. Beltwide. Prod. Res. Conf. San Antonio, Texas,* January 2–6 1983: 218-220.

Zhang, G. F.,Wan, F. H., Murphy, S. T., Guo, J. Y. and Liu, W. X. 2008. Reproductive biolop, of two nontarget insect species, *Aphis gossypii* (Homoptera: Aphididae) and *Orius sauteri* (Hemiptera: Anthocoridae), on Bt and non-Bt cotton cultivars. *Environ. Entomol.*37:1035–1042.

Zhang, L. G., Shi, S.B. and Zhang Q.D. 2001. IPM in Hubei. *Hubei Plant Protection* 3: 3

Zhang, L. G., Zhang, Q.D. and Zhou, W.K. 2002. Extend IPM and facilitate sustainable agricultural development. *Plant Protec. Technol. Ext.*22(4): 38.

Zhao, J.H., Ho, P. and Azadi, H. 2011. Benefits of Bt cotton counterbalanced by secondary pests? Perceptions of ecological change in China. *Environ. Monit. Assess.*173:985–994.

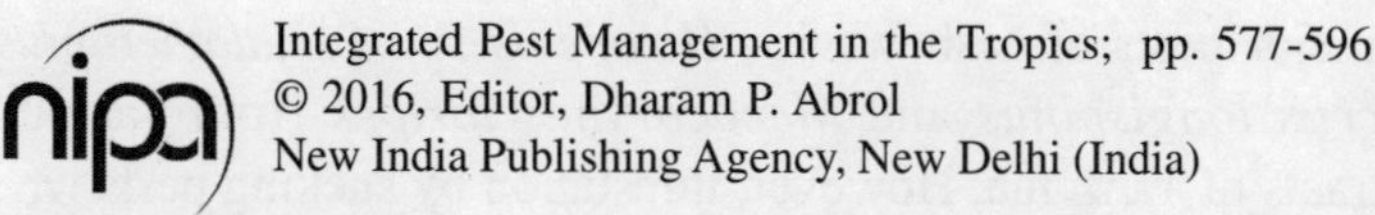
Integrated Pest Management in the Tropics; pp. 577-596
© 2016, Editor, Dharam P. Abrol
New India Publishing Agency, New Delhi (India)

Chapter - 18

Integrated Pest Management in Plantation Crops – Cashew and Rubber

P. Rethinam and Chandrika Mohan
Coconut Development Board, Kochi, Bhagireth 18, Lakshmi Nagar, S.N. Palayam Coimbatore 641007, Tamil Nadu

Plantation crops are high-value commercial crops of greater economic importance which play a pivotal role in the world economy. These include coconut, arecanut, tea, coffee, rubber, cocoa, oil palm, cashew. In this chapter only Cashew and Rubber are dealt with since the other plantation crops have been dealt in the earlier chapters.

18.1 Cashew

Cashew *(Anacardium occidentale* L.), native to northeast Brazil is grown mostly in arid, coastal areas of India, Southeast Asia, East and West Africa. The tree is cultivated in 2.7 million ha. and the major producers are India, Brazil, Nigeria, Indonesia, Guinea-Bissau, Benin, Vietnam and Tanzania (FAO, 2003). Cashew is attacked by around 180 species of insect and non- insect pests in India resulting in substantial yield loss (Ayyar, 1942; Sundararaju, 1993b; Maruthadurai *et al.*, 2012; Bhat and Rupa, 2012). The most important pests which limit the production are the cashew stem and root borer and tea mosquito bug (Bhat and Raviprasad, 2011). In addition, leaf and blossom webbers, shoot tip caterpillar, apple and nut borers *etc.*, cause damage to cashew. Although fifty insect species are reported as pests of cashew from Benin, *Acrocercops syngramma* (*Eteoryctis gemoniella), Apate terebrans* and *Helopeltis sp.* were documented as the three key species (ICIPE, 2013). Moses Olotu *et al.* (2012)

gave account of major pests of cashew viz., *Helopeltis* sp., *Pseudotheraptus wayi*, *Pseudococcus longispinus* and *Mecocorynus loripes* from various cashew growing tracts of Tanzania. However, infestation by sucking pests *viz.*, *Helopeltis anacardii*, *H. schoutedeni* and *Pseudotheraptus wayi* were considered as the key production constraints. Dwomoh *et al.* (2012) surveyed and reported 170 insect species occuring on cashew in Ghana.

18.1.1 Cashew stem and root borer *(Plocaederus ferrugineus* L.*)* (Coleoptera: cerambycidae)

Cashew stem and root borers (CSRB) are serious pests in all major cashew growing areas of India causing 1-19% damage every year (Abraham, 1958; Sundararaju and Bakthavatsalam, 1990; Mohapatra and Mohapatra, 2004). *Plocaederus ferrugineus* L. is the primary species infesting cashew. *P. obesus* G. and *Batocera rufomaculata* DeG. were the other species reported (Abraham, 1958; Uthaiah *et.al.*, 1989 and Rai, 1984). Cashew trees severely infested by *P. ferrugineus* die within two years resulting in considerable tree loss and up to 40% infestation is recorded in the endemic areas (Maruthadurai *et al.,* 2012). Anilwe *et al.* (2007) reported *Spondias mombin* (purple mombin) as a favoured alternate host of *P. ferrugineus* and also stated that this pest is accountable for quick mortality of cashew trees in Nigeria. Asogwa *et al.* (2009) reported 36.3% infestation rate by this pest.

18.1.1.1 Damage symptoms

Occurrence of small borer holes at the collar region, extrusion of coarse dust power like frass through the holes and discharge of gum at the basal region of the tree are some of the characteristic symptoms of pest identification. The newly hatched grubs tunnel in to the sub epidermal and vascular tissues of the stem and root region through the bark. This prevents release of nutrient supply to the plant parts resulting in yellowing and shedding of leaves and finally death of the tree. Severe injury to the roots also affect stand of the tree which may tilt to one side (Maruthadurai *et al.*, 2012))

18.1.1.2 Biology

CSRB has an annual life cycle. The reddish-brown long horned beetle is medium sized (25-40 mm). Fecundity is 60-90 eggs. Older trees with rough bark or injuries are preferred for oviposition and fissures in loose bark tissues on the stem or uncovered portion of roots above the soil are the favourite oviposition sites. Eggs measuring 3mm, whitish and ovoid, hatch out in 4-7 days. Grubs tunnel into fresh bark tissues and complete grub stage in 4-7 months. Final instar grubs are off- white coloured and measure a length of 7-10 cm. Positive correlation of pest incidence with maximum temperature was reported by Mohapatra and Mohapatra (2004) from Orissa, India.

18.1.1.3 Integrated Pest Management Techniques

Phytosanitation is an important component of CSRB infestation. To achieve this, the dead trees and trees which are beyond recovery (those having more than 50% of bark circumference damage and / or showing yellowing of the canopy) should be uprooted and removed from the plantation before and after monsoon, since they serve as natural inoculum for multiplication and spread of the pest (Raviprasad *et al.*, 2009; Sundararaju and Bakthavatsalam, 1990). Curative treatment of the infested trees in initial stages of infestation can save the tree. All the trees in the plantation are to be examined for symptoms of infestation by CSRB at the tree bases at monthly interval especially during harvest (January- May). In case external infestation symptoms are noticed, the CSRB grubs should be mechanically removed by careful chipping the bark. The injured portion as well as base of the tree and exposed root should be swabbed/drenched with chlorpyriphos 0.2 per cent solution (Raviprasad *et al.*, 2009; Sundararaju and Bakthavatsalam, 1994; Bhat and Rupa, 2012). The newly planted grafts should be trimed to have branching at a height of 0.75m to 1.00m from ground level for facilitating better inspection and adopting pest management techniques effectively. Anikwe *et al.* (2007) reported from Nigeria that prophylactic treatment of healthy trees with coal tar–kerosene mixture (1:2 ratio) gave 100% efficiency for twelve months. They also recorded 98% success in pest management by using the chemical Propoxur (carbamate) and physical management by poking of bore- holes of the beetle using a wire and killing the insect.

Bhat and Raviprasad (1996) reported mycosis of grubs of CSRB with entomopathogens *viz.*, the green muscardine fungus *Metarhizium anisopliae* and white muscardine fungus *Beauveria bassiana.* Mixing the spawn with organic matter like FYM, neem cake and cashew apple are practiced to enhance the spore load under the field condition. The eggs of CSRB were parasitized by *Avetiniella batocerae* (Bhat and Raviprasad, 2011).

18.1.2 Tea mosquito bug

The tea mosquito bugs (TMB), *Helopeltis* spp. are the most serious foliage and fruit pests of cashew in India causing 20- 60 per cent yield loss to the crop. Extensive review of literature containing information on species distribution, nature and extent of damage, biology, natural enemies, host plants and management of the pest are published by Devasahayam and Nair (1986); Sundararaju and Bakthavatsalam (1994); Sundararaju (1996) and Sundaraju and Sundarababu (1999). Out of the 41 species of *Helopeltis* recorded from tropics, only three species (*H. antonii,* Sign., *H. theivora* Wat. and *H. bradyi* Wat.) are confined to India. *H. antonii* occurs in southern and eastern India, whereas, *H. theivora* exists in West Coast, Western Ghat and North East India. *H. bradyi* is found in West Coast and Western Ghat regions. Cashew and cocoa are damaged by all the three species. *H. antonii* also infests guava, ber and drumstick (Sundararaju and Sundarababu, 1999). Damage by *H. antonii* was first reported on Cashew in India by Ayyar (1932). Stonedahl (1991) and Sundararaju (1996) reported occurrence of *H. bradyi* on cocoa, guava and cashew and Stonedahl (1991) documented literature on oriental species of *Helopeltis.* Under west coast condition, all these three species of *Helopeltis* (*H. antonii*, *H. theivora* and *H. bradyi*) infest cashew, whereas in other regions, only one species (*H. antonii*) was reported on cashew. *H. antonii* is the dominant species infesting cashew causing severe damage and is distributed in most of cashew growing regions of Kerala, Karnataka, Maharashtra (Pillai *et al.,* 1976); Tamil Nadu (Abraham, 1958; Pillai *et al.,* 1976); Goa (Sundararaju, 1979) and Orissa (Jena *et al.,* 1985). ICIPE (2013) reports indicated that among the biological constraints causing sizeable yield loss in cashew in Western Africa, infestation by *Helopeltis* sp. is a major factor which also depends on the variety, location and season.

18.1.2.1 Nature of damage

On cashew, the nature of damage caused by *H. antonii* has been described by various workers (Abraham, 1958; Pillai and Abraham, 1975; Sathiamma, 1978, Ambika and Abraham, 1979; Pillai, 1979; Devasahayam and Nair, 1986; Satapathy, 1993; Sundraraju, 1996). Both nymphs and adults suck sap from tender shoots, panicles and immature nuts and apples. The typical feeding damage symptom is the formation of necrotic lesion around the point of stylet insertion by the bug. The lesions on shoots and panicles coalesce and ultimately result in shoot blight or blossom blight. During outbreak situation, the entire flush dries up and the trees present a scorched appearance. Even though this pest is capable of inflicting cent percent loss in yield, losses in nut yield of 25 to 50 per cent have been reported from Karnataka, Goa, Kerala and West Bengal (CCRS, 1966; Desai *et al.,* 1977; Abraham and Nair, 1981; Chatterjee, 1989a).

18.1.2.2 Biology

Like other tropical mirids, *Helopeltis* spp. exhibit a more or less continuous cycle of generations throughout the year (Stonedahl, 1991; Sundararaju, 1996). The adult *H.antonii* is a reddish-brown bug measuring 6-8mm with distinct colouration of head (black), thorax (red) and abdomen (black and white). Tender tissues of new shoots and soft tissues of inflorescence branches are the preferred oviposition sites of female bugs. Chorionic threads seen outside the tissues indicate the presence of eggs inside, which are reniform and creamy-white in colour. Female lays about 50 eggs and the incubation period is 5-7 days (Ambika and Abraham, 1979; Satapathy, 1993; Sundararaju, 1996). The five nymphal instars are completed in 5-10 days. Longevity of male is more (9-10 days) than female (7 days). Many trees are reported as alternate hosts of *Helopeltis* spp. *viz.*, guava, neem, cocoa, cotton, mahogany, apples, cinchona, grapes, drumstick, Jamun and black pepper. Stonedahl (1991) and Sundararaju (1996) reported an average life cycle of 25-32 days for *Helopeltis* spp.

18.1.2.3 Seasonal incidence

Pest population build up starts from October- November coinciding with the emergence of new flushes in cashew and attains its peak when the trees are in full blossom stage during January -February. Young trees are more infested due to the availability of luscious growth throughout the year. (Pillai *et al.,* 1984; Rai, 1984; Sundararaju, 1984; Stonedahl, 1991; Satapathy, 1993; Sundararaju, 1996).

18.1.2.4 Varietal resistance

All the recommended cashew varieties are susceptible to this pest. However, the mid season/late season flowering cashew varieties are able to escape from the severity of the pest to certain extent. Cashew accessions *viz.* VTH 153, Kunthur 24, Goa 11/6, VTH 153/1, 9/78 and 51 different cashew types in Karnataka, accession No. 665 in Kerala and BLA - 39-4 in West Bengal were reported as least susceptible to *H. antonii* (Ambika *et al.,* 1979; Ghosh and Chatterjee, 1987; NRCC, 1988; Uthaiah *et al.,* 1989; Hiremath, 1991). The cashew accession, Goa 11/6 with yield of 2 t/ha under unsprayed situations is reported to have only moderate level of pest incidence. This accession has been released as a variety "Bhaskara" from the Directorate of Cashew Research, Puttur, Karnataka, India (Sundararaju *et al.*, 2006). Beevi and Mohapatro (2007) and Mohapatro (2008 a) had published field screening methodology of cashew genotypes against TMB with a 0-4 scoring, which is based on the number and type of lesions caused by TMB.

18.1.2.5. Integrated Pest Management techniques

18.1.2.5.1. Natural enemies : *Erythmelus helopeltidis* Gahan. (Hymenoptera: Mymaridae) *Telenomus* sp. (Laricis group) (Scelionidae) *Chaetostricha* sp. (Trichogrammatidae) and *Gonatocerus* sp. nr. *bialbifuniculatus* are the egg parasitoids reported on this pest from West Coast regions while, *Ufens* sp. is the egg parasitoid reported from East Coast of India. Pest population buildup is naturally regulated by these egg parasitoids (Devashayam, 1989; Sundararaju, 1993a; Sundararaju, 1996). It is also reported that activities of parasitoids are favoured naturally by increased minimum temperature and relative humidity during the susceptible period of November - February. The solitary nymphal endoparasitoids, *Leiophron* spp. (Hymenoptera: Braconidae) appear to be most promising in Indonesia and Africa (CIBC, 1983), but the activities of hyper parasitoids limit their potentiality in biological control. *Crematogaster wroughtonii* Forel (Formicidae) has been recorded as a predator on nymphs of *Helopeltis* spp. Several species of spiders (*Hyllus* sp., *Oxyopes sehireta*, *Phidippus patch* and *Matidia* sp) and five species of reduviid bugs (*Sycanus collaris* (Fab), *Sphedanolestes signatus* Dist., *Endochus inornatus* Stal., *Irantha armipes* Stal. and *Occamus typicus* Dist.) were recorded as predators of the tea mosquito bug. Use of *Chrysoperla* sp. as predator of *H. theivora* has been reported by Somchoudhury *et al.* (1993). *Aspergillus flavus* and *A. tamarii* were reported as pathogenic to *H. antonii* (Ambika and Abraham, 1979; Devashayam and Nair, 1986 and Sundararaju, 1984).

Cashew farmers in Masasi, Tanzania have successfully used the African weaver ant to combat insect pests (Moses Olotu *et al.,* 2012; Masasi, 2013). Effectiveness of the weaver ant *Oecophylla longinoda* in protecting tropical trees by vigorously patrolling tree canopies and preying up on or deterring potential pests was reported by Paul Van Mele (2008). Potential of weaver ant, *O. longinoda* for pest management of cashew in Ghana was popularized among farmers through farmer field schools (Dwomoh *et al.*, 2009). They also reported that population buildup and control efficiency of the weaver ants are checked by the ferocious boundary fights between colonies.

18.1.2.5.2. Chemical pesticides : Insecticides *viz.* carbaryl (0.1-0.15%), quinalphos (0.05%), dimethoate (0.05%), phosalone (0.07%), phosphomidan (0.03%) were reported to be effective against *H. antonii* (Muthu and Bhaskaran, 1979; Nair and Abraham, 1982, 1984; Samiayyan and Palaniswamy, 1984; Singh and Pillai, 1984, Sundararaju, 1984; AICRPC, 1989; Chatterjee, 1989b; Godse, *et al.* 1991; 1993). Dust formulations of carbaryl (5 and 10%), phosalone (4%), malathion (5%) and quinalphos (1.5%) were also found to be effective against *H. antonii* (Nair and Abraham, 1983; Hiremath, *et al.,* 1987; Bakthavatsalam, *et al.* 1993; Mohapatro, 2008b)

Raviprasad *et al.* (2005) reported that, even though many insecticides and plant products (botanicals) were effective, their ovicidal action is limited. In the endemic areas it is recommended to spray three times with any of these insecticides during most vulnerable periods of crop coinciding with flushing, flowering and fruiting stage based on pest incidence. Spraying of recommended insecticides will be remunerative if the trees are giving economical yield (>2.0 kg/tree). Although, cashew is an insect pollinated crop, spraying of these insecticides during flowering season did not influence the fruit set (Pillai *et al.*, 1984; Rai, 1984; Sundararaju and Sundarababu, 1999). Proper surveillance during flushing, flowering and fruiting period is essential for the management of TMB. Pest management treatment has to be initiated within 5 days when 5-10% of the flushes show pest incidence. Spraying lambda cyhaolothrin (0.003%) is recommended at flushing stage followed by L-cyhalothrin (0.003%) or carbaryl (0.1%) at flowering stage and third spray of carbaryl (0.1%) should be given at full fruit set stage to protect cashew from TMB (Bhat and Raviprasad, 2011). The Juvenile hormone analogues (farnesyl methyl ether @0.5 ug/nymph and graded doses of ZR 512 and MV 678) cause abnormalities on *H. antonii* when applied on fifth instar nymphs (Ambika and Abraham, 1984).

18.1.3 Leaf and blossom webbers

The cashew shoots bearing fresh flushes and flowers get attacked by two species of leaf and shoot webbing caterpillars, *Lamida* (= *Macalla*) *moncusalis* Wlk. (Lepidoptera: Pyraustidae) and *Orthaga exvinaceae* Hamps. (Lepidoptera: Noctuidae). The symptoms of infestation are the presence of webs on terminal portions with clumped appearance and drying of webbed shoot/ inflorescences. The caterpillars feed on the fresh leaves and flowers hiding inside the silken gallery web reinforced with plant scrapes. Presence of these galleries and larval castings on the terminal portion of new shoots and blossom indicate incidence of caterpillars. Yellowish green coloured eggs are laid solitary or in small groups of 3-5 on leaves. The dark green coloured caterpillars bear yellow longitudinal bands and pinkish dorsal line. After completing five larval instars, they pupate in the leaf webs or in soil and complete the total life cycle in about 37 days. (Rao *et al.,* 2002).

18.1.4 Shoot tip caterpillars

The tiny yellowish to greenish brown larvae of the moth, *Hypotima* (=*Chelaria*) *haligramma* M. (Lepidoptera: Gelechidae) damage the shoot tips and inflorescences. The tender shoot tips are infested occasionally up to 25 to 35 mm leading to drying up of the shoot tips. This pest is regularly reported from the East Coast tracts of India (Mohapatra *et al.*, 1998).

18.1.5 Apple and nut borers

Several lepidopteran species have been recorded as apple and nut borers of cashew. The extent of damage by the most common species, *Thylocoptila paurosema* M. (Pyralidae) has been reported in different cashew growing tracts.

These borers tunnel near the joint of apple and nut and the damaged fruits can be easily located as the infested fruits have frass at the damaged portion.

Attack by the caterpillar on fruits results in shrinking and premature nut fall. The medium sized adult moth has dark forewings and pale dark hind wings. Larval period is completed in 15-33 days covering 5 instars. Pupation takes place in the soil for about 8-10 days in earthen cocoons. Maruthadurai *et al.* (2012) reported 10% yield loss due to *Thylocoptila paurosema* in certain cashew growing areas during severe infestation. Spraying carbaryl 50WP 0.1% @ 2g/L during fruit setting and removal and proper disposal of dead and dried inflorescence during pre-flowering period are recommended for effective management of the pest. *Panerotoma* sp. (Braconidae) and *Trathala* sp. (Ichneumonidae) have been recorded as hymenopteran larval parasitoids of *T. paurosema*.

18.1.6 Thrips

Leaf and flower infesting thrips of cashew are exceedingly polyphagous allowing steadiness of thrips population. Leaf infesting thrips spp *viz.*, *Rhipiphorothrips cruentatus* (Hood), *Selenothrips rubrocinctus* (Giard) and *Retithrips syriacus* (Mayet) tends to exhibit infestation pattern in accordance to physiological state of plants like water stress and nutrient distribution. Ananthakrishnan (1985) reported substantial premature flower shedding due to infestation by flower thrips *viz.*, *Scirtothrips dorsalis* (Hood), *Thrips hawaiiensis* (Morgan) and *Haplothrips ganglbaueri* (Schmutz). Thrips lay solitary eggs in the leaf epidermis and incubation period is about 10 days. Colonies comprising of adults and nymphs are seen on the abaxial leaf surface sucking sap from the leaves, inflorescence, nuts and apples. Infestation by the thrips results in yellowing of leaves, scab formation on flower branches, apples and nuts. Severe infestation by this pest results in leaf shedding and stunted growth of trees.

18.1.7 Mealy bugs

Among the mealy bugs infesting cashew *Ferrisia virgata* (Pseudococcidae: Heteroptera) is the major species recorded from most of the cashew growing tracts. *Planococcus lilacinus*, *Planococcus citri* and *Phenococcus solenopsis* are the other mealy bugs reported from cashew. Mealy bugs colonize tender leaves, flower panicles and fruit peduncles of cashew. They suck sap from these tender parts and affects normal growth which leads to substantial yield loss in pest outbreak conditions. Additionally, the honey dew secreted by mealy bugs invites sooty mould growth affecting normal photosynthetic efficiency of the trees leading to indirect crop loss. Constant scouting for detection of infestation at an early stage itself is essential for taking up timely pest management procedures. Pruning and destroying pest infested plant parts and keeping the plantation weed- free helps in pest management. Further pest spread can be prevented by collecting and burning pest infested fallen leaves under the tree canopy. Pesticide spraying *viz.*, dichlorvas 76 WSC 0. 2% (@ 2.5ml / L) or dimethoate 30 EC 0.05% (@ 1.75ml / L) in combination with fish oil resin soap @ 20 g per litre of water is recommended to combat the pest. It is better to avoid repetitive usage of the same insecticides for more effectiveness. Care should be taken to cover the entire lower surface of leaves, twigs and branches where the immature crawler stages are seen in throngs. Predators viz., *Chrysoperla carnea, Menochilus sexmaculatus, Coccinella septumpunctata* and *Scymnus coccivora* are effective in mealy bug management. Conservation of predator fauna in and around cashew plantations was recommended (Maruthadurai *et al.,* 2012).

18.2 Rubber

Hevea brasiliensis Muell. Arg., the rubber tree, is the main source of natural rubber. This evergreen tree belonging to the family Euphorbiaceae and is one of the major economic plants grown in the tropical regions mainly between 15°N and 10°S. *H. brasiliensis*, native of tropical America, grows luxuriously in the Amazon river basin. Natural rubber is not cultivated widely in its native continent of South America due to the existence of South American leaf blight, and other natural biotic stresses of the rubber tree. Asia is the main source of natural rubber and the three largest producing countries, Thailand, Indonesia, and Malaysia, together account for around 72% of all natural rubber production. Because of the latex in the bark, rubber is not a preferred host plant for many common polyphagous insect pests and generally pest attacks are rather sporadic. Scale insect, thrips, mealybugs, termites, cockchafer grubs, mites, slug and snail, bark feeding caterpillars and rats are the major pests reported from rubber plant in India.

18.2.1 Scale insects

The scale insect, Saissetia nigra Nietn. is a polyphagous species that are seen generally in young plantations and nurseries in almost all rubber areas. Severe infestation by the sap feeding scale insects on leaflets, leaf stalks and developing soft leaf parts result in drying up of shoots. Jayasinghe (1999) reported scale insect spp *viz., Saissetia nigra, Pulvinaria maxima, Lepidosaphes cocculi, Saissetia* spp., *Aspidiotus destructor, Hemiberlesia cyanophylli, Hemiberlesia palmae, Parlatoria proteus, Phenacaspis dilatata, Pinnaspis* sp. and *Pinnaspis theae* infesting rubber from Sri Lanka. Natural enemies like insect parasites and entomopathogenic fungi keep this pest in check. Spraying organophosphorus insecticides like Malathion at 0.05% concentration is recommended in case of severe pest out break (Rubber Board, 2013; KAU, 2013).

18.2.2 Mealybugs

The common polyphagous striped mealy bug, *Ferrisia virgata* Ckll (Hemiptera: Pseudococcidae), infesting rubber trees enjoys a wide distribution. *Planococcus citri, Rastrococcus iceryoides, Dysmicoccus* sp. and *Pseudococcus maritimus* are the other scale insects infesting rubber trees in Sri Lanka (Jayasinghe, 1999). Mealy bugs colonize tender plant parts mostly in nurseries in all rubber areas and suck sap from tender plant parts. Wilting and drooping of leaves with the presence of white mealy growth along the underside of leaf stalks are symptoms of mealy bug infestation in rubber trees. Spraying insecticides like malathion 0.1% (2 ml/L) or quinalphos 0.075% (Rubber Board, 2013) or fish oil rosin soap (KAU, 2013) are recommended for pest management.

18.2.3 Termite

Odontotermes obesus (Rambur) Occurs in dry regions of Central Kerala (Thrissur & Palakkad) and non-traditional areas like Dapchari in Orissa. Severe infestation by termites sometimes leads to drying of young plants. Mud tunnels of termites on the tapping panel and collection cups of older trees create operational difficulties in tapping. Soaking the soil at the base of infested rubber plants with chlorpyriphos 0.1% solution is recommended for termite control (Rubber Board, 2013).

18.2.4 Cockchafer Grub (White grubs)

Holotrichia serrata F., *H. rufoflava* (Brenske), *Anomala varians* (Ol). are the major white grubs reported from rubber. Nehru and Jayaratnam (1993) reported that in rubber plantations, white grubs are common in loose soils near forest areas on roots of nursery seedlings and juvenile plants resulting in wilting of the plants. Jayasinghe, (1999) reported various white grub species *viz.*, *Lachnosterna (Holotrichia) bidentata, Holotrichia insularis, Psilopkolis vestita, Leucopholis rorida, Leucopholis nummicudens, Leucopholis tristis, Exopholis hypoleuca* and *Lepidiota stigma* infesting rubber from Sri Lanka. Once it was abundant throughout Europe and a major pest in the periodical years of "massflight", it had been nearly eradicated in the middle of the 20th century through extensive use of pesticides and has even been locally exterminated in many regions. However, with an increase in regulation of pest control in the 1980s, its numbers have started to grow again. Insecticides *viz.*, carbaryl 5D or carbaryl 4G @ 100 kg/ha were recommended for *Holotrichia serrata* on rubber seedlings in India (Nehru and Jayaratnam, 1988; 1993). Entomopathogens *viz., Beauveria brongniartii, B. bassiana, Bacillus popilliae* (as Doom) were found promising for the management of *H. serrata* on rubber seedlings plants and among the chemical insecticides, the best control was reported by isofenphos (Nehru *et al.*, 1991; Nehru and Jayarathnam, 1993).

18.2.5 Mites

Sporadic incidence of the mite, *Hemitarsonemus dorsalis* on young rubber plants in nurseries reported as a minor pest. HDRA (2012) and Bastos *et al.* (1981) reported *Tetranychus cinnabarinus*, the red spider mite, as a key pest of rubber in South America. Mite infested leaves develop yellow spreading parches on the upper surface, slowly crinkle, wither and shed due to drainage of plant sap by the pest. Effectiveness of *Phytoseilus riegeli*, a predatory mite, as biological control agent for pest management was also documented. Feres (1992) reported *Calacarus heveae* and *Tenuipalpus heveae* as the two major predominant species of plant feeding mites infesting rubber in Brazil. *C. heveae*, reaches its peak during February to April in the northwestern region of the State of São Paulo resulting in substantial premature fall of the leaves. Hernandes and Feres (2006) had published an inclusive review about mites (Acari) infesting rubber trees in Brazil. Dusting or spraying sulphur 0.2% (Sulfex 80 WP 2.5g/L) were reported effective in mite management.

18.2.6 Slug and Snail

The terrestrial pulmonate slug, *Mariaella dussumieri* Gray and snail *Crytozona (Xestina) bistralis* Beck are reported as pestiferous on rubber tree from different parts of the Kerala state (India) *viz.*, Kanjirappally, Thrissur and Calicut areas. Nursery seedlings are more prone to infestation. They occasionally feed on the rubber latex by scratching the young leaves and buds affecting normal growth of buds. Slugs also feed on latex from the tapped ends and collection cups. Jose *et al.* (1989) recommended metaldehyde 2.5% bait pellets application at the bottom of affected plants and incorporation in to the seedling beds to manage the snails. Brushing Bordeaux paste (10%) on the stem at 30 cm above the bud union part was also suggested to repel the slugs and snails.

18.2.7 Bark feeding Caterpillars

Aetherastis circulata Meyr. and *Ptochoryctis rosaria* Meyr. are the major bark feeding caterpillars infesting rubber. The incidence of bark feeding caterpillar on rubber was sporadic up to 1968, but become extensive by 1980's and many workers (Nehru *et al.*, 1983; 1987; Nehru and Jayaratnam, 1987; Jayaratnam *et al.*, 1991) reported infestation of rubber tree by *A. circulata.* The pest was recorded from various locations of Kerala *viz.,* Nedumangad, Punalur, Thrissur and Nagercoil. The caterpillars infesting rubber trees make a type of gallery cages reinforced with faecal matter and silk around the tree trunk and branches. Jose *et al.* (2000) and Rubber Board (2013) reported that this pest generally infests dead bark and occasionally feeds on live bark tissues leading to exudation of latex where a profound feeding mark could be observed. Bark feeding caterpillars *viz., Homodes brachteigutta (Euproctis subnotata)* and *Acanthopsyche snelleni* reported on rubber from Sri Lanka (Jayasinghe, 1999).

During severe pest infestation, power dusting of insecticides *viz.*, carbaryl 5% (Sevin) at the rate of 10 kg hectare^{-1} or fenvalarate dust 0.4% (Fenval 4D) at the rate of 7 kg hectare^{-1} are recommended by Rubber board (2013) for pest management. Spraying fenvalerate 0.02% (Tatafen/Arfen 20 EC, 1ml/L) on to the trunk is also effective. Thankamony (2012) reported suppression of *A. circulata* with neem oil 0.5% which induce 79.31 per cent mortality of caterpillars followed by Azadirachtin 0.03% (65.55%) compared to 96% control by the insecticides fenvalerate (0.02%) and carbaryl (0.01%).

18.2.8 Thrips

Thrips infest under surface of leaves leading to silvery-grey coloured scarring growth which in turn curl up and drop. Pesticide application is resorted only in case of severe infestation (Rubber Board, 2013).

18.2.9 Leaf eating & mining caterpillars

Jayasinghe (1999) reported *Amsacta lactinea, Tiracola plagiata, Miltochrista (Orgyia turbata), Crematopsyche pendula, Mocis undata, Thosea sinensis, Hyposidra talaca, Adoxophyes privatana, Euproctis suhnbtata and Acrocercops* sp. as the important leaf eating caterpillars infesting rubber trees in Sri Lanka.

18.2.10 Lace bug: *Leptopharsa heveae* Drake & Poor (Hemiptera: Tingidae)

Santos, Rodrigo Souza, (2010) reported the lace bug, *Leptopharsa heveae* as one of the most important insect pests infesting rubber in South America. *L. heveae* infests leaves of nursery seedlings as well as juvenile and older plants (Lara and Tanzini, 1997). The adults suck the sap and damage the leaves and reduce photosynthesis. Heavy infestation defoliates the tree and repeated defoliation reduces plant growth and latex yield. Chemical insecticides (deltamethrin, triclorfon, metamidophos) were reported effective in managing the pest. Santos and Freitas (2008) reported egg parasitism by the mymarid, *Erythmelus tingitiphagus* (Soares) in rubber tree plantation in Brazil.

18.2.11 Vertebrate pests

The major rodents affecting rubber cultivation are *Bandicota indica* Bech, *Bandicota bengalensis* Gray and *Rattus meltada* Gray. They devour seed kernels in germination beds and roots of juvenile plants. Baiting with the single doze anticoagulant rodenticide, Bromodiolone 0.005%, or poison mixed food baits are recommended by Rubber Board (2013) for controlling rodents.

Monkeys (*Presbytis melalophos nobilis*) and wild pigs (*Sus scrofa*) were reported as the major vertebrates damaging rubber plants especially the establishing plantations. They dig out and devour tap root or debark basal portion of juvenile trees. Williams *et al.* (2001) and Joshi *et al.* (2002) observed that even if the young plants are not totally uprooted by the attack by pigs, they are often broken resulting in retarded growth.

18.2.12 Rubber wood destroying Beetles

Being a nondurable wood, rubber wood is very susceptible to fungi, wood borers, and termites (CIRAD, 2003; Wong *et al.*, 2005). Insect borers such as Bostrichidae, Curculionidae: Platypodinae and Scolytinae damage the wood at all phases from cut log to seasoned wood and refined final products. Browne (1961), Hussein (1981) and Ho and Hashim (1997) reported sixteen species of ambrosia beetles from Malaysia. Eight species belonging to the subfamilies Scolytinae and Platypodinae attack cut down trees and processed rubber wood. Nine species of powder post beetles (Bostrichidae) and one species of ambrosia beetle (Scolytinae) were reported from processed sawn timber. Nair (2007) and Mathew (1982) reported six powder post beetles with prevalence of Bostrichidae attcking stored rubber wood and sawn timber from India. In Thailand also wood destroying beetles were reported (Wisut Sittichaya and Roger Beaver, 2009) boring in the wood of felled trees including ten species of powder post beetles (Bostrichidae) and eleven species of bark and ambrosia beetles (Platypodinae and Scolytinae). *Sinoxylon unidentatum* (F.) and *Sinoxylon anale* Lesne (Bostrichidae) were recorded as the predominant species in air-dried and seasoned rubber wood sawn timber, while *Euplatypus parallelus* (Fabricius) (Platypodinae) was the dominant species observed in stacked rubber wood logs. *Lyctoderma coomani* Lesne and *Lyctus tomentosus* Reitter (Bostrichidae: Lyctinae) were also reported from Thailand (Wisut Sittichaya and Roger Beaver, 2009).

18.3 Conclusion

The plantation crops ecosystem is very vital on account of the multi species cropping system being adopted in many crops for maximum utilization of unit area. The pest scenario in plantation crops is dynamic. The perennial nature of the crops coupled with conducive cropping system in most cases provides an ideal substrate for the build of several pest species. For the implementation of Integrated Pest Management (IPM) a bouquet of technologies are available for the farmers to choose and adopt. Crop habitat manipulation like shade regulation and field sanitation are important aspects in IPM. Even though biological control using parasitoids, predators and pathogens are also well utilized in pest management of this sector, pesticides continued to play a vital role in suppression of pests. Application of chemicals for pest suppression in these perennial crops induces ill effects on beneficial flora and fauna which in turn changes the pest scenario by way of new pest out breaks. As most of the plantation crops are export oriented, biological method of pest suppression has to be popularized as a long term strategy to curb the pest problem in these perennial crops on account of the pesticide residues in the products and also to have a green and clean environment.

18.4 Future Research Thrust

Cashew

Development of eco-friendly IPM strategies for management of major insect pests, standardization of mass rearing of Cashew Stem and Root Borer; flower and fruit pests and their natural enemies are the future priority areas. Since organic cashew has a preference, there is need to develop an eco-friendly pest management system. In order to address concerns of food quality standards, analysis of pesticide residues in cashew produce and cashew production ecosystem also warrants research.

Rubber

Developing cost effective chemical, biological and integrated insect and non insect pest management system as well as the development of insecticide application techniques need to get priority. Identification of molecular bases for pest resistance/tolerance to pest in *Hevea* germplasm and to develop newer rubber clones are yet another priority areas in rubber.

References

Abraham, C.C. and Nair, G.M. 1981. Effective management of the tea mosquito bugs for breaking the yield barriers in cashew. *Cashew Causerie,* 3(1): 6-7.

Abraham, E.V. 1958. Pests of Cashew (*Anacardium occidentale*) in South India. *Indian. Journal of Agricultural Sciences*. 28: 531-544.

AICRPC 1989. All India Co-ordinated Research Project on Cashew (1989) National Research Centre for Cashew, Puttur, Karnataka, India (Mimeographed) pp. 88.

Ambika, B. and Abraham, C.C. 1979. Bioecology of *Helopletis antonii* Sign. (Miridae: Hemiptera) infesting cashew trees. *Entomon* 4: 335-342.

Ambika, B. and Abraham, C.C. 1984. In: *Cashew Research and Development.* (Eds). RAO, E.V.V.B. and KHAN, H.H. Indian Society for Plantation Crops, CPCRI, Kasaragod, India, p111-115.

Ambika, B. Abraham, C.C. and Vidyadharan, K.K. 1979 *Proceedings Plantation Crops Symposium PLACROSYM* II: p513-516.

Ananthakrishnan, T.N. 1985. Host relationship and damage potential of thrips infesting cashew. *Acta Hort.* (ISHS) 108:131-134 http://www.actahort.org/books/ 108/108 24.htm (accessed April, 2013).

Anikwe, C., Okelana, F.A., Otuonye, H.A., Hammed, L.A., Aliyu, O.M. 2007. The integrated management of an emerging insect pest of cashew: A case study of the cashew root and stem borer, *Plocaederus ferrugineus* in Ibadan, Nigeria. *Journal of Agriculture, Forestry and the Social Sciences.* ISSN: 1597-0906. Vol. 5, No 1.

Asogwa, E.U., Anikwe, J. C., Ndubuaku, T.C.N., Okelana, F.A. and Hammed, L.A. 2009. Host plant range and morphometrics descriptions of an emerging insect pest of cashew, *Plocaederus ferrugineus* L. (Coleoptera: Cerambycidae) in Nigeria: a preliminary report. *Int. J. Sustain. Crop Prod.* 4(3), 27-32.

Ayyar, T.V.R. 1932. Bulletin -cashew *Agricultural Department of Madras* No. 27. Pp 95.

Ayyar, T.V.R. 1942. Insect enemies of cashew nut plant *(Anacardrum occidentale)* in South India. *Madras Agric. J.,* 30, 223-26.

Bakthavatsalam, N., Sundarababu, D. and Bhat, P.S. 1993 *The Cashew.* **7**(3): 12-13.

Bastos, J. A. M., Flechtmann, C. H. W., Figueiredo, R. W. de 1981. Control of the complex of *Tetranychus bastosi* T.B. & S. *Tetranychus cinnabarinus* (Boisduval), with some synthetic organic acaricides, on 'manicoba'. *Fitossanidade*. Vol. 4 (1): 11-13.

Beevi, S.P. and Mohapatro, G.K. 2007. A new field scoring methodology for cashew genotypes against tea mosquito bug, *Helopeltis antonii* Sign. *Journal of Plantation Crops* 35(3): 139-145.

Bhat, M.G. and Rupa, T.R. 2012. Current trends in cashew research in India and future focus pp19-27. In: Souvenir, PLACROSYM XX (Radhakrishnan, B., Rajkumar, R., Palani, M.N., Premkumar. R., Ilango, R.V., Kumar, M.S. and Chandrashekara, K.N. Eds) UPASI Tea Research Foundation, Tamil Nadu, India 37p.

Bhat, P.S. and Raviprasad, T.N. 1996. Pathogenicity of entomopathogenic fungi against cashew stem and root borer *Plocaederus ferruginues* L. (Coleoptera : Cerambycidae). *Journal of Plantation Crops* 24:265-271.

Bhat, P.S. and Raviprasad, T.N. 2011. Management strategies for the pest and diseases of cashew in India. Pp 95-106. In: *Health care in plantation Crops*. Souvenir *Phytophthora* -2011. International workshop, seminar and exhibition on *Phytophthora* diseases of plantation crops and their management. 12-17 September 2011, Rubber Research Institute of India, Kottayam.

Browne, F.G. 1961. The Biology of Malayan Scolytidae and Platypodidae. *Malayan Forest Records* No.22, Forest Department, Kuala Lumpur, Malaysia.

CCRS 1966. Central Cashew Research Station, Annual Progress Report 1965-66. Ullal, Karnataka.

Chatterjee, M. L. 1989a. *The Cashew* 3(3): 19-20.

Chatterjee, M.L. 1989b. A study on cashew flower thrips infestation and their control by different insecticides. *The Cashew,* 3(4): 12.

CIBC 1983. Commonwealth Institute of Biological Control. 1983. *Biocontrol News and Information* 4: 7-11.

CIRAD 2003. Rubber Wood General Properties. Tropix 5.0. French Agricultural Research Centre for International Development, France.

Desai, M.A., Kulkarni, S.V. and Rodrigues, A. 1977. *Report on the Marketing Survey on cashew in Goa.* Government of Goa, Daman and Diu. pp. 11-17.

Devasahayam, S. 1989. *Erythmelus helopeltidis* Gahan (Hymenoptera: mymaridae) a new parasite of *Helopeltis antonii* Sign. on cashew . *J. Bombay Nat. Hist. Soc.* 86: 113.

Devasahayam, S. and Nair, C.P.R. 1986. The mosquito bug, *Helopeltis antonii* Sign. on cashew in India. *Journal of Plantation Crops* 14:1-10.

Dwomoh, E.A., Afun, J.V., Ackonor, J.B. and Agene, V.N. 2009. Investigations on *Oecophylla longinoda (*Latreille) (Hymenoptera: Formicidae) as a biocontrol agent in the protection of cashew plantations. *Pest Management Science* 65(1) : 41-46.

Dwomoh, E. A,, Ackonorand, J. B and Afun, J. V. K. 2012. Survey of insect species associated with cashew (*Anacardium occidentale* Linn.) and their distribution in Ghana. *Journal of Agricultural Research and Fisheries* Vol. 1 (1): 006-016.

FAO 2003. Food and Agriculture Organization of the United Nations. *FAOSTAT statistics database.* Rome: UN Food and Agriculture Organization. http://apps.fao.org. (accessed April 2013.)

Feres, R.J.F. 1992. A new species of *Calacarus* Keifer (Acari, Eriophyidae, Phyllocoptinae) from *Hevea brasiliensis* Muell. Arg. (Euphorbiaceae) from Brazil. *Int. J. Acarol.* 18:61-65.

Ghosh, S.N. and Chatterjee, M.L. 1987. *The Cashew* 1(1): 21-23.

Godse, S.K, Dumbre, R.B. and Kharat, S.B. 1991. *The Cashew* 5(2): 14-15.

Godse, S.K, Dumbre, R.B. and Kharat, S.B. 1993. *The Cashew* 7(2): 13-14.

HDRA 2012. South America Tropical Advisory Service, December 2002 HDRA - the organic organization, Ryton Organic Gardens Coventry, CV8 3LG, UK, Tree Species No. TTS15.*http://www.rainforestrescueinternational.org* (accessed April 2013).

Hernandes, F.A., Feres, R.J.F. 2006 Review about mites (Acari) of rubber trees (*Hevea* spp., Euphorbiaceae) in Brazil Biota Neotrop. vol.6 (1): http://dx.doi.org/10.1590/ S1676-06032006000100005 (accessed April 2013).

Hiremath, I. G. 1991. *Cashew Bull.* 28(7): 9-13.

Hiremath, I.G., Khan, M.M. and Kumar, D.P. 1987. *The Cashew* 1(2): 8-9.

Ho, Y.F. and Hashim, S. 1997. Wood-boring beetles of rubber wood sawn timber. *Journal of Tropical Forest Products*. 3:15-19.

Hussein, N.B. 1981. A preliminary assessment of the relative susceptibility of rubber wood to beetle infestation. *The Malaysian Forester*. 44: 482-487.

ICIPE 2013. Integrated Management of Major Insect Pests and Diseases of Cashew in East and Western Africa, Fact sheet, The International Centre of Insect Physiology and Ecology (ICIPE)http://www.giz.de/Themen/en/dokumente/giz2011-1-en-factsheet-icipe-integrated-management-insect-pests.pdf (accessed April, 2013).

Jayarathnam, K, Nehru, C.R., and Jose, V.T. 1991. Field evaluation of some newer insecticides against bark feeding caterpillar *Aetherastis circulata* infesting rubber. *Indian Journal of Natural Rubber Research* 4(2):131-133.

Jayasinghe, C.K. 1999. Pests and diseases of *Hevea* rubber and their geographical distribution. *Bulletin of the Rubber Research Institute of Sri Lanka*, 40:1-8.

Jena, B.C. Patnaik, N.C. and Satapathy, C.R. 1985. Insect pests of cashew. *Cashew Causerie* 7 (3): 10--11.

Jose, V.T., Nehru, C, R and Jayarathnam K, 1989. Effect of Bordeaux paste as a repellent of slugs (*Mariaella dussumieri* Gray) infesting rubber plants. *Indian Journal of Natural Rubber Research* 2(1): 70-71.

Jose, V.T., Thankamany, S., and Kothandaraman, R. 2000. Control of bark feeding caterpillar *Aetherastis circulata* (Meyr) (Ypnomutidae: lepidoptera) infesting rubber with insecticidal dust. *Advances in Plantation Crops Research*. Pp 352-354.

Joshi, L., Wibawa, G., Vincent, G., Boutin, D., Akiefnawati. R., Manurung, G., Noordwijk, M.V. and Williams, S. 2002. Jungle Rubber: a traditional agro forestry system under pressure, International Centre for Research in Agroforestry Southeast Asia Regional Research Programme Jl. CIFOR, Situ Gede, Sindang Barang, Bogor, Indonesia. ISBN 979-3198-04-4.

KAU 2013. Centre for E learning Kerala Agricultural University. *http://www.celkau.in/Crops/Plantation%20Crops/ Rubber/pests.aspx* (accessed April, 2013).

Lara, M.F. and Tanzini, M.R. 1997. Non preference of the Lace Bug *Leptopharsa heveae* Drake & Poor (Heteroptera: Tingidae) for Rubber Tree clones. *An. Soc. Entomol. Brasil* 26(3): 429-434.

Maruthadurai, R., Desai, A.R., Prabhu, H.R.C. and Singh, N.P. 2012. Insect Pests of Cashew and their Management. *Technical Bulletin No. 28*, ICAR Research Complex for Goa, Old Goa.

Masasi 2013. Masasi farmers use weaver ants to combat insect pests in cashew *http://masasi-farmers.webs.com* (accessed April, 2013).

Mathew, G. 1982. A survey of beetles damaging commercially important stored timber in Kerala, Research Report, Kerala Forest Research Institute No. 10, pp. 93.

Mohapatro, G.K. 2008a. A refined damage rate scoring method for the tea mosquito bug infestation in cashew *Indian Journal of Entomology* 70(3), 273-277.

Mohapatro, G.K. 2008b. Evaluation of insecticides for control of tea mosquito bug, *Helopeltis antonii* and other insect pests of cashew. *Indian Journal of Entomology* 70(3): 217-222.

Mohapatra, L.N and Mohapatra, R.N. 2004. Distribution, intensity and damage of cashew stem and root borer (*Plocaederus ferrugineus* L.) in Orissa. *Indian Journal of Entomology* 66(1): 4-7.

Mohapatra, L.N., Behara, A.K. and Satapthy, C.R. 1998. Influence of the environmental factors on the cashew nut shoot tip caterpillar, *Hypotima haligramma* Meyr. *Cashew Bulletin* 35: 17-18.

Moses I. Olotu, Hannalene du Plessis, Seguni, S.Z., Maniania1, K.N. 2012. Efficacy of the African weaver ant *Oecophylla longinoda* (Hymenoptera: Formicidae) in the control of *Helopeltis* spp. (Hemiptera: Miridae) and *Pseudotheraptus wayi* (Hemiptera: Coreidae) in cashew crop in Tanzania. *Pest Management Science* 2012 DOI: 10.1002/ps.3451.

Muthu, M. and Baskaran, P. (1979) *Proc. PLACROSYM* II. p474-483.

Nair, G.M. and Abraham, C.C. 1982. *Agric. Res. J. Kerala* 20(2): 41-48.

Nair, G.M. and Abraham, C.C. 1983. *Agric. Res. J. Kerala* 21(1): 21-26.

Nair, G.M. and Abraham, C.C. 1984. *Agric. Res. J. Kerala.* 22: 118-123.

Nair, K.S.S. 2007. Tropical Forest Insect pests: Ecology, Impact, and Management. Cambridge University Press, Cambridge, 404 pp.

Nehru, C.R. and Jayarathnam, K. 1993. Evaluation of biological and chemical control strategies against the white grubs (*Holotrichia serrata*) infesting rubber seedlings. *Indian Journal of Natural Rubber Research* 6(1-2): 159-162.

Nehru, C.R. and Jayarathnam, K. 1987. Field incidence of bark-feeding caterpillar *Aetherastis circulata* (Meyr.) on different alternative host plants and its control. *Pesticides* 21(7): 39-40.

Nehru C.R. and Jayarathnam, K. 1988. Control of white grub (*Holotrichia serrata* F.) attacking rubber at the nursery stage in India. *Indian Journal of Natural Rubber Research* 1988 Vol. 1 No. 1 pp. 38-41.

Nehru C.R., Jayarathnam, K. and Karnavar, G.K. 1991. Application of entomopathogenic fungus *Beauveria brongniartii* for management of chafer beetle of the white grub *Holotrichia serrata* infesting rubber seedlings. *Indian Journal of Natural Rubber Research* 4(2): 123-125.

Nehru C.R., Jayarathnam, K. and Pillai, P.N.R. 1983. Incidence of bark-feeding caterpillar *Aetherastis circulata* Myer on rubber (*Hevea brasiliensis* Muell. Arg.). *Indian Journal of Plant Protection* 11: 1-2, 150.

Nehru, C.R., Jayarathnam, K. and Thankamany, S. 1987. Field incidence of bark-feeding caterpillar *Aetherastis circulata* (Meyr) on different alternative host plants and its control. *Pesticides* 21(7): 39-40.

NRCC 1988. (National Research Centre for Cashew) *Annual Report for 1987*, Puttur, Karnataka, India. pp. 41.

Paul van Mele 2008. A historical review of research on the weaver ant *Oecophylla* in biological control. *Agricultural and Forest Entomology* 10: 13-22.

Pillai, G.B. 1979. Pests. pp. 55-72. In Cashew *(Anacardium occidentale)* (Nair, M.K., Rao, E.V.V.B., Nambiar, K.K.N. and Nambiar, M.C., Eds.). Central Plantation Crops Research Institute, Kasaragod, India.

Pillai, G.B. and Abraham, V.A. 1975. *Indian Cashew J.* 10(1): 5-7.

Pillai, G.B., Dubey, D.P. and Vijay Singh. 1976. Pests of cashew and their control in India -A review of current status. *Journal of Plantation Crops* 4: 37-50.

Pillai, G.B. Singh, Vijay, Dubey, O.P. and Abraham, V.A. 1984. Seasonal abundance of tea mosquito bug, *Helopltis antonii* on cashew in relation to meteorological factors. In: *Cashew research and development* (Eds. E.V.V.B. Rao and H.H. Khan) Indian Society for Plantation Crops, CPCRI, Kasaragod, India, pp.103-110.

Rai, P.S. 1984. Handbook of cashew pests. Researchco Publications, Delhi, India.

Rao A. R., Naidu V. G., and Prasad P. R. 2002. Studies on the Biology of Cashew Shoot and Blossom Webber (*Lamida moncusalis* Walker) *Indian Journal of Plant Protection* 30(2): 167-171.

Raviprasad, T.N. Bhat, P. S. and Sundararaju, D. 2009. Integrated pest management approaches to minimize incidence of cashew stem and root borers (*Plocaederus* spp.) *Journal of Plantation Crops* 37(3), 185-189.

Raviprasad, T.N. Sundararaju, D. and Bhat P.S. 2005. Efficacy of botanicals against *Helopeltis antonii* Sign. infesting cashew. *The Cashew* XIX(4): 9-14.

Rubber Board 2013. http://rubberboard.org.in/RubberCultivation.asp. Rubber Board, India. (assessed April 2013).

Samiayyan, K. and Palaniswamy, K.P. 1984. *Cashew Causerie* 6(4): 9-11.

Santos, Rodrigo Souza 2010. Biological Control of Rubber Tree Lace Bug by Endophytic Parasitoid in Brazil *Entomology Papers from Other Sources.* Paper 114. *http://digitalcommons.unl.edu/entomologyother/114* (assessed April 2013).

Santos, RS and Freitas, F.D. 2008. Parasitism of *Erythmelus tingitiphagus* (Soares) (Hymenoptera: Mymaridae) in *Leptopharsa heveae* Drake & Poor (Hemiptera: Tingidae) eggs, in rubber tree plantation (*Hevea brasiliensis* Müell. Arg.) *Neotropical Entomology* 37(5): 571-6.

Satapathy, C.R. 1993. Bioecology of major insect pests of cashew *(Anacardium occidentale* Linn.) and evaluation of certain pest management practices. *Ph.D. Thesis*, U.A.S., Bangalore, pp. 224.

Sathiamma, B. 1978. Occurrence of insect pests on cashew. *Cashew Bulletin* 15(4): 9-10.

Singh, V., and Pillai, G.B. 1984. In: *Cashew Research and Development* (Eds Rao, E.V.V.B., and Khan, H.H.) Indian Society for Plantation Crops, CPCRI, Kasaragod, India. pp. 301.

Somchoudhury, A.K., Samanta, A. and Dhar, P. 1993. *Proc. International Symposium on tea science and human health* pp. 330-338. (Teatech,1993). Tea Research Association, India.

Stonedahl, G.M. 1991. *Bull. Entomol. Res.*81: 465-490.

Sundararaju, D. 1979. A note on major pest problem of cashew, coconut and arecanut and their control in Goa. *Proceedings of PLACROSYM II.* pp. 513-16.

Sundararaju, D. 1984. Studies on cashew pests and their natural enemies in Goa *Journal of Plantation Crops* 12: 38-46.

Sundararaju, D. 1993a. Studies on the parasitoids of the mosquito bug, *Helopeltis antonii* Sign. (Hymenoptera: mymaridae) on cashew with special reference to *Telenomous* sp. (Hymenoptera: Scelionidae). *Journal of Biological Control.* 7: 68.

Sundararaju, D, 1993b. Compilation of recently recorded and some new pests of cashew in India. *The Cashew* 7: 15-19.

Sundararaju, D. 1996. Studies on *Helopletis* spp. with reference to *Helopeltis antonii* Sign.in Tamil Nadu. Ph.D. thesis. T.N.A.U., Coimbatore, India.

Sundararaju, D.and Bakthavatsalam, N. 1990. Cashew pest management for coastal Karnataka. *The Cashew* 4: 3-6.

Sundararaju, D.and Bakthavatsalam, N. 1994. Pests of cashew. In: *Advances in horticulture, Vol.10* (eds. K.L.Chadha and P.Rethinam) Mathotra Publishing House, New Delhi, India, pp 759-785.

Sundararaju, D and Sundarababu, P.C. 1999. *Helopeltis* spp. (Heteroptera: Miridae) and their management in plantation and horticultural crops of india. *Journal of Plantation Crops* 27(3): 155-174.

Sundararaju, D., Yadukumar, N., Bhat, P.S., Raviprasd, T.N., Venkattakumar, R. and Sreenath Dixit, 2006. Yield performance of "Bhaskara" cashew variety in coastal Karnataka. *Journal of Plantation Crops* 34(3): 216-219.

Thankamony, S. 2012. Evaluation of Bio Pesticides and Insecticides on the control of bark feeding caterpillar, *Aetherastis circulata* attacking Rubber Trees. International Rubber Conference, 28-31, October 2012, Kovalam, Kerala, India. Abstracts p149-150. *http://clinic.rubberboard.org.in/PDF/abstract.pdf* (accessed April, 2013).

Uthaiah, B.C., Rai P.S., Khan, M.M., Hiremath, I.G., Kumar, D.P and, K.B. 1989. Pre bearing performance of some cashew type in coastal Karnataka, *The Cashew* 3: 9-11.

Williams, S. E., Noordwijk, M.V., Penot, E., Healey, J.R., Sinclair, F.L. and Wibawa, G. 2001. On farm evaluation of the establishment of clonal rubber in multi-strata agro forests in Jambi, Indonesia. *Agroforestry Systems*, 53(2): 227-237.

Wisut Sittichaya, and Roger Beaver 2009. Rubber wood destroying beetles in the eastern and gulf areas of Thailand (Coleoptera: Bostrichidae, Curculionidae: Scolytinae and Platypodinae) *Songklanakarin J. Sci. Technol.* 31 (4): 381-3879.

Wong, A.H.H., Kim, Y.S., Singh, A.P. and Ling, W.C. 2005. Natural Durability of Tropical Species with Emphasis on Malaysian Hardwoods - Variations and Prospects. The International Research Group on Wood Preservation, the 36th Annual Meeting, Bangalore, India, April 24-28, 2005, 33 pp.

Integrated Pest Management in the Tropics; pp. 597-625

New India Publishing Agency, New Delhi (India)

CHAPTER - 19

Insect Sex Pheromones in Integrated Pest Management

Prasad, AR., Durairaj, C[1]., Prasuna, A.L., Jyothi, K.N. and Sambathkumar, S[1]
Indian Institute of Chemical Technology, Hyderabad, Telangana State, India
[1]Tamil Nadu Agricultural University, Coimbatore, Tamil Nadu, India
email: c_durairaj@yahoo.com

19.1. Introduction

A wave of research on insect pheromones has been initiated in early 1960's in response to public concern and alarm about chemical pesticides to develop an elegant and ecologically safe technology for the control of insect pests. Pheromones, a class of semiochemicals are considered as 'Biochemical Pesticides', belong to the subdivision of Biorational pesticides which include "Microbials" (Viruses, Bacteria, Pathogens *etc.*) and "Biochemicals" (insect plant growth regulators, semiochemicals *etc.*). The term biorational (biological + rational) pesticides can be defined as the use of specific and selective chemicals, often with a unique modes of action, that are compatible with natural enemies and the environment, with minimal effect on non-target organisms. Biorational control is based on a diversity of chemical, biological and physical approaches for controlling insect pests which results in minimum risk to man and the environment (Horowitz *et al.*, 2009).

Chemical messages that trigger various behavioural responses in insects are collectively referred to semiochemicals. The term pheromone is used to describe semiochemicals that operate intra-specifically (Karlson and Luscher,

1959). Allelochemical is the general term for inter specific effectors (Whittaker and Feeny 1971). First, pheromones are designated as either releaser pheromones, whereby reception of the pheromone triggers an immediate behavioral response (*e.g.*, a male insect being immediately attracted by a female's volatile sex pheromone), or primer pheromones, which induce a delayed response which may be manifested as a physiological change as in the caste determination in termites (Kaib, 1999) and honeybees (Pettis *et al.*, 1999; Keeling *et al.*, 2004).

Insect Pheromones are potentially of high interest in Integrated Pest Management (IPM) as these are naturally occurring, eco-friendly and used at concentrations close to those found in nature. Pheromones/semiochemicals are volatile and perceived through olfaction. The versatility and potential of pheromones in the pest management has been well reviewed by many researchers (Pickett *et al.*, 1989; Agelopoules *et al.*, 1999; Aldrich *et al.*, 2003). Pheromones are simple organic molecules secreted to the outside by insects themselves which evoke specific behavioural responses on reception, in the individuals of the same species. It can evoke many responses including mating (sex pheromone), congregation for food or reproduction (aggregation pheromones), dispersal on attack of predators (alarm pheromones), deterrence towards oviposition sites (epideictic pheromones) and trail marking (trail pheromones). Pheromones dissipate rapidly, leave no residues and are non toxic to humans and other living organisms (Witzgall *et al.*, 2010). Their use in IPM programmes proved mainly to rationalize the use of conventional insecticides with the reductions of up to 50% (Baker and Heath, 2004). Insect pheromones with their unique mode of action provide appropriate and effective management for pests targeting the adult insects. The potential of sex pheromones particularly of insects that belong to Lepidoptera, a very important economic insect order, has been well exploited globally in pest management practices. Sex pheromone based technology is now internationally recognized as green plant protection technology marked with ecological security and sustainability. Pheromone application is found highly suitable and only possible practice for the control of insect pests like leaf miners, webbers, fruit borers, dwellers and storage pests. By virtue of the ecology of these pests it is practically difficult to control their immature stages (larvae) the most damaging stage of the pest as they protect themselves from the insecticide sprayings by remaining inside the leaf mines or webs.

Development and implementation of pest management tactics based on semiochemicals have progressed considerably ever since the first detection of sex pheromone in silk moth *Bombyx mori* in 1959 (Butenandt *et al.,* 1959). During the last four decades pheromone systems of many agricultural pests of the order Lepidoptera and coleoptera have been identified, the chemicals are synthesized, formulated as lures using various dispensers and are being used in pest management. A database on insect pheromones and related attractants has been created with hundreds of chemicals (Arn *et al.*, 2000; El-Sayed, 2013). To date the study of pheromones/semiochemicals which deals with the interactions within and between the individuals has been recognised as an important section within the discipline of Chemical Ecology. This chapter will summarize the state of the art regarding the practical use of synthetic pheromones in crop pest management focusing mainly on the work carried out in India.

19.2. Pheromones in Pest Management

The various approaches of pheromones/ semiochemicals in pest management are Monitoring, Mass trapping , Mating, Disruption Attract & Kill and Push-Pull.

19.2.1. Pheromones in monitoring

Monitoring is the corner stone of any pest management program. Pest monitoring provides data about the presence and density of the target pest species. Monitoring helps in determining when and how often to time control actions. Pest monitoring minimizes environmental pollution while undertaking any pest management practices by avoiding the unnecessary pesticide applications. For monitoring pest populations a small number of traps (3 to 5 per ha) loaded with sex pheromone of a specific pest are placed in the field to attract the opposite sex. The traps act as specific and sensitive indicators regarding the targeted pest. Lepidopteran pest populations in agricultural systems are monitored by collecting males in traps baited with synthetically – produced female pheromones (Mitchell, 1986; Roelofs and Carde, 1977). One of the first widely used monitoring system was established for pea moth, *Cydia nigricana* for which an action threshold was established based on trap catches (Wall *et al.*, 1987). Monitoring helps to evaluate the effectiveness of mating disruption to detect the presence of pests in quarantine or survey programmes and aids in assessing the levels of insecticide resistance. An equally important and rapidly growing use of pheromone-baited traps is for the detection and interception of invasive species, Semiochemically-baited traps form the cornerstone of detection for notorious invasive pests such as the gypsy moth *Lymantria dispar* (L.), the Japanese beetle *Popillia japonica* Newman, the boll weevil *Anthonomus grandis* Boheman, the pink bollworm *Pectinophora gossypiella* (Saunders), and the Mediterranean [*Ceratitis capitata* (Wiedeman)] and related fruit flies (Ridgway *et al.*, 1992; Howse *et al.*, 1998). In addition to deploying pheromones for detection of pest insects, there is increasing interest in using pheromones for sampling of beneficial insects also (Suckling *et al.*, 1999, 2000, 2006; Stanley *et al.*, 2000; Cossé *et al.*, 2004).

19.2.2. Pheromones in mass trapping

This technique is an extension of the use of pheromone baited monitoring traps, with the aim of reducing or eradicating target pests by capturing as many individuals as possible. Success is depending on the number of traps installed, ecology and life cycle of the insect, reproductive behaviour of the insects and the type of pheromone used in the optimally efficient trap. Examples of mass trapping for pest control are fairly plentiful (Arnon Shani ,1993) and classic case studies have been reviewed extensively by El-Sayed *et al.* (2006; 2009) and Witzgall *et al.* (2010). Mass trapping technique is particularly successful against Coleopteran pests wherein the aggregation type of pheromones is employed for trapping both sexes of target pests. Aggregation pheromone is similar to sex pheromone but attractive to both sexes and thus help in locating food sources, hibernation sites, oviposition and also to some extent mating as it causes con specifics to gather at one site (Ali and Morgan 1990). Aggregation pheromones offer excellent potential for direct control in pest management when incorporated into traps along with host plant volatiles and currently being used to control forest pests (Macaulay *et al.*, 1985; Kogan, 1998). However, the most effective and largest scale use of pheromone-based mass trapping has been in the management of large tropical weevils such as the palm weevils *Rhynchophorus palmarum* Linnaeus and *R. ferrugineus* Olivier (Oehlschlager *et al.*, 2002; Faliero, 2006) and related species, and the banana weevil *Cosmopolites sordidus* Chevrolat.

19.2.3. Pheromones in mating disruption

Mating disruption is used typically with female sex pheromone of Lepidoptera since long range sex pheromones in most of the moth species have been shown to be produced by females to attract the males (Byers, 2007;2012). The technique involves the permeation of the air around the crop with large quantity of synthetic pheromone to prevent the male moth to locate the female successfully for mating. Pheromone-based mating disruption, which is the most commonly used approach, may work via a number of overlapping mechanisms that interfere with mate location and reproduction. Sanders (1996), Cardé and Minks (1995) and Miller *et al.* (2006a,b) have reviewed different mechanisms of mating disruption as

- Sensory fatigue, representing either adaptation or habituation
- Competition between the pheromone dispensers and the calling insect (false trail)
- Masking of the natural pheromone concentration.

The first mating disruption product for control of an insect pest *Pectinophora gossypiella* was registered in 1978. Additional registrations came slowly in the 1980's, but increased dramatically over the past two decades. Now more than 50 years later, mating disruption technology is used worldwide to control many insect pests in both agriculture and forestry. The area under mating disruption has increased exponentially from the 1990's and worldwide coverage was 770,000 ha in 2010 (Witzgall *et al.*, 2010). Currently, 30 mating disruption pheromone based products are registered by the US-EPA for the control of eleven lepidopterous pest species. The three species with the highest land area under mating disruption include gypsy moth in North America, codling moth in apple and pear trees world wide and the grapevine moth in grape in the EU and Chile (Witzgall *et al.*, 2010). Mating disruption is also being attempted with insects other than moths such as for example, a pheromone formulation to vine mealy bug *Planococcus ficus* (Signoret) in vineyards came on the market in 2008 (Suterra LLC., Bend, Oregon, USA).

19.2.4. Pheromones in attract and kill technology

This is a promising safe strategy for pest management that improves the synthetic pesticides usage. The strategy is to attract the adult insects by using semiochemicals to a site (trap, trap crop or a selected location in the field *etc.*) and kill them using a chemical insecticide or pathogen or any biocontrol method. A combination of the sex pheromone which attracts males and an aggregation pheromone or food lure with a toxicant which attracts females can be used to get the desired results. To date, attracticide technology has been most widely applied for the control and eradication of invasive tephritid fruit flies such as the Mediterranean fruit fly *Ceratitis capitata* (Millar, 1995), the oriental fruit fly *Dacus dorsalis* Hendel, the melon fly *D. cucurbitae* Coquillet, the Mexican fruit fly *Anestrapha ludens* (Loew), and the olive fruit fly *D. oleae* (Gmelin) (Metcalf and Metcalf, 1992). For most of these species, the best formulations consist of food baits combined with a toxicant, but pheromone-based attracticides have now been developed for control of olive fruit fly (*e.g.*, Mazomenos *et al.*, 2002).

19.2.5. Pheromones in push-pull strategy

Push pull strategies involve the behavioural manipulation of insect pests and their natural enemies through integration of stimuli which make the protected resource unsuitable to the pest (push) while luring them towards the attractive bait (pull) from where they can be removed or destroyed successfully. Combination of repellent and attractant semiochemicals may find use in push-pull tactics. Push-pull strategy has been established as a reliable IPM system reducing the pesticide input. Semiochemical-based push-pull strategy was found appropriate in controlling stem borers of maize and sorghum (Khan *et al.*, 2008).

Ever since the first pheromone product was made available for commercial purpose in 1991, the utility of pheromone-mediated field application technology has been amply demonstrated in several instances (Ridgway *et al.*, 1992). An initial survey until 2000 indicated that worldwide 1.3 – 2.0 million ha of variety of crops are treated with pheromones through all three techniques as part of IPM strategy. Commercially, sales of semiochemicals (pheromones) represent a very small fraction of the current global market for pest control products. However, the estimated annual growth rate in pheromone sales is about five times that of conventional pesticides and one estimate predicts that pheromone sales will rise from approximately 1% of global pesticide market to 20% by 2015. At present, about half of the total sales of semiochemical-based products are in the form of pheromone lures and traps for monitoring purposes, while most of the remaining are for mating disruption products. The estimated annual production of lures for monitoring and mass trapping is on the order of tens of millions, covering at least 10 million hectares. Insect populations are controlled by air permeation and attract-and-kill techniques on at least 1 million hectares (Witzgal *et al.*, 2010). Globally, North America has the largest regional market for pheromones used in IPM programs accounting for over 50.0% of the total market.

19.3. Development and Use of Pheromone–Based Pest Management Strategies in India

Monitoring the activity of insects using pheromone or semiochemical baited traps has become an integral part of many pest management programmes in several parts of the world. Over the last 50 years, chemists have extracted and analyzed natural insect pheromones from over 1,500 different insect species and developed synthetic routes to prepare most of these chemicals in large quantities to use them in the fields. This has made Pheromone Application Technology (PAT) to establish rapidly as one of the major eco-friendly pest management practice in the developed countries. However, in developing countries like India progress in the application of pheromones for pest control has been rather slow and is still negligible. A few of the studies recorded in this field are presented below.

19.3.1. Early studies on insect pheromones in India

The Natural Resources Institute (NRI), UK has extensively worked in the Indian Sub continent to utilise insect pheromones for the control of a wide range of insect pests (Cork and Hall, 1998). Most of the work conducted on pheromones of cereal pests particularly on rice pests. The effect of trap type, trap height, position in the crop and age of the lure on trap catch of sex pheromone of major insect pest of rice, the yellow stem borer, *Scirpophaga incertulas* were optimised by Ganeswara Rao and Krishnaiah (1994; 1995). Initially pheromone mediated mass trapping of yellow stem borer of rice was successfully demonstrated by Directorate of Rice Research (DRR) Hyderabad by procuring pheromone components from Agrisense, UK and by impregnating onto locally available dispensers. Their trials demonstrated significant reduction in YSB infestation in pheromone treated plots with significant increase in yield. DRR in collaboration with Natural Resources Institute (NRI) UK has also demonstrated the feasibility of mating disruption of YSB with imported pheromone formulations (Cork *et al*., 1996; 1998). A PVC-resin formulation of sex pheromone was tested for mating disruption of yellow rice stem borer, *S. incertulas*, in replicated 1 ha plots in India over two seasons. Even at an application rate of 10 g a.i./ha/ season there was a significant reduction in damage relative to that in plots treated with insecticides. However, application rates of 30–40 g a.i./ha/season are recommended for general use by farmers (Cork *et al*., 2008).

Several researchers in India have monitored the incidence of Pink Bollworm, *Pectinophora gossypiella,* with its sex pheromone blend of 1:1 mixture of (Z,Z) and (Z,E) 7-11, hexadecadienyl acetate (Sharma and Singh, 1982; Sidhu and Dhawan, 1985). Taneja and Jayaswal (1981 and 1983) reported a significant correlation between the moth catches of pink bollworm in traps and larvae in flowers and bolls of cotton at Hissar, Haryana. Ranga Rao *et al.* (1991a and b) conducted intensive sex pheromone monitoring of tobacco caterpillar, *Spodoptera litura*. Their studies indicated 10 annual generations of *S.litura* in Hyderabad and 7 generations in Pant Nagar, Uttarakhand during 1985-87.

The development of a trapping system for monitoring *Heliothis armigera* male moths using a synthetic sex pheromone attractant was carried out by International Crops Research Institute for the Semi Arid Tropics (ICRISAT), Patancheru, India. A white funnel trap was most efficient and was adopted as the ICRISAT standard trap. ICRISAT standard traps are being used in a network to study the populations of *H. armigera* and their movements across the Indian Sub-continent (Pawar *et al*., 1988). Patel *et al.* (1985) used 1000 traps over 175ha of cotton in Gujarat in a mass trapping campaign for *H.armigera* and *S.litura* and recorded a reduction in insecticide applications by 3 to 6 sprays and increased yields. In black gram, Krishnaiah (1986) conducted monitoring and mass trapping studies for *S. litura* and obtained excellent results recording suppression of pest population, reduction in number of insecticide sprays and increase in grain yield to the extent of 71 kg per ha. There is also a report on successful mass trapping of codling moth, *Cydia pomonella* by Tuhan and Pawar (1980) who felt that if continued for several generations could eradicate the problem.

Studies on sex pheromone of Diamond Back moth (DBM), *Plutella xylostella*, were reported by Reddy and Urs in 1995 and also reported the relative performance of various types of traps in cole crops and successful mass trapping of DBM in cabbage fields using synthetic pheromone. (Reddy and Urs 1996; 1997). Reddy and Guerrero (2000) successfully evaluated Pheromone – based IPM to control DBM in cabbage fields. The IPM programme, based on the pheromone trap catch threshold of eight moths per trap per night confirmed the potential use of synthetic pheromone as a monitoring and management tool for DBM. Mass trapping of Red Palm Weevil (RPW) *Rhynchophorus ferrugineus* Olivier was successfully carried out in 45ha of weevil infested coconut gardens with lures procured from Costa Rica (USA), pherobank (Holland) and indigenously produced pheromones from Central Plantation Crops Research Institute (CPCRI), Kasaragod, Kerala were evaluated along with the food bait and among them lure synthesized by CPCRI Kasargod, Kerala, although 50 % less efficient than imported lures, is found to be economical (Faleiro, 2006).

Male-based sex pheromone in Coffee bean weevils (CBW) that attracts females was reported by Sreedharan and Karan Singh (1987). Hall *et al.* (1998) identified (S)-2-hydroxy-3-decanone as the major component of the male sex pheromone of CBW from India. According to Hall *et al.* (2006) and Venkatesha and Dinesh (2012), further studies on trap design and additional pheromone components CBW are still needed.

The female sex pheromone of sugarcane Inter node borer, *Chilo sacchariphagus indicus* is identified as a mixture of Z_{13} -octadecenyl acetate and Z_{13} – octadecen 1- ol in the ratio of 7:1 respectively at Sugarcane Breeding Institute with the help of Overseas Development and Natural Resource Institute (ODNRI), London (David *et al.*, 1981). Small scale field trials at Pugalur, Tamilnadu with above lures gave encouraging results (David *et al.*, 1985). Successful attempts have been recorded for the management of Early shoot borer - *Chilo infuscatellus* and Internode borer-*Chilo sacchariphagus* indicus (Narasimhan , 1995). Lack of indigenous technology, good quality pheromone components and economic viability hindered further progress and development of pheromone application for management of sugarcane pests (Easwara murthy *et al.*, 2003).

19.3.2. Constraints for popularization of pheromone technology in India

The major constraints/ drawbacks that can be attributed for less popularization of pheromone application are :

- Lack of indigenous technologies to produce pheromones of high purity in large quantities for regular supply.

- Most of the field trials are conducted with imported pheromone lures which led to the lack of continuous availability of material resulting in inconsistent data generation.
- Improper implementation of the technology due to poor knowledge of the targeted insects' behavior.
- Mostly, research groups (Entomologists and Chemists) working in isolation at laboratory levels.
- Deployment of spurious lures in the Indian market.

19.3.3. Pheromone research at Indian Institute of Chemical Technology, Hyderabad (CSIR-IICT)

The study on semiochemicals especially insect sex pheromones and the associated application technology which includes the chemical identification, preparation of synthetic mimics, development of devices for field release and the application methodologies can be collectively referred as Pheromone Application Technology (PAT). This kind of study needs an interdisciplinary collaborative approach between chemists, entomologists, agronomists and finally the crop producers. Recognizing the versatility of insect pheromones in Integrated Pest Management, Council of Scientific and Industrial Research -Indian Institute of Chemical Technology (CSIR-IICT) with its strong base in synthetic organic chemistry entered this field of Green Chemistry way back in 1995 (Yadav and Reddy, 1998). Centre for Semiochemicals at CSIR-IICT, Hyderabad with its multidisciplinary group has initiated the work on Pheromone R&D as an ongoing activity and further supported by funding agencies like DBT, DST, ICAR, APNL/ BTU etc. The centre perhaps is the only place in India, through its international collaborative activities has acquired adequate expertise and established the minimum required state of art facilities like Electroantennogram (EAG) and EAG coupled Gas chromatograph (GC-EAD), GC coupled Mass spectroscopy.

19.4. Common Structures of Pheromones

Structurally, the sex pheromones of Lepidoptera are multi component aliphatic linear unsaturated compounds from 10 to 20 carbon atoms being the most frequent 12, 14 and 16 with alcohol, ester, aldehyde, ketone or epoxide functionality at the end of carbon chain. They have one or more double bonds at various carbon numbers. The chemical structures of insect pheromones are quite diverse and most of them fall into one or the other following category of compounds:

(a) Unsaturated Straight Chain Aliphatic compounds

OAc

Z-13-Octadecenyl acetate

CHO

Z-9-Hexadecenal

(b) Intramolecular-linked ketal (Beetles)

(-)frontalin Exobrevicomin

(c) Spiroketals (wasp, beet, olive fly)

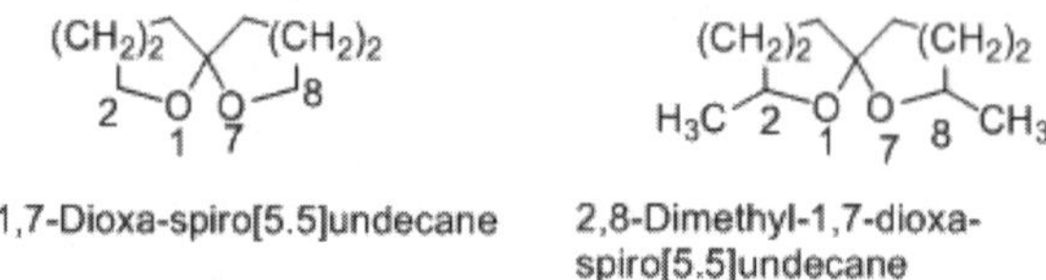

1,7-Dioxa-spiro[5.5]undecane 2,8-Dimethyl-1,7-dioxa-spiro[5.5]undecane

(d) Oxygen & Nitrogen Heterocycles (Fire Ants)

N $(CH_2)_6CH_3$

$H_3C(H_2C)_{10}$ N

2-Methyl-6-undecyl-piperidine

O

O

2-Ethyl-6-methyl-2,3-dihydro-pyran-4-Hepialone

(e) Terpenoids (Bollweevils, anthropods, termites etc.)
Grandisol, Ipsenol, several sesqui and diterpenes

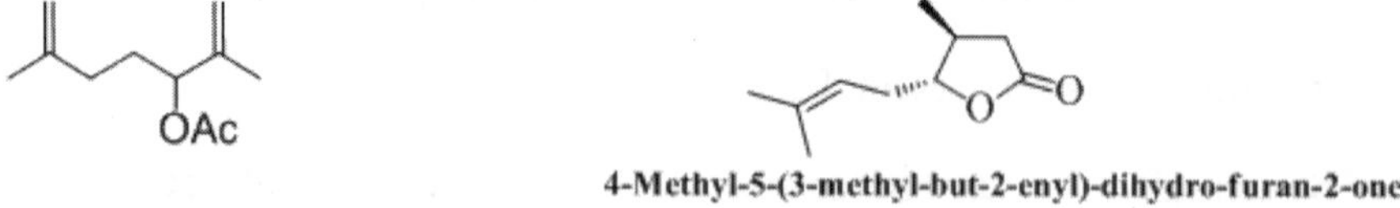

4-Methyl-5-(3-methyl-but-2-enyl)-dihydro-furan-2-one

Structurally, the sex pheromones of Lepidoptera are multi component compounds. They comprise of mostly aliphatic linear carbon chains ranging from 10 to 20 carbon atoms with one or more double bonds frequently at 12, 14 and 16 carbons. Insects achieve species specificity for the pheromones through 1. Basic chemical structure, 2. Fuctional groups (alcohol, acetate, aldehyde, lactone *etc.*) 3. Position, configuration of double bonds & Optical isomerism and 4. Defined blend ratio of Pheromone components.

19.5. Field Evaluation of Synthetic Pheromones

Field evaluation of synthesized components is the final and important step to bring the developed technology to the actual application for the benefit of the end users. To achieve this goal, IICT has developed strong linkages with State Agriculture Universities, Indian Council of Agriculture Research (ICAR) Institutes and Non- Governmental Organizations (NGO's) for popularization of PAT. Several village level on farm demonstrations in collaboration with agriculture research scientists involving the local farmers under participatory approach, proved the efficacy of indigenously synthesized pheromone blends. All the concerned farmers informed that they are benefited through pheromone application technology. During this field implementation programmes the farmers, agriculture extension personnel, local NGO's were educated and trained on all aspects of PAT. The following are the insect pests for which large scale village level field evaluations were carried out at different locations (Table 19.1).

Table 19.1 : Details of field demonstrations with indigenously developed pheromone lures

S. No.	Insect Pest	Crop	Location (India)	Approach/ Method
1.	Yellow Stem Borer, *(Scipophaga incertulas)*	Rice	Andhra Pradesh	Monitoring and Mass trapping
			Tamil Nadu	Monitoring
			Orissa	Monitoring
2.	Groundnut leaf miner, *Aproaerema modicella*	Groundnut	Andhra Pradesh	Monitoring and Mass trapping
			Tamil Nadu	Monitoring and Mass trapping
			Gujarat	Monitoring
3.	Brinjal shoot and fruit borer, *Leucinodes orbonalis*	Brinjal	Andhra Pradesh	Monitoring and Mass trapping
			Tamil Nadu	Mass trapping
			Orissa	Monitoring
4.	Sugarcane inter node borer, *Chilo sacchariphagus indicus*	Sugarcane	Tamil Nadu	Monitoring and Mass trapping
5.	Top shoot borer, *Scirpophaga excerptalis*	Sugarcane	Tamil Nadu	Monitoring and Mass trapping
6.	Sugarcane top borer, *Chilo infuscatellus*	Sugarcane	Tamil Nadu	Monitoring and Mass trapping
7.	American bollworm, *Helicoverpa armigera*	Cotton and vegetables	Andhra Pradesh Maharashtra	Monitoring and Mass trapping

8.	Pink bollworm, *Pectinophora gossypiella*	Cotton	Andhra Pradesh Maharashtra	Monitoring and Mass trapping
9.	Spiny bollworm, *Earias vitella*	Cotton	Andhra Pradesh Maharashtra	Monitoring Mass trapping
10.	Diamondback moth, *Plutella xylostella*	Vegetables	Andhra Pradesh	Monitoring Mass trapping

The group has created facilities for the study of insect behavioural biology, electrophysiology and related aspects of chemical ecology unique to each insect. The Centre also provides service in semiochemical research to Agricultural based Universities, organizations and State Agriculture department in isolation and identification of semiochemical compounds and to determine the target insects activity to the synthetic pheromones by employing EAG, GC-EAD, GC-MS, wind tunnel and olfactometers *etc.* (Durairaj *et al.*, 2015; Jyothi *et al.*, 2014; 2002; Rajasekhara Rao *et al.*, 2010; 2013; Padmaja *et al.*, 2012; Kumar *et al.*, 2011; Ganesh *et al.*, 2011; Prasad and Subhaharan, 2010; Prasuna *et al.*, 2008; Varshney *et al.*, 2005; Mamata *et al.*, 2010; Krishna Kumari *et al.*, 1998; Persoons *et al.*, 1993). Information on PAT, which includes the pheromone blend composition, suitable dispenser, trap type along with the recommendations for implementation is available as package at IICT for crop pests like rice yellow stem borer, groundnut leaf miner and sugarcane borers and vegetable pests like diamondback moth, brinjal shoot and fruit borer *etc.* (Prasad *et al.*, 2010, 2012 and 2013; Yadav, 2007; Jyothi *et al.*, 2014; Radhika *et al.*, 2014). Various parameters standardized for implementation of PAT for direct control using mass trapping strategy is as follows (Table 19.2).

Table 19.2 : Details of PAT developed for various insect pests

Name of the insect	Type of Trap	Number of traps/hectare		Duration of lure replacement	Dispenser Type	Pheromone dose/ lure	Trap height
		Monitoring	Mass trapping				
Groundnut leaf miner (GLM)	Delta sticky trap	5-7	16-18	21 days	Plastic vial	3 mg	5 cm above the crop canopy
Brinjal shoot and fruit borer	Sleeve trap	5-7	18-22	21 days	Plastic vial	3-5 mg	Canopy level
Sugarcane inter node borer	Water trap	5-7	18-20	21-24 days	Rubber sepa	3 mg	3 feet from ground up to 5th week
Rice Yellow stem borer	Sleeve trap	5-7	18-20	21 days	Rubber septa	5 mg	Canopy level
American bollworm	Funnel trap	5-7	18-20	21 days	Rubber septa	5 mg	1mt above the canopy
Diamondback moth	Sticky trap	5-7	18-20	21days	Rubber septa/ Plastic vial	3 mg	Canopy level
Pink bollworm	Delta sticky/ Sleeve trap	5-7	18-20	21- 24 days	Plastic vial	3 mg	5 cm above canopy level
Spotted bollworm	Sleeve funnel trap	5-7	18-20	21 days	Green rubber septa/ Plastic vial	3 mg	5 cm above canopy level

19.6. Case Studies on Sex Pheromone Based IPM

19.6.1. Rice Yellow Stem Borer (YSB), *S. incertulas*

Rice (*Oryza sativa* L.) is the important and staple food crop for more than two third of the population of India. One third of all food grain produced in India comes from rice, approximately 74 million tonnes from 42 million ha. In India, the major pre harvest loss is associated with *S. incertulas* which completes between 4 and 5 generations a year in South Indian conditions. The pheromone system of YSB comprises of Z-11-Hexadecenal & Z-9-Hexadecenal in the ratio of 3 : 1 as identified by Cork *et al.* (1985). Feasible methodologies for the two components were developed and field experiments were conducted on village level basis with indigenously synthesized pheromone blend impregnated on to locally available dispensers.

CHO CHO

Z_{11}-Hexadecenal (3) : Z_9-Hexadecenal (1 part)

Mass trapping technique was demonstrated to the farmers through on-farm multi location trials, area/village wide/ compact-block front line demonstrations. Directorate of Rice Research (DRR), Hyderabad collaborated with IICT in conducting the field experiments. Eight different pheromone dispensers were prepared and evaluated for their efficacy and persistence in the fields. YSB damage was reduced by 50 to 65% and avoided the grain yield loss of 0.5 to 1.5 ton / hectare at the end of the trial. For mass trapping, white rubber septum or rubber tube impregnated with 10 mg of pheromone was proved persistent and provided season long control of YSB without lure replacement (Krishnaiah *et al.*, 1998). The technique was found suitable and promising even in small farm holdings of up to one hectare.

19.6.2. Groundnut Leaf Miner (GLM) *A. modicella*

Groundnut, *Arachis hypogea* L. is one of the important oilseed crops in India. Although, India ranks first in area under groundnut cultivation, the productivity is quite low (1432Kg / ha) compared to USA (3000 kg /ha) China (2600 Kg / ha) Argentina (2100 Kg / ha) due to pest and disease problems. Among the pests that cause severe damage to groundnut crop are leaf miner, *A. modicella* and Tobacco caterpillar, *S. Litura* are considered as major defoliators of in India. The larvae of GLM feed on foliage mining into the leaf lets and subsequently seize the plant growth. Existence of female based sex pheromone system in GLM was first established by Lalitha Kumari and Reddy (1991) and Nanadagopal (1995). Sex pheromone of GLM constitutes three components: Z-7,9-decadienylacetate (I) E-7 decenyl acetate (II) and Z-7-decenyl acetate (III) in 100:20:14 ratios as identified by Hall *et al.* (1994).

OAc OAc OAc

Z-7, 9-decadienylacetate (10) : E-7 decenyl acetate (2) : Z-7-decenyl acetate (1.4)

The use of synthetic sex pheromone blend as an effective monitoring tool for GLM populations was first reported in India by Whiteman *et al.* (1995) using synthetic sex pheromone from NRI, UK. Subsequently, basic studies conducted in the laboratory using electrophysiological and behavioural equipment like EAG, GC-EAD and wind tunnel confirmed the laboratory efficacy of the indigenously synthesized pheromone blend (Yadav *et al.*, 2007; Jyothi *et al.*, 2008).

A series of trials (1999-2006) were conducted by IICT in association with the collaborating Agriculture Institute/University to optimize trap design, pheromone dispenser, pheromone lure dosage, lure longevity etc., which resulted in the development of an efficient and easy to use trapping system for GLM. During these trials IICT collaborated with Directorate of Groundnut Research Junagadh; Acharya N.G. Ranga Agriculture University (ANGRAU) Hyderabad, voluntary organizations (KrishiVignan Kendra (KVK), Gaddipally Nalgonda; Society for Development of Drought Prone Area (SDDPA)–Wanaparthy, Mahboobnagar. Results obtained confirmed that 3 mg dose of the 3-component blend of GLM was found effective under field conditions.

19.6.2.1. Large scale village based demonstration of PAT in Ananthapur district of Andhra Pradesh, India by adopting Mass Trapping as a direct control Strategy

The climatic factors such as high temperatures and low humidity (33° C and RH <40) in Ananthapur district made GLM as an important and predominant pest occurring regularly. Highly encouraged by the potential benefits of pheromone mediated mass trapping trials conducted in Nalgonda and Mahaboobnagar districts of Andhra Pradesh, large scale field demonstrations covering 5000 ha of groundnut crop were proposed for a 5 year period with financial support from CSIR.

PMMT was found highly successful and effective in reducing the GLM infestation from 25.5% to 8.25% by the end of 3rd year. In farmers' practice control plots with insecticide treatment, reduction in pest population was insignificant where as in control plots without insecticides an increase pest damage by 2 % were recorded. Installation of pheromone traps in the groundnut fields not only resulted in the reduction in pest incidence/damage but also the no. of insecticide sprays. Adoption of PAT in place of pesticides generated substantial economic benefits to the farmers. Estimated benefits will be even more when we consider the environmental and health aspect due to reduction in pesticide to the environment and society as whole.

Results obtained during mass trapping experiments conducted successively for three years are presented in the table 19.3 below.

Table 19.3 : Details of Mass trapping of GLM in comparison to Insecticide treatment plots

Period	*Number of moths trapped plot	Per cent damage			Yield Kg/hectare		
		Mass trapping plot	Control with insecticide	Control without insecticide	Mass trapping insecticide	Control with insecticide	Control without
2007-08	7080 ± 1336	25.50	28.66	35.50	1930	1780	1460
2008-09	3940 ± 580	11.55	25.50	34.00	1950	1820	1500
2009-10	1285 ± 340	8.25	26.15	36.33	1985	1710	–

19.6.3. Sex Pheromone-based control for sugarcane borers

Sugarcane (*Saccharum officinarum* L.), an important cash crop in India is attacked by several pests in various growth stages. Among them sugarcane borers cause severe damage to the cane production in all the sugarcane growing regions. The borer pests cause shoot death at young age as well as cause damage at the joints of the mature canes. The estimated sugar loss due to borer pests reaches upto 30-60 kg/ha. An indigenous pheromone technology was developed for three of the borer pests of sugarcane (Table 19.4). Detailed studies were conducted on optimization of effective dispenser, lure dosage and lure longevity with indigenously synthesized pheromone blend of sugarcane inter node borer by Yadav *et al.,* 2007. Large scale village based field demonstrations were carried out by Rajashree sugars in their sugarcane fields at Theni, Coimbatore. PMMT strategy using 20 traps/ hectare were optimized and was proved effective to reduce populations of Sugarcane Inter node borer significantly. The trials provided good evidence that mass trapping could significantly reduce pest damage with improvement in cane yield as depicted in (Table 19.5).

Table 19.4: Pheromone systems with their respective blend ratio of sugarcane borers for which preparative methodologies are available

S. No.	Pest	Pheromone Blend Ratio
1.	Sugarcane internode borer, *C. sacchariphagus indicus*	(Z)-13-Octadecenyl Acetate (7) : (Z)-13-Octadecenol (1)
2.	Top shoot borer, *S. excerptalis*	(E)-11- Hexadecenal (7) : (Z)-11- Hexadecenal (3)
3.	Early shoot borer, *C. infuscatellus*	(Z) -11-Hexadecenol

Table 19.5 : Yield loss estimation in control plot (without pheromones) and in treatment plots (with use of pheromones)

S. No.	Parameter	Treatment plot (with pheromones)	Control plot (without pheromones)
1.	Percent incidence of dead hearts (1)	3.74	14.18
2.	Average single cane Weight of dead heart affected canes (Kgs)	0.64	0.38
3.	Average single cane Weight of unaffected healthy canes (Kgs)	1.22	1.34
4.	Loss in individual cane weight as a result of INB incidence and dead heart formation (Kgs)	0.58	0.96
5.	Yield loss per acre (MT's)	0.80	5.44

The results indicated a significant difference in percent incidence of dead hearts between pheromone treated (3.74%) and untreated control plots (14.18%). Though the loss in single cane weight due to inter node borer incidence and resultant dead heart formation in treated and control plots were not significant, the higher percent incidence level in control plot resulted in significant yield loss in control plot (5.44 MT's) over pheromone treated plot (0.80 MT's) (Table 19.5). An average yield loss of 4.64 MT's per acre can be mitigated by application of pheromone technology. Considering the significant yield losses in control plot and efficacy of pheromone in treated plots, it can be inferred that the use of pheromones is cost effective and an efficient control measure that can be adapted for control of internode borer. The average losses in recovery between pheromone treated and control plots are significant indicating that pheromones are effective in minimizing "in-factory" quality losses due to borer pests. An average yield loss of 4.64 MT's per acre can be mitigated by the application of pheromone technology. M/s. Rajashree Sugars, Coimbatore successfully implemented PMMT in 5000 hectares of sugarcane on community basis by utilizing the transferred expertise of IICT.

19.6.4. Sex Pheromone-based IPM developed for cotton pests

Cotton is one of the principal commercial crops. It is cultivated in an area of 9 million hectares. But our productivity is still below the average world productivity of countries like Australia, China, USA and Turkey. Cotton production in India is constrained by low yields per hectare due to frequent and occasionally disastrous attack of "Bollworm complex". These include *H. armigera,*, *P. gossypiella* and *Earias* sp. Farmer's community has extensively used pesticides in order to minimize the losses due to these pests. Due to this indiscriminate use of pesticides, pests have developed resistance to insecticides, increased cost of cotton cultivation and damage to environment.

In order to popularize the Sex Pheromone based Application Technology in IPM and / to clear the general apprehensions / limitations amongst the Indian farming commun ity with regard to its Implementation, IICT in collaboration with College of Agriculture, MPKV, Dhule has demonstrated the feasibility of PMMT. The field demonstrations were successfully carried out in 8 different villages in and around Dhule, Maharashtra using indigenously developed pheromone lures of all the three bollworm species (Table 19.6). Also, developed and identified the locally available dispensers as the best suitable for all the three species of bollworms in comparison with the imported dispensers. PMMT strategy was found effective in trapping all the three species of bollworm complex as evidenced by reduction in bollworm damage & increase in yield in comparison to control plots. The trapped adults of all the three bollworm species are depicted in Fig 1a, 1b and 1c.

Table 19.6 : Pheromone systems with their respective blend ratio of Cotton bollworms

S.N.	Pest	Pheromone components in their blend ratio
1.	American Bollworm, *H. armigera*	Z-11-Hexadecenal (97) : Z-9-Hexadecenal (3
2.	Spiny Bollworm, *E. vittella*	(E,E)-10, 12 – Hexadecadienal (10) (Z)-11-Octadecenal (2) (Z)-11-Hexadecenal (2)
3.	Pink bollworm, *Pectinopora gossypiella*	(Z,Z)-7,11-Hexadecadienylacete (1) (Z,E)-7,11-Hexadecadienylacetate (1)

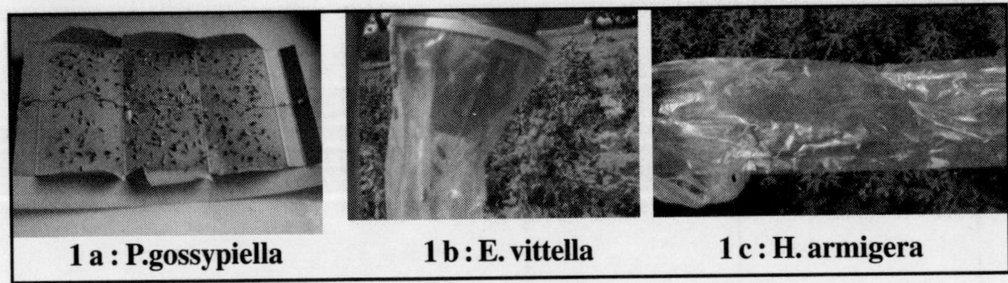

1 a : P.gossypiella 1 b : E. vittella 1 c : H. armigera

Fig. 19.1 : Trapped adults of bollworm species

The type of pheromone trap, dispenser with the developed synthetic methodologies for pheromone components in required blend ratios for all the three species of bollworm complex along with the methodology of implementation is available at IICT as a package.

19.6.5. Pheromone-based IPM developed for insect pests of vegetables and fruits

Synthetic methodologies developed at IICT for the pheromone blend components of the vegetable and orchard pests are depicted in (Table 19.7).

Table 19.7 : Insect pests of vegetable and fruit crops and their pheromone systems

Sl. No	Name of the Insect Pest	Host	Pheromone Components & Blend ratio
1.	American bollworm, *H. armigera*	Chillies, cotton, pigeon pea *etc.*	CHO Z-11-Hexadecenal (97) : CHO Z-9-Hexadecenal (3)
2.	Brinjal shoot and fruit borer, *L. orbonalis*	Brinjal	OAc (E)- 11-Hexadecenyl acetate (100) : OH (E)-11-Hexadecenol (1)
3.	Spiny Bollworm, *E. vittella*	Cotton and Okra	CHO (E,E)-10, 12 – Hexadecadienal (10) . CHO (Z)-11-Octadecenal (2) : CHO (Z)-11-Hexadecenal (2)
4.	Diamondback moth *P. xylostella*	Cabbage and cole crops	CHO (Z)- 11-Hexadecenal(10) : OH (Z)-11-Hexadecenol (0.1) : (Z)- 11-Hexadecenylacetate (10)
5.	Sweet potato weevil, *C.formicarius*	Sweet potato	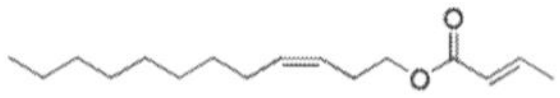 O O (Z)-dodec-3-en-1-yl (E)-2-butenoate
6.	Citrus Leaf Miner, *P. citrella*	Citrus	O (7Z,11Z,13*E*)-hexadecatrienal O (7Z,11Z)-hexadecadienal

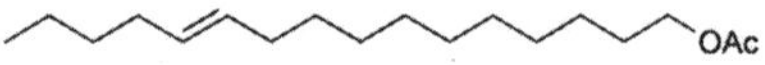

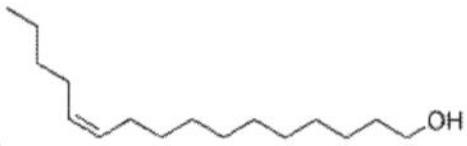

19.6.5.1. Brinjal shoot and fruit borer (BSFB), *L. orbonalis*

Brinjal is one of the most popular vegetables in India which is cultivated year-round all over the country. The area under cultivation is estimated to be around 5 lakh ha with a total production of 82 lakh metric tonnes. It is mainly grown in small plots as a cash crop by marginal and poor farmers and the average yields are reported to be around 200-350 quintals per ha.

Brinjal shoot and fruit borer (BSFB), *Leucinodes orbonalis* Guenee, is a serious pest of brinjal or eggplant (*Solanum melongena*L.) causing severe yield losses up to 70%. Zhu *et al.* (1987) identified E-11-hexadecenyl acetate as the pheromone and subsequently Attygalle *et al.* (1988) identified E-11- hexadecenol as the minor component. A pheromone based mass trapping technique was developed by Cork *et al.* (2001, 2003 and 2005) wherein blend, optimisation, dose, dispenser and trap placement in the field were standardized. IICT developed feasible methodologies for the synthesis of two component blend of BSFB- E-11-hexadecenyl acetate (100) and E-11- hexadecenol (1). PMMT strategy was successfully implemented in 2000 ha of brinjal crop over a period of five successive years (2007-2012). Agricultural Research Station (ARS), Ananthapur of Acharya N.G. Ranga Agricultural University (ANGRAU), Andra Pradesh state and Tamil Nadu Agricultural University (TNAU), Coimbatore, Tamil Nadu states of India collaborated for field evaluations through farmers' participation.

Mass trapping technique was demonstrated to the farmers through on-farm multi location trials, village wide and front line demonstrations in selected villages of Ananthapur (Andhra Pradesh) and Coimbatore (Tamil Nadu). The percent incidence of damaged shoots/fruits, yield and quality losses were significantly higher in insecticide plots without pheromones than pheromone treated plots. These village based demonstrations covering approximately 100 villages both in Tamil Nadu and Andhra Pradesh demonstrated that mass trapping can be successfully employed for the control of BSFB for the benefit of the small farm holders. PMMT resulted in decline of BSFB infestation from 15% to 1% in pheromone treated plots while it increased to 25% in control plots with pesticide treatment. The installation of pheromone traps in Brinjal crop resulted not only in the reduction of pest incidence but also in the reduction of insecticide sprays by 60 %. Adoption of PAT in place of pesticides generated substantial economic benefits to the farmers. Brinjal farmers were benefited by saving to a tune of Rs. 24,000/ acre through implementation of PAT (Table 19.8).

Table 19.8 : Economics of PAT in pheromone treated and control plots

Particulars	Cost of cultivation (Rs)	Cost of Plant Protection	Gross returns	Net to PAT farmer returns (Rs)	Additional benefit
PAT farmer	76,000	19,500	2,40,000	1,64,000	24,000
Non-PAT farmer	22,500	22,500	2,24,000	1,40,000	-

19.6.5.2. Diamondback moth (DBM), *P. xylostella*

IICT scientists conducted Mass trapping trials in 50 acres of cabbage crop in the farmer's fields of Rayannagudem village of Nalgonda District of AP. Delta sticky traps per acre baited with 3 mg of the 3 - component pheromone blend along with equal quantity of antioxidant (Butylated Hydroxy Toluene) in rubber septa dispensers were installed in the fields and monitored the trapped population of DBM adults on every alternate day. Percent damage in pheromone treated crop was brought down from 30% to 5%. The results indicated that mass trapping can be successfully employed to protect cabbage from DBM.

19.6.5.3. Sweet potato weevil, *C. formicarius*

An IPM schedule using synthetic sex pheromone (Z 3 dodecenol E 2 butenoate) as the principal component against sweet potato weevil has been evaluated for the first time in India by Central Tuber Crops Research Institute (CTCRI) with indigenously developed pheromone lures. Installation of pheromone traps from the first day of planting at 10 m distance (one trap/100 m^2) to destroy the male weevils; The continuous elimination of the male population by pheromone traps in the treatment has resulted in a drastic decline in the population build up of the weevil and consequent reduction in the damage, leading to increased production of marketable tubers over 53% (Pillai *et al.*, 1993).

19.6.5.4. Citrus leaf miner (CLM), *P. citrella*

Citrus is one of the important fruit crops in India having a production of 75.7 lakh tons from 8.43 lakh hectares with the productivity of 9.0 tons/ha at National level compared to 30 tons/ha in advanced citrus producing countries (NHB, 2008). The citrus leaf miner, *P. citrella* is a major pest of *Citrus* spp. throughout the world. Attraction of male *P. citrella* to (Z,Z)-7,11-hexadecadienal was accidentally discovered by Ando and collaborators while conducting field screenings of potential attractants for species in the Gelechiidae family, and caught males of the Japanese populations of the citrus leaf miner (Ando *et al.*, 1985). In the recent past the sex pheromone of *P. citrella* was reported by a host of researchers (Leal *et al.*, 2006; Moreira *et al.*, 2006). A 3:1 blend of (Z,Z,E) -7,11,13-hexadecatrienal and (Z,Z)-7,11-hexadecadienal is highly attractive to males. IICT has successfully synthesized the two component pheromone blend of CLM indigenously. Silicon septa impregnated with CLM sex pheromone blend gave excellent trap catches of 320/trap/night during field evaluation studies conducted by NRCC, Nagpur (Fig: 19.2a and b).

Fig. 19.2a : CLM Adult Trapping

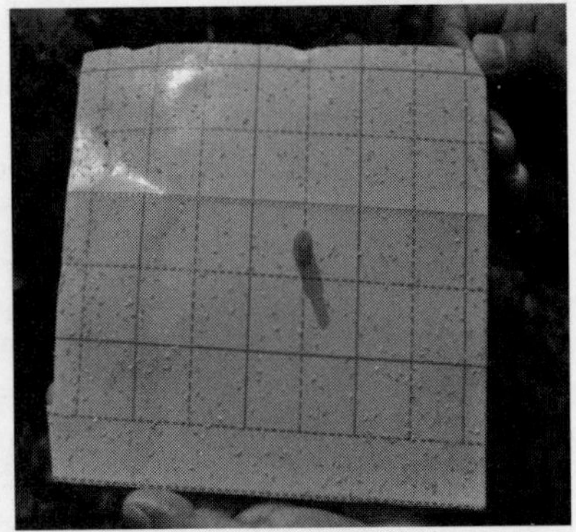

Fig. 19.2b : Trapped CLM Male Moths (One Night Record 320/trap)

19.7. Conclusions

The idea of controlling insect populations through species-speciûc manipulation of sexual communication, without adversely affecting non target organisms, has been a driving force for pheromone research worldwide. Our work on insect sex pheromones /semiochemicals and their applications for control of crop pest insects has shown that synthetic sex pheromone components in their natural blend ratio proved as significant alternatives to the conventional chemical insecticides.

In India, after initial constraints, awareness has been created in the farming community about the utility of pheromones in IPM. Mass trapping as a stand-alone control method is demonstrated for the first time with indigenously synthesized pheromone lures. The large number of trapped moths in the present case studies of GLM, YSB, BSFB and SIB successfully demonstrated that PAT as a reliable and efficient control tool in pest management. Based on the feedback and earnest requests from farmers for the continuation of the programme with pride and satisfaction it can be concluded that the farming community in India is slowly gaining the confidence and realizing the benefits of PAT. Furthermore, it is clear that multidisciplinary research work at one place in the area of pheromones like in Centre for Semiochemicals certainly provided the necessary impetus for implementation and popularization of pheromone application as promising tool for control of economically important crop pests in Indian agriculture.

As environmental awareness is increasing and countries are becoming increasingly interdependent on trade, harmonization of regulations and restrictions will help to promote greater use of pheromones in both developed and developing countries. Our efforts on popularization of PAT indicates that PAT will undoubtedly play a key role in developing global competitiveness of the country for environmentally and ecologically friendly pesticide free agro-produce.

References

Agelopoulos N, Birkett MA, Hick AJ, Hooper AM, Pickett JA (1999) Exploiting semiochemicals in insect control. *Pestic. Sci.,* 55: 225-235.

Aldrich JR, Bartelt RJ, Dickens JC, Knight AL, Light DM, Tumilson JH (2003) Insect chemical ecology research in the United States Department of Agriculture–Agricultural Research Service. *Pest Mgmt. Sci.,* 59: 777-787.

Ali MF, Morgan ED (1990). Chemical communication in insect communities: a guide to insect pheromones with special emphasis on social insects. *Biol. Rev. Cambridge Phil. Soc.,* 65: 227-247.

Ando T, Taguchi, K Y, Uchiyama M, Ujiye T, Kuroko H (1985) (7Z,11Z)-,11Hexadecadienal: sex attractant of the citrus leafminer moth, *Phyllocnistis citrella* Stainton (Lepidoptera: Phyllocnistidae). *Agric. Biol. Chem.,* 49:3633–3635.

Arn H, Toth M, Priesner, E. (2000) List of sex pheromones of female Lepidoptera & related male attractants. Available: http://phero.net/pherolist.

Arnon Shani (1993) Role of pheromones in integrated pest management (Survey conducted in July – November 1990). in Insect Pheromones , *IOBC WPRS Bull.,* 16 (10) : 359-369.

Attygalle AB, Schwarz J, Gunawardena NE (1988) Sex pheromone of brinjal shoot and fruit borer *Leucinodes orbonalis* Guen'ee (Lepidoptera: Pyralidae: Pyraustinae). *Z. Naturforschung,* 43:790–792.

Baker TC, Heath JJ (2004) Pheromones-function and use in insect control. In LI Gilbert, K Iatro S S Gill (eds.). *Mol. Insect Sci.,* 6: 407-460.

Butenandt A, Beckmann R, Stamm D, Hecker E (1959) Über den Sexual-Lockstoff des Seidensspinners Bombyx mori. Reindarstellung und Konstitution. *Z Naturforsch.* 14b: 283 – 284.

Carde RT, Minks AK (1995) Control of moth pests by mating disruption: successes and constraints. *Annu. Rev. Entomol.,* 40:559-585.

Cork A, De Souza K, Hall DR, Jones OT, Casagrande E, Krishnaiah K, Syed Z (2008) Development of PVC resin-controlled release formulation for pheromones and use in mating disruption of yellow rice stem borer, *Scirpophaga incertulas. Crop Prot.,* 27:248-255.

Cork A, Alam SN, Rouf FMA, Talekar NS (2005) Development of mass trapping technique for control of brinjal shoot and fruit borer, *Leucinodes orbonalis* (Lepidoptera:Pyralidae) *Bull. Entomol. Res.,* 95: 589-596.

Cork A, Alam SN, Rouf FMA, Talekar NS (2003) Female sex pheromone of the brinjal fruit borer, *Leucinodes orbonalis* (Lepidoptera: Pyralidae): Trap optimization and application in IPM trials. *Bull. Entomol. Res.,* 93: 107–113.

Cork A, Alam SN, Das A, Das CS, Ghosh GC, Farman DI, Hall DR, Maslen NR, Vedham K, Phythian SJ, Rouf FM, Srinivasan K. (2001) Female sex pheromone of brinjal fruit and shoot borer, *Leucinodes orbonalis* blend optimization. *J. Chem. Ecol.,* 20:1867-77.

Cork A, Hall DR (1998) Application of Pheromones for Crop Pest Management in the Indian Sub-Continent. *J. Asia-Pac. Entomol.,* 1 (1) : 35-49.

Cork A, Souza K D, Krishnaiah K, Kumar D V S S R, Reddy A A, Casagrande E and De-Souza K (1996) Control of yellow stem borer, *Scirpophaga incertulas* (Lepidoptera: Pyralidae) by mating disruption on rice in India: Effect of unnatural pheromone blends and application time on efficacy. *Bull. Entomol. Res.,* 86 (5): 515-524.

Cossé AA, Bartelt RJ, Zilkowski BW, Bean DW, Petroski RJ (2004) Identification of the aggregation pheromone of *Diorhabda elongata*, a biological control agent of saltcedar (Tamarix spp.). *J. Chem. Ecol.,* 31: 657–670.

David H, Easwaramoorthy S, Nesbitt B F, Nandagopal V (1981) Sex attraction studies in sugarcane internode borer, *Chilo sacchariphagus indicus* (K.) with synthetic pheromone. *Pestology,* 5: 9-11.

David H., Nesbitt B F., Easwaramoorthy, S and Nandagopal, V (1985) Application of sex pheromones in sugarcane pest management. *Proc. Indian Acad. Sci. (Animal),* 94: 333-339.

Durairaj C, Rajesh Kumar J, Prasad AR (2015) Studies on the impact of climate change on pheromone trap catches of Gram pod borer, *Helicoverpa armigera* and Brinjal shoot and fruit borer, *Leucinodes orbonalis* under vegetable ecosystem. In: *Proceedings of International Conference on İnnovative Insect Management Approaches for Sustainable Agro Ecosystems (IIMASAE).* Jan. 27-30, Tamil Nadu Agricultural University (TNAU), Madurai, India.

Easwaramurthy S, Mukunthan N, Singaravelu B (2003) Pheromones in the management of sugarcane (Lepidoptera: Gelechidae) internode nborer *Chilo sacchariphagus indicus* (Kapur) *In: Proceedings of South Indian Sugarcane and Sugar Technology*. Association Convention. Hyderabad, p. 49-51.

El-Sayed, A M (2013) The Pherobase: database of pheromones and semiochemicals. [Online]. Available: www.pherobase.com [Accessed: 6 June, 2013]

EL-Sayed, A.M, Suckling DM, Wearing CH, Byers J (2006) Potentail of mass trapping for long term pest management and eradication of invasive species. *J. Econ. Entomol.,* 99: 1550-1564.

Faliero JR (2006) A review of the issues and management of the red palm weevil *Rhynchophorus ferrugineus* (Coleoptera: Rhynchophoridae) in coconut and date palm during the last one hundred years. *Int. J. Trop. Insect Sci.,* 26: 135–154.

Ganesh P, Bhat P, Sudha K, Kanaujia KR, Jyothi KN, Prasuna AL (2011) Studies on electrophysiological, olfactometric response and chemical analysis of groundnut extracts against groundnut bruchid, *Caryedon serratus* (Oliver). *J. Agric. Tech.,* 7: 1265-1273.

Ganeswar Rao A, Krishnaiah K, Cork A, Hall DR (1994) Disruption of sex pheromone mediate communication in yellow stem borer, *Scirpophaga incertulas* Walker with a controlled release pheromone formulation. *J. Entomol. Res.,* 18: 287-288.

Ganeswar Rao A, Krishnaiah K. (1995) Factors influencing yellow stem borer, *Scirpophaga incertulas* (Walker) moth captures in pheromone traps. Pheromone in IPM. In: National Workshop on pheromone in IPM, Hyderabad, 1995. Directorate of Rice Research, Rajendranagar, Hyderabad. pp. 40-42.

Hall DR, Beevor PS, Campion DG, Chamberlain DJ, Cork A, White R, Almetre A, Henneberry TJ, Nandagopal V, Wightman J, Ranga Rao GV (1994) Identification and synthesis of new pheromones. *IOBC WPRS Bull.,* 16: 1-10 .

Hall DR, Cork A, Phythian SJ, Chittamuru S, Jayarama BK, Venkatesha MG, Sreedharan K, Kumar PKV, Seetharama HG, Naidu R (2006) Identiûcation of components of male-produced pheromone of coffee white stemborer, *Xylotrechus quadripes*. *J. Chem. Ecol.,* 32: 195–219.

Hall DR, Cork A, Phytian S, Sumathi C,Venkatesha M G, D'Souza M V, Naidu R. (1998) Studies on the male sex pheromone of the coffee white stemborer, *Xylotrechus quadripes* Chevrolat (Coleoptera: Cerambycidae), pp. 50–52. *In Proceedings of the 2nd International Symposium on Insect Pheromones, 30 March–3 April 1998. WICC-International Agriculture Centre, Wageningen.*

Horowitz AR, Ellsworth PC, Ishaaya I (2009) Biorational Pest Control – An overview pp 1-20, *In* : Ishaaya I, Horowitz AR (eds) *Biorational Control of Arthropod Pests: Application and Resistance Management.* Springer Dordrecht Heidelberg, London, New York.

Howse PE, Stevens IDR, Jones OT (1998). *Insect Pheromones and Their Use in Pest Management.* Chapman and Hall, London. 369 pp.

Jyothi KN, Prasuna AL, Sighamony S, Prasad AR, Yadav JS (2002) Electroantennogram responses of *Apanteles obliquae* (Hym., Braconidae) to various infochemicals. *J. Appl. Entomol.,* 126 : 175-181.

Jyothi KN, Prasuna AL, Prasad AR, Kabre GB (2014) Development of indigenous pheromone technology for monitoring of spotted bollworm Earias vittella Fab (Lepidoptera: Noctuidae) on cotton *J. Agric. Tech.,* 40: 503-513.

Jyothi KN, Prasuna AL, Prasad AR (2014) Electrophysiological and behavioural responses of groundnut seed beetle *Caryedon serratus* (Olivier) (Coleoptera: Chrysomelidae) to conspecfic and groundnut seed chemical cues. *Afric. Entomol.,* 22: 505-511.

Jyothi KN, Prasuna AL, Prasad AR, Yadav JS (2008) Electrophysiological responses of both sexes of groundnut leaf miner, *Aproaerema modicella* (Lepidoptera :Gelechiidae) to synthetic female sex pheromone blend. *Curr. Sci.,* 94(5): 629-633.

Kaib M (1999) Termites. Pp. 329–353. *In* : Hardie J, Minks AK (eds.). *Pheromones of Non-Lepidopteran Insects Associated with Agricultural Plants.* CABI Publishing, Wallingford, UK.

Karlson P, Luscher M., (1959) "Pheromones": a new term for a class of biologically active substances. *Nature,* 183: 53-56.

Keeling CI, Plettner E, Slessor KN (2004) *Hymenopteran Semiochemicals.* pp.134–177.

Khan ZR, Midega CAO, Amudavi DM, Hassanali A, Pickett J A (2008) On-farm evaluation of the 'push-pull' technology for the control of stemborers and striga weed on maize in western Kenya. *Field Crops Res.,* 106: 224-233.

Kogan M (1998) Integrated Pest Management: Historical Perspectives and Contemporary Developments *Annu. Rev. Entomol.,* 43: 243-270.

Krishnaiah K (1986) Studies on the use of pheromones in the control of *Spodoptera litura* Fab. on black gram grown in rice fallows. *Indian J. Plant Prot.,* 14: 43-46.

Krishnaiah, K.; Zainalabeuddin, S.; Ganeswara Rao, A.; Kumar, D.V.S.S. and Rama Gopalavarma, N. (1998) Pheromone monitoring systems of rice yellow stem borer, *Scirpophaga incertulas* Walker. *Indian J. Plant Prot.,* 26: 99-106.

Krishnakumari B, Prasuna A L, Jyothi KN, Valli MY, Sighamony S, Prasad AR, Yadav JS (1998) Behavioural and electrophysiological responses of Achaea janata Linn. (Lepidoptera: Noctuidae) males to synthetic female-produced sex pheromone components. *J. Entomol. Res.,* 22: 197-202.

Kumar PA, Srinivas TN, Prasad A R (2011) Identification of Fruity Aroma-Producing Compounds from *Chryseobacterium* sp Isolated from the Western Ghats, India. *Curr. Microbiol.,* 63: 193.

Lalita Kumari VL, Reddy DRR (1991) Sex pheromone in the groundnut leaf miner, *Aproaerema modicella. Int. Arachis Newslett.,* May,15-16.

Leal WS, Parra-Pedrazzoli AL, Cossé AA, Murata Y, Bento JMS, Vilela EF (2006) Identification, synthesis and field evaluation of the complete sex pheromone from the citrus leafminer, *Phyllocnistis citrella. J. Chem. Ecol.,* 32: 173-186.

Macaulay EDM, Etheridge P, Garthwaite DG, Greenway AR, Wall C, Goodchild RE (1985) Prediction of optimum spraying dates against pea moth, *Cydia nigricana* (F.), using pheromone traps and temperature measurements. *Crop Prot.,* 4: 85-98.

Mamata Latwal, Ganesh Panday, Om Prakash, Sudha Kanaujia, Kanaujia R, Prasuna AL, Jyothi KN (2010) Ovipositional deterrence by egg extracts of *Caryedon serratus* (Oliver). *J. Biopestic.,* 3 (Special issue): 400-404.

Mazomenos B, Pantazi-Mazomenou A, Stefanou D (2002) Attract and kill of the olive fruit fly Bactrocera oleae in Greece as a part of an integrated control system. *In:* Witzgall P, Mazomenos B, Konstantopoulou M (eds.). Proceedings of the meeting""Pheromones and Other Biological Techniques for Insect Control in Orchards and Vineyards", Samos (Greece), 25–29 September 2000. *Bulletin International Organization for Biological Control, Working Group "Pheromones and Other Semiochemicals in Integrated Control"* 25: 137–146.

Millar JG (1995) An overview of attractants for Mediterranean fruit fly. Proceedings of a Workshop on the Med fly Situation in California: Defining Critical Research, Riverside, CA, 11–13 November 1994, pp. 123–144. College of Natural and Agricultural Sciences, Riverside CA.

Mitchell, ER (1986) Pheromones: as the glamour and glitter fade– the real work begins. *Fl. Entomol.,* 69: 132–139.

Moreira JA, McElfresh JS, Millar JG (2006) Identification, synthesis and field testing of the sex pheromone of the citrus leafminer, *Phyllocnistis citrella*. *J. Chem. Ecol.,* 32: 169-194.

N.H.B.(2008) Indian Horticulture Database. *Agricultural Research Data Book,* ICAR New Delhi.

Nandagopal V, Reddy PS (1990) Report of sex pheromone in groundnut leaf miner, *Aproaerema modicella* Deventer. *Int. Arachis Newslett.,* 7: 5-6.

Narasimhan S. Application of pheromone technology in the control of Chilo infuscatellus (snell)(Lepidoptera:Pyralidae)in sugarcane. Eds Narasimhan NS, Narasimhan S, Suresh G, Madhavan S. *In: Proceedings of the Indo-British Workshop on Current Approaches to Pheromone Technology.* SPIC Science Foundation, India, pp.61-64.

Oehlschlager AC, Chinchilla C, Castillo G, Gonzalez L. (2002) Control of red ring disease by mass trapping of *Rhynchophorus palmarum* (Coleoptera: Curculionidae). *Fl. Entomol.,* 85: 507–513.

Padmaja PG, Madhusudhana R, Patil JV, Prasuna A L, Jyothi KN, Prasad AR. (2012) Role of Plant Volatiles in Resistance to Shootfly *Atherigona soccata* Rondani in Sorghum. In: *International Symposium on Plant Biotechnology held National Agriculture Science Centre*, New Delhi 21-24 Feb.

Patel RC, Yadava DN, Joshi AD, Patel AK (1985) Impact of mass trapping on *Heliothis armigera* and *Spodoptera litura* on cotton with pheromones. Eds. Raghupathi A and Jayaraj S. *In: Proc. National Seminar on Behavioural and Physiological Approaches for the Management of Crop Pests.* Tamil Nadu Agricultural Universily, Coimbatore, India 120-122.

Pawar CS, Sithanantham S, Bhatnagar VS, Srivastava CP, Reed W (1988) The development of sex pheromone trapping of *H.armigera* at ICRISAT, India. *Trop. Pest Mgmt.,* 34: 39-43.

Persoons CJ, Vos JD, Yadav JS, Prasad AR, Sighmony S, Jyothi KN, Prasuna AL(1993) Indo-Dutch Cooperation on Pheromones of Indian Agricultural Pest Insects: Sex Pheromone Components of *Diacrisia obliqua, (Arctiidae), Achaea janata* (Noctuidae) and *Amsacta albistriga* (Arctiidae). *IOBC/WPRS Bull.,* 16: 136-140.

Pettis J, Pankiw T, Plettner E (1999) Bees. Pp. 429–450. *In:* Hardie J, Minks AK (eds.). *Pheromones of Non-Lepidopteran Insects Associated with Agricultural Plants*. CABI Publishing, Wallingford, UK. 466.

Pickett JA, Wadhams LJ, Woodcock CM (1989) Chemical ecology and pest management: Some recent insights. *Insect Sci. Applic.,* 10: 741-750.

Pillai KS, Rajamma P, Palaniswami MS (1993) New technique in the control of sweet potato weevil using a synthetic sex pheromone in India. *Int. Pest Mgmt.,* 39: 84-89.

Prasad AR, Jyothi KN, Prasuna AL, Yadav JS (2013) Semiochemicals: Tools for Pest Management. *Proc. Andhra Pradesh Acad. Sci.,* 15(2): 15-18.

Prasad AR, Prasuna AL, Jyothi KN, Yadav JS (2012) Mass Trapping As Control Method for Groundnut Leaf Miner *Aproaerema modicella* (Gelechiidae: Lepidoptera) with Synthetic Sex Pheromone in India. *Semiochemicals: The Essence of Green Pest Control IOBC/Wprs Pheromones and Other Semiochemicals Working Group* Bursa, Turkey, October 1-5.

Prasad AR (2010) Pheromone Application Technology (PAT) in Pest Management : Efforts of Indian Institute of Chemical Technology (IICT), In: *Proceedings of National symposium on Emerging Pest Management Strategies Under Changing Climate Scenario*, Department of Entomology, College of Agriculture, Orissa University of Agriculture and Technology, Bhubaneswar.

Prasad AR, Subhaharan K (2010) *National Symposium on Plantation Crops*, Rubber Research Institute, Robber Board, Kottayam, Kerala, 7-10 December.

Prasuna AL, Jyothi KN, Prasad AR, Yadav JS (2008) Olfactory Responses of Banana Pseudostem Weevil, *Odoiporus longicollis* (Coleoptera: Curculionidae) to Semiochemicals from Conspecifics and Host Plant. *Curr. Sci.,* 94(7): 896 to 900.

Radhika P, Prasad AR, Reddy BS Padmalatha Y, Prasuna AL, Jyothi KN (2014) Pheromone Mass Trapping: An efficient technique to manage *Aproaerema modicella* Deventer in Groundnut under Low rainfall Zone of Andhra Pradesh. *Annu. Plant Prot. Sci.,* 22: 18-21.

Rajasekhara Rao K, Swagatika M, Naskar SK, Bhaktavatsalam N, Prasad AR, Khushboo S, Jayaprakas CA Mukherjee A (2013) Plant volatile organic compounds as chemical markers to identify reistance in sweet potato against weevil *Cylas formicarius*. *Curr. Sci.,* 105: 1247-1253.

Rajasekhara Rao K, Naskar SK, Prasad AR, Prasuna AL, Jyothi KN (2010) Differential volatile emission from sweet potato plant: mechanism of resistance in sweet potato for weevil *Cylas formicarius* (Fab.). *Curr. Sci.,* 99: 1597-1601.

Ranga Rao GV, Wightman JA, Ranga Rao DV (1991a) Monitoring *Spodoptera litura* (F) (Lepidoptera: Noctuidae) using sex attractant traps: Effect of trap height and time of the night on moth catch. *Insect Sci. Appl.,* 12: 443-447.

Ranga Rao GV, Wightman JA, Ranga Rao DV (1991b) The development of a standard pheromone trapping procedure for *Spodoptera litura* (F) (Lepidoptera:Noctuidae) population in groundnut (*Arachis hypogea*) crops. *Trop. Pest Mgmt.,* 37: 37-40.

Reddy, GVP and Guerrero A (2000) Pheromone based integrated pest management to control the diamondback moth *Plutella xylostella* in cabbage fields. *Pest Mgmt Sci.,* 56: 882-888.

Reddy, GVP, Urs KCD (1996) Studies on the sex pheromone of the diamond back moth, *Plutella Xylostella* (Linnaeus) (Lepidoptera:Yponomeutidae) in India. *Bull Entomol. Res.,* 86: 585-590.

Reddy, GVP, Urs KCD (1997) Mass trapping of diamond back moth *Plutella xylostella* in cabbage fields using synthetic pheromones. *Int. Pest Cont.,* 39: 125-126.

Ridgway RL, Silverstein RM, Inscoe MN (1992) (eds). Behavior-modifying chemicals, for insect management. In: *Applications of Pheromones and Attractants*. Marcel Dekkar Inc. New York. 761p.

Roelofs WL, Carde' RT (1977) Responses of Lepidoptera to synthetic sex Pheromone chemicals and their analogues *Annu. Rev. Entomol.,* 22: 377-399.

Sharma HC, Singh OP (1982) Sex pheromones for the control of pink bollworm *Pectinophora gossypiella*, their feasibility under Indian conditions. *Cott. Develop.,* 11(4): 35-38.

Sreedharan K, Karan-Singh (1987) Studies of the pheromone system of the coffee white stem-borer, *Xylotrechus quadripes* Chev. *J. Coffee Res.,* 17: 129.

Suckling DM, Gibb AR, Gourlay H, Conant P, Hirayama C, Leen R, Szöcs G (2000) Sex attractant for the gorse biocontrol agent *Agonopterix ullicetella* (Oecophoridae). *N. Z. Plant Prot.,* 53: 66–70.

Suckling DM, Gibb AR, Johnson TT, Hall DR (2006) Examination of sex attractants for monitoring weed biological control agents in Hawaii. *Biocont. Sci Technol.,* 16: 919–927.

Suckling DM, Hill RL, Gourlay AH, Witzgall P (1999) Sex attractant-based monitoring of a biological control agent of gorse. *Biocont. Sci Technol.,* 9: 99–104.

Taneja SL, Jayaswal AP (1981) Captire threshold of pink bollworm moths on Hirsutum cotton. *Trop. Pest Mgmt.,* 27: 318-324.

Taneja SL, Jayaswal AP (1983) Factors affecting pink bollworm moth catches in gossyplure baited traps. *Indian J. Plant Prot.,* 11: 78-83.

Tuhan NC, Pawar AD (1980) Monitoring of Codling moth *Cydia pomonella* in Kashmir valley by sex pheromone traps and fruits examination. *Indian J. Plant Prot.,* 7: 185-188.

Varshney VK, , Rameshwar Dayal R, Bhandari S, Jyothi KN, Prasuna AL, Prasad AR,Yadav JS (2005) Behavioral response of the borer beetle Hoplocerambyx spinicornis to volatile compouds of the tree *Shorea robusta. Chem. Biodiver.,* 2: 785-791.

Venkatesha MG, Dinesh AS (2012) The coffee white stemborer *Xylotrechus quadripes* (Coleoptera: Cerambycidae): bioecology, status and management. *Int. J. Trop. Insect Sci.,* 32(4): 177–188.

Wall C G, Arthwaite D G B, Lood S, Myth J A,Herwood SA (1987) The efficacy of sex-attractant monitoring for the peamoth, *Cydia nigricana* in England, 1980-1985. *Ann. Appl. Biol.,* 110: 223-229.

Whittaker RH, Feeny PP (1971) Allelochemics: Chemical interactions between species. *Science,* 171: 757-770.

Wightman JA, Ranga Rao GV, Lakshmi Narayana M (1995) Insect pheromones in India: Review and requirements for oilseed crops. (Eds) Krishnaiah K, Divakar *In: Proceedings of the National Workshop on Use of Insect Pheromones in Integrated Pest Management.* BJ. DRR, Hyderabad, India, pp.53-60.

Witzgall. P, Kirsch , Cork. A. (2010) Sex Pheromones and Their Impact on Pest Management. *J. Chem. Ecol.,* 36: 80-100

Yadav JS, Reddy ER (1988) Synthesis of insect sex pheromones. *Curr. Sci.,* 57: 1321- 1330.

Yadav JS, Prasad AR, Prasuna AL, Jyothi KN (2007) Potential of sex pheromone for the management of groundnut leaf miner *Aproaerema modicella*, Deventer (Gelechiidae: Lepidoptera). *J. Appl. Zool. Res.,* 18(1): 9-14.

Zhu P, Kong F, Yu S, Yu Y, Jin S, Hu X, Xu J (1987) Identiûcation of the sex pheromone of egg plant borer, *Leucinodes orbonalis* Guen'ee (Lepidoptera: Pyralidae). *Z. Naturforschung.,* 42:1347–1348.

Integrated Pest Management in the Tropics; pp. 627-685

New India Publishing Agency, New Delhi (India)

CHAPTER - 20

Chemical Ecology in Pest Control

N. Bakthavatsalam and Kesavan Subaharan
ICAR-National Bureau of Agricultural Insect Resources
P. B. No 2491, H. A. Farm Post, Bengaluru
Karnataka, 560024, India
email: nbakthavatsalam@yahoo.com

Chemical ecology is the study of ecological interactions between organisms mediated by chemical produced by those organisms. In short it is considered as the ecological functions of the chemicals. An insect, in various stages of its life history, performs various activities like host plant location, host plant selection, feeding and oviposition, mating, escape from its natural enemies etc. and the chemicals play a crucial role than visual and audio signals. Chemical ecology encompasses the studies of these chemicals based interactions.

Several books/reviews were written by earlier workers. (Lenoir *et al.*, 2001; Ayasse and Jarau, 2014; Carde and Belle, 1995; Carde and Miller, 2004; Eisner and Meinwald, 1995; Miller and Haynes, 1998; Müller-Schwarze, 2001; Wajnberg and Colazza, 2013; Dicke and Takken, 2006; Miller and Haynes 1998; Schult, 2005; Romeo, 2005; Bell and Carde, 1985, Nishida, 2014).

Chemical ecological studies have been given foremost importance in insects due to compelling demand for using them in the management of insect pests. Of all the total number of articles on chemical ecology published more than 80% are on insect chemical ecology. Chemical ecology of almost all the orders of insects had been studied and documented. The maximum families studied were Coleoptera, followed by Lepidoptera and Diptera. Least number of families were studied in Mecoptera, Chemicals were broadly classified as pheromones (intra specific) and allelochemicals (interspecific) depending on their role they mediated. Allelochemicals are again divided into allomones (favourable to the emitter), kairomone (favourable to the receiver) and synomones (favourable to both the emitter and receiver).

The chemicals involved in the interactions are further classified as alarm pheromone, aggregation pheromone, contact sex pheromone, brood pheromone, trail pheromone and sex pheromone, chiefly depending on the function they elicit. In this chapter we will study mainly various types of semiochemicals involved in alarm, aggregation and trailling behaviours, the various chemicals involved in those behavioural interactions with a few examples from certain important orders along with the application of these semiochemicals in insect pest management.

Alarm pheromone

Alarm pheromones are the chemicals produced by individuals in distress to alert the other individuals of the same species to elicit some behavioral response. Alarm pheromones were studied for several species of insects such as grass-cutting ant, *Atta bisphaerica* (Forel). The mandibular secretions of this species ants contain a chemical, 4-methyl-3-heptanone, this chemical warns other individuals to either move away or change their trailing tracts (Hughes *et al.*, 2002). There are considerable variations in the mandibular secretions of species belonging to the same genus, for eg 2-phenylpropenal and 3-decanone were found in all three species. The *Crematogaster nigriceps* (Emery) secret 3-methyl-2-pentanone, 3-methyl-2-pentanol, 3-methyl-2-hexanol, 3-methyl-2-heptanone, 3-methyl-2-heptanol, 2-phenylethanol and 2-methylheptanoic acid while *C. sjostedti* (Mayr) secretes butanoic acid, 3-heptanol, 2-methylpentanoic acid, glycerol, 6-methyl-3-octanol, 2-pyrrolidinone and 3-nonanol and nonanal in *C. mimosae* Santschi (Wood *et al.*, 2002). The alarm pheromones were also released by disturbed female mirid bugs, *Lygocoris pabulinus* (L.) (hexyl butanoate) from their metathoracic glands in higher amounts than the undisturbed females (Groot, *et al.*, 2001) and *Acyrthosion pisum* (Harris) (β-farnesene) , *Chrysoperla carnea* (Stephens) (Z-4-tridecene, Zhu-JunWei *et al.*, 2000), *Frankliniella occidentalis* (Pergande) (dodecyl acetate) (Kirk *et al.*, 1998), *Neanura muscorum* (Templeton) (1,3-dimethoxybenzene) (Messer *et al.*, 1999) *Megaponera foetens* (Fab.) (Actinidine) (Janssen *et al.*, 1995), *Oxytrigona mediorufa* (Cockerell) (2-heptanone) (Cruz-Lopez *et al.*, 2007), *Camponotus obscuripes* (n-undecane) (Fujiwara-Tsujii *et al.*, 2006), *Cephalonomia stephanoderis* (skatole) (Gomez *et al.*,2005) and *Prorhinotermes canalifrons* (Sjöstedt) (E,E)-alpha -farnesene (Sobotnik *et al.*, 2008).

The production of alarm pheromone in quantities dependant on their life stages, for *e.g.* first and second instar of *Acyrthosion pisum* produce comparatively more farnesene than late instars and adults (Mondor *et al.,* 2000).

The alarm pheromone often acts as a kairomone for a variety of predators and species such as *Coccinella septempunctata* L. and *Aphidius uzbekistanicus* Luzhetzki (Micha and Wyass, 1996) were often attracted to β-farnesene, the alarm pheromone of aphids.

Although, broadly the term used for referring to the release of chemicals of certain species of insects to warn off their other individuals of the colony, they also act as kairomone, releasing the valuable information favorable to the receiver. Best examples of alarm pheromones exist in the worker ants of *Atta bisphaerica.* Polyethism is a behavioural response where the individuals of colony respond to differently to their alarm pheromones, for *e.g.* when the crushed mandibles of leaf cutting ant *Atta capiguara* Goncalves were applied to their colonies, the majority of the minor workers showed alarm response but some workers did not show any response (Hughes and Goulson, 2001).

Genetic engineering has been used to produce experimental plants with E beta f synthase genes to produce sufficient quantities of β-farnesene which interacted in two trophic level, both repelling the pest *Myzus persicae* and attracting the parasitoid, *Diaeretiella rapae* (M'Intosh) (Beale *et al.,* 2006).

The staphylinid adult beetles of the genus *Pella* mimic producing the alarm pheromones, undecane and 6-methyl-5-hepten-2-one (sulcatone) produced by the ant, *Lasius fuliginosus* (Latreille) in an apparent attempt to avoid predation by *L. fuliginosus* (Stoeffler *et al.,* 2007).

Brood pheromones

Brood pheromones are those which showed releaser activity in the regulation of broods, however, some authors consider them as kairomones. In honeybees, the brood pheromone consists of 10 fatty acids, whose main function is colony regulation and they are involved in the inhibition of ovaries resulting in the reproductive division of labour, and two components methyl palmitate and ethyl oleate are involved in the increased activity of hypo-phrayngeal glands which inturn affect nursing of the larvae by workers (Mohammedi *et al.,* 1998; Le Conte *et al.*, 2000). Yun-Feng *et al.*, (2010) observed that methyl palmitate increased the weight of larvae inside the queen cells in *Apis cerana cerana* F. while, ethyl oleate significantly depressed weight and ovary of pheromones. A review on queen pheromone have been presented by Hu-FuLiang and Xuan-Hong Zhuan (2004).

Contact Sex pheromones

Several sex pheromones of beetles, especially belonging to the Cerambycidae act in long range, short range or contact pheromone. Sugeno *et al.*, (2006) identified 9-methylheptacosane, 11-methylheptacosane, 9-methylnonacosane and 11-methylnonacosane, as contact sex pheromone in Chrysomelid, *Gastrophysa atrocyanea* Mot. *Tetropium fuscum* (Fab.) and *T. cinnamopterum* Kirby also utilize (S)-11-methyl-heptacosane at 40 µg/ female elicited mating behavior in the males (Silk *et al.*, 2011). Males of *Aphytis melinus* De Bach also rely on the contact sex pheromone for locating their partners (Bernal and Luck, 2007). Cuticular hydrocarbons also play a major role in the mate finding in the case of male *Ellychnia corrusca* (L.) (Ming-QingLei; and Lewis, 2010).

Defense substances

Most of insects feed on obnoxious substances produced by their host plants and which in turn are expressed in their body, probably playing the role of defense. *Uresiphita reversalis* (Guenee) when fed on *Cytisus monspessulanus* L. assimilate quinolizidine alkaloids, (+)-2,3-dehydro-10-oxo- alpha -isosparteine which act as defense substances from their natural enemies (Nihei *et al.*, 2006).

Marking pheromone

The ovipositing females usually identify the presence of immature stages or brood of their species through the specific pheromones, which are termed as marking pheromones (Nufio and Papaj, 2001). Sometimes, males after copulation often mark the females as observed in the aphids, *Cinara cuneomaculata* (Del Guercio) *C. laricis* (Borner) (on larch) and *Lachnus roboris* (L.) probably preventing the females from other males (Dagg and Scheurer, 1998), *Schistocerca gregaria* (Forskai) (Seidelmann, 2006), several tephritid flies such as *Anastrepha ludens*, *A. obliqua* (Macquart) and *A. serpentina* (Wiedemann) utilize marking pheromones to indicate oviposition by their females which are intraspecific in nature (Aluja, Diaz-Fleischer, 2007). Citronellol, 2,3-dihydrofarnesol, and (S)-isomer of (E)-2,3-dihydrofarnesol have been known to be the major host marking pheromones of bumble bees and cuckoo bumble bees (Luxova *et al.*, 2004; Valterova *et al.*, 2007). Several predatory insects belonging to Hemiptera, Coleoptera, Diptera and Neuroptera use marking pheromones, mediating in foraging, intraguild predation and cannibalism (Mu-LiLi; Li-GuoQing, 2007).

Recognition pheromones

Recognition of nest mates is achieved in several species of honeybees and ants through the use of recognition pheromones (Gamboa *et al.*, 1996; Akino, *et al.*, 1999; Breed *et al.*, 2007). These recognition pheromones are often mimicked by insects such as *Maculinea rebeli* (Hirschke) to avoid predation from ants.

Trail pheromones

Trail pheromones are used in colonial insects which allow the conspecifics to follow their trails. The trail pheromones are best exploited in ants and termites and poison glands, rectal glands, hind gut, Dufours gland, tibial gland and Pavan gland secrete the trail pheromones. Several functional groups such as aldehyde, alcohol, amines, ketones and coumarin substances were known to act as trail pheromones (Table. 1).

Table. 1 : Some examples of trail pheromones used by several species of insects

Species	Compound	Source of secretion	Reference
Aphaenogaster rudis (Enzmann)	anabasine, anabaseine and 2,3'-bipyridyl N-isopentyl-2-phenylethylamine	poison-gland	Attygalle, *et al.*, 1998
Linepithema humile (Mayr)	(Z)-9-hexadecenal		Suckling, *et al.*, 2010
Atta vollenweider (Forel) *Atta sexdens*	4-methylpyrrol-2-carboxylate and 2-ethyl-3, 6-dimethylpyrazine		Kleineidam, *et al.*, 2005
Camponotus balzani and C. *sericeiventris* (Linnaeus)	3,4-dihydro-8-hydroxy-3,5,7-trimethyliso-coumarin	hindgut	Kohl, *et al.*, 2003
C. rufipes Emery	3,4-dihydro-8-hydroxy-3,7-dimethylisocoumarin	hindgut	Ubler, *et al.*, 1995
Camponotus atriceps (Smith) and *C. floridanus* (Buckley)	3,5-dimethyl-6-(1'-methylpropyl)-tetrahydropyran-2-one and nerolic acid	hind gut	Haak, *et al.*, 1996
Camponotus castaneus (Latreille)	3,5-dimethyl-6-(1'-methyl-propyl)-tetrahydro-2H-pyran-2-one	hindgut	Kohl, *et al.*, 2003
C.pennsylvanicus	3,4-dihydro-3,5-dihydroxy-6-methyl-pyran-4-one	rectum	Hillery, & Fell, 2000
C.pennsylvanicus	n-undecane n-tridecane	Dufour's gland	Hillery, and Fell, 2000
C. pennsylvanicus	Palmitic acid	poison gland	Hillery, and & Fell, 2000
Camponotus silvicola (Forel)	3,4-dihydro-8-hydroxy-3,5,7-trimethylisocoumarin	hindgut	Ubler, *et al.*, 1995

Contd...

Camponotus socius Roger	(2S,4R,5S)-2,4-dimethyl-5-hexanolide and 2,3-dihydro-3,5-dihydroxy-6-methylpyran-4-one	hindgut	Kohl, *et al.*, 2001
Coptotermes formosanus (Shiraki)	(3Z,6Z,8E)-3,6,8-dodecatrien-1-ol ((Z,Z,E)-DTE-OH and (Z,E,E)-dodecatrien-1-ol		Tokoro, *et al.*, 1994
Coptotermes formosanus and *Reticulitermes* sp.	2-phenoxyethanol		Jian-Chen *et al.*, 1998
Crematogaster castanea (Smith)	(R)-2-dodecanol	tibial glands of the hind legs	Morgan, *et al.*, 2004
Dolichoderus thoracicus (Smith)	(Z)-9-octadecenal	Pavan gland	Attygalle, *et al.*, 1998
Gnamptogenys striatula (Mayr)	2E)-3,4,7-trimethyl-2,6-(octadien-1-ol (4-methylgeraniol), and (2E)-3,4,7-trimethyl-2,6-nonadien-1-ol	Dufour's gland	Blatrix, *et al.*, 2002; Schulz, *et al.*, 2002
L. *fuliginosus*	mellein		Veit , 2001
L. *fuliginosus*	3,4-Dihydro-8-hydroxy-3-methylisocoumarin (mellein) and 2,3-dihydro-3,5-dihydroxy-6-methyl-4H-pyran-4-one	hindgut	Kern, *et al.*, 1997
Lasius fuliginosus (Latreille)	n-hexanoic acid and n-heptanoic acid	hind gut	Akino, and Yamoka, 1996
M. madagascarica (Gregg)	methyl pyrrole-2-carboxylate	poison gland secretions	Holldobler, *et al.*, 2002
Mastotermes annandalei (Silvestri), *M. barneyi* (Light) *Odontotermes hainanensis* (Light) and *O. maesodensis* (Ahmad)	(E)-dodec-3-en-l-ol	sternal glands	Peppuy, *et .al.*, 2001
Mastotermes darwiniensis (Froggatt) *Termopsidae Porotermes adamsoni* (Froggatt) *Stolotermes*	(E)-2,6,10-trimethyl-5,9-undecadien-1-ol	sternal glands	Sillam-Dusses, *et al.*, 2007

Contd...

victoriensis Hill			
Megaponera foetens	N, N-dimethyluracil and	poison gland	Janssen, *et al.,* 1995
Pachycondyla (Paltothyreus) tarsata (Fabricius)	9-heptadecanone	sternal glands	Janssen, *et al.*, 1999
Pogonomyrmex barbatu (Smith) *P. maricopa* (Wheeler) *P. occidentalis* (Cresson) and *P. rugosus* (Emery)	2,5-dimethylpyrazine, trimethylpyrazine and 3-ethyl-2,5-dimethylpyrazine (EDMP)	poison-gland secretions	Holldobler, *et al.*, 2001
Pogonomyrmex vermiculatus (Emery)	alkylpyrazines, 2,5-dimethylpyrazine, 2,3,5-trimethylpyrazine, and 3-ethyl-2,5-dimethylpyrazine	poison gland	Torres-Contreras *et al.*, 2007
P. vermiculatus	tridecane to heptadecane	Dufour gland Contreras	Torres- *et al.*, 2007
Reticulitermes hesperus (Holmgren)	(3Z,6Z,8E)-dodecatrien-1-ol	sternal gland	Saran, *et al.*, 2007
Solenopsis invicta (Buren)	(E,E)- alpha - farnesene	worker whole body, head, thorax, and abdomen	Chen, and Fadamiro, 2007
S. *invicta*	Z,E- alpha - farnesene		Suckling, *et al.*, 2012
Trigona recursa (Smith)	Hexyl decanoate	mandibular glands	Jarau, *et al.*, 2006
Trigona spinipes (Fabricius)	octyl octanoate	mandibular glands	Schorkopf, *et al.*, 2007

Aggregation pheromone

Aggregation pheromones, mostly secreted from the male prothoracic glands of coleopterans and hemipterans facilitate aggregations of male and female and many of them were identified. Insect species belonging to Diptera, Lepidoptera, and Coleoptera produced aggregation pheromone though the chemicals differ signficantly Table 2.

Table 2 : Aggregation pheromone identified for some insect species

Species	Compound	Reference
Acalymma vittatum (F.)	(3R,4R)-3-methyl-4-(1,3,5,7-tetramethyloctyl)oxetan-2-one	Morris, *et al.*, 2005
Anthonomus rubi (Herbst)	(Z)-2-(3,3-dimethylcyclohexylidene) ethanol, (cis)-1-methyl-2-(1-methylethenyl) cyclobutaneethanol, and (2-(1-methylethenyl)-5-methyl-4-hexen-1-ol (lavandulol))	Innocenzi *et al.*, 2001
Bactrocera papayae (Drew & Hancock)	2-allyl-4,5-dimethoxyphenol (DMP) and (E)-coniferyl alcohol (CF)	Khoo *et al.*, 2005
Biprorulus bibax (Breddin)	((3R,4S,1'E)-3,4-bis(1'-butenyl) tetrahydro-2-furanol, linalool, farnesol and nerolidol)	James *et al.*, 1994
Carpophilus dimidiatus (F.)	(3E,5E,7E,9E)-6,8-diethyl-4-methyl-3,5,7,9-dodecatetraene	Bartelt *et al.*, 1995
Conotrachelus nenuphar. (Herbst)	(+)-(1R,2S)-1-Methyl-2-(1-methylethenyl)cyclobutaneacetic acid (trivial name, grandisoic acid)	Eller and Bartelt, 1996
Cydia pomonella (L.)	(E)-2-octenal, (E)-2-nonenal, sulcatone and geranylacetone	Jumean *et al.*, 2004
Dendroctonus ponderosae (Hopkins)	trans-verbenol and exo-brevicomin	Borden *et al.*, 2008
Diorhabda elongata (Desbrochers)	(2E,4Z)-2,4-heptadienal (1) and (2E,4Z)-2,4-heptadien-1-ol	Cosse *et al.*, 2005
Drosophila mulleri and *D. busckii,* Coquillett,	(S)-2-tridecanyl acetate and (S)-2-pentadecanyl acetate	Kamezawa *et al.*,1994
Dryocoetes affaber (Mann.)	(+)-exo-brevicomin and (+)-endo-brevicomin ((+)EXOB and (+)ENDOB)	Camacho *et al.*, 1994
Dryocoetes confusus Swaine *and D. autographus* (Ratzeburg)	(+)- or (+or-)-exo-brevicomin	Williams, & Borden, 2004
Erysarcoris lewisi (Distant)	(2Z,6R)-2-methyl-6-(4'-methylenebicyclo [3.1.0]hexyl)hept-2-en-1-ol	Mori, 2007
Euschistus conspersus (Uhler)	methyl (2E,4Z)-decadienoate	Krupke & Brunner, 2003
Eysarcoris lewisi (Distant)	(Z)-2-methyl-6-(4-methylenebicyclo[3.1.0]hex-1-yl)hept-2-en-1-ol.	Takita, *et al.*, 2008
Frankliniella occidentalis (Pergande)	(R)-lavandulyl acetate and neryl (S)-2-methylbutanoate	Hamilton, *et al*, 2005

Contd...

Galerucella calmariensis (L.) and *Galerucella pusilla.* (L.)	dimethylfuran lactone, 12,13-dimethyl-5, 14-dioxabicyclo[9.2.1]tetradeca-1(13), 11-dien-4-one	Bartelt, *et al.*,2006
Gnatocerus cornutus (Fabricius)	(1S,4R,5R)-(+)- alpha -Acoradiene (1,8-dimethyl-4-(1-methylethenyl)spiro [4.5]dec-7-ene)	Tashiro, *et al.*, 2004
Halyomorpha halys (Stål)	methyl (E, E, Z)-2,4,6-decatrienoate	Lee-KyuChul *et al.*, 2002 Khrimian, *et al.*, 2008
Harmonia axyridis (Pallas)	(-)- beta -caryophyllene	Brown, *et al.*, 2006
Ips duplicatus (Sahlberg)	Ipsdienol (Id), E-myrcenol (EM), and Amitinol (At)	Chen-GuoFa *et al.*,2009
Ips nitidus (Eggers)	2-methyl-3-buten-2-ol, approx. 74%-(-)-ipsdienol, and (-)-cis-verbenol	Zhang, *et al.*,2009
Ips shangrila (Cognato and Sun)	2-methyl-3-buten-2-ol and 99%-(+)-ipsdienol	Zhang, *et al.*,2009
Ips spp.	ipsdienol (2-methyl-6-methylene-2,7-octadien-4-ol) and ipsenol (2-methyl-6-methylene-7-octen-4-ol).	Sandstrom, *et al.*,2008
Ips typographus (L.)	(-)- alpha -pinene..	Erbilgin, *et al.*, 2007
	cis-verbenol (cV) and 2-methyl-3-buten-2-ol (MB)	Sun-XiaoLing *et al.*,2006
	Vertenol contains methyl butanol (1500 mg), (-)(cis)-verbenol (70 mg) and a terpenoid alcohol (trans)-7-methylene-3,8-nonadien-5-ol (10 mg),	Lebedeva, *et al.*,2003
Ips typographus (L.) and *Pityogenes chalcographus* (L.)	PheropraxReg. or Chalcoprax Reg.	Niemeyer and Watzek, 1996
Leptinotarsa decemlineata (Say)	(S)-3,7-dimethyl-2-oxo-oct-6-ene-1,3-diol [(S)-CPB I]	Kuhar, *et al.*, 2006; Babu, & Chauhan,, 2009
Metamasius hemipterus (Olivier)	4-methyl-5-nonanol (1) 2-methyl-4-heptanol (2) or 2-methyl-4-octanol (3)	Ramirez-Lucas, *et al.*, 1996
Metamasius hemipterus sericeus (Olivier)	5-methyl-nonan-4-ol and 2-methyl-heptan-4-ol (8:1 ratio) 'metalure'	Giblin-Davis, *et al.*,1996

Contd...

Monochamus scutellatus scutellatus (Say)	2-(undecyloxy)-ethanol (monochamol)	Fierke, *et al.*,2012
Murgantia histrionica. (Hahn)	4-[3-(3,3-dimethyloxiran-2-yl)-1-methylpropyl]-1-methylcyclohex-2-en-1-ol	Zahn, *et al.*,2008
Neoclytus acuminatus acuminatus (F.)	(2S,3S)-hexanediol (R)-3-hydroxyhexan-2-one	Lacey, *et al.*,2004; 2007
Oncopeltus fasciatus (Dallas)	hexyl, (E)-2-hexenyl, and (E)-2-octenyl acetates	Zhang, & Aldrich, 2003
Oryctes elegans	4-methyloctanoic acid (1: major component)	Rochat, *et al.*, 2004
(Prell.)	and ethyl 4-methyloctanoate (2), occasionally mixed with minor components: 4-methyloctanyl acetate (3), methyl 4-methyloctanoate (4), 4-methyloctanol (5), and nonanyl acetate (6)	
Oryctes monoceros (Olivier)	ethyl 4-methyloctanoate (1), 4-methyloctanoic acid (2),	Allou, *et al.*,2006
Oryctes rhinoceros (L.)	Ethyl 4-methyloctanoate	Morin, *et al.*,1996; Hallett, *et al.*,1995
Oryctes sp.	(+or-)-4-methyloctanoic acid	Ragoussis, *et al.*,2007
Oulema melanopus (Linnaeus)	(E)-8-hydroxy-6-methyl-6-octen-3-one	Rao, *et al.*,2003
P. quercivorus (Ciesla)	trans-1-methyl-4-methyl-ethyl-2-cyclohexen-1-ol	Nakashima, *et al.*,2005
Pityophthorus carmeli (Swaine) and *Pityophthorus setosus* (Blackman)	(E)-(+)-Pityol	Dallara, *et al.*,2000
Platypus quercivorus	(1S,4R)-p-menth-2-en-1-ol	Kashiwagi, *et al.*,2006
(Murayama)	trans-1-methyl-4-methyl-ethyl-2-cyclohexen-1-ol	Nakashima, *et al.*,2005
Prostephanus truncatus (Horn)	Trunc-call 1 (T1) and Trunc-call 2 (T2).	Hodges, *et al.*,2004
Rhynchophorus cruentatus	(5S,4S)-5-Methyl-4-octanol (cruentol),	Giblin-Davis, R.M
(Fabricius), *R phoenicis* (Fabricius) and *R palmarum* (L.)	(3S,4S)-3-methyl-4-octanol (phoenicol) and (4S,2E)-6-methyl-2-hapten-4-ol (rhynchophorol)	*et al.*,1996
R. ferrugineus, R.	(4S,5S)-4-Methyl-5-nonanol (ferrugineol)	Giblin-Davis,

Contd...

vulneratus (Olivier) *R. bilineatus* (Olivier), *M. hemipterus* (L.) *and D. borassi* (F.) and *R. palmarum* (L.)		*et al.*,1996 Gunawardena, *et al.*,1995 Gunawardena, & Bandarage, 1995
Rhynchophorus bilineatus (Montr.)	(4S,5S)-4-methyl-5-nonanol	Oehlschlager, *et al.*,1995
Rhynchophorus palmarum L.	6-methyl-2 (E)-hepten-4-ol (rhynchophorol	Duarte, *et al.*,2003
Rhyzopertha dominica (F.)	dominicalure 1 ((S)-(+)-1-methylbutyl-(E)-2-methyl-2-pentenoate) and dominicalure 2 ((S)-(+)-1-methylbutyl-(E)-2,4-dimethyl-2-pentenoate),	Razkin, *et al.*,1996 Selitskaya, & Shamshev, 1994
Riptortus clavatus (Thunberg)	(E)-2-hexenyl (E)-2-hexenoate, (E)-2-hexenyl (Z)-3-hexenoate (E2HZ3H), and tetradecyl isobutyrate, octadecyl isobutyrate	Mizutani, *et al.*, 1999; Mizutani, 2006; Endo, *et al.*, 2005; Yasuda, *et al.*, 2007; Huh-HyeSoon *et al.*,2009; Huh-Wan & Park-Chung Gyoo, 2005
Rosalia funebris (Mots)	(Z)-3-decenyl (E)-2-hexenoate	Ray, *et al.*,2009
Scolytus amygdali (Geurin-Meneville)	4-methyl-3-heptanol	Zada, A *et al.*, 2004
Scolytus multistriatus (Marsh.).	racemic beta -multistriatin	Berens, & Scharf, 1995
Scyphophorus acupunctatus (Gyllenhal)	2-methyl-4-heptanol and 2-methyl-4-octanol	Ruiz-Montiel, *et al.*,2003
Sitophilus granarius (L.)	(2S,3R)-sitophilate ((2S,3R)-3-hydroxy-2-methylpentanoate	Chambers, *et al.*,1996
Sphenophorus levis. (Vaurie) stink bugs	2-methyl-4-octanol	Zarbin, *et al.*,2003
	methyl (2E,4Z)-decadienoate	Leskey and Hogmire, 2005
Tribolium castaneum (Herbst)	(4R,8R)- and (4S,8R)-4,8-dimethyldecanal 4,8-Dimethyldecanal	Santangelo, *et al.*, 2006 Kim-JunHeon *et al.*,2005
Trypodendron lineatum, (Olivier) *Ips typographus* (L.)	Linoprax and Pheroprax in black flight barrier traps	Babuder, *et al.*,1996

Sex Pheromones

Sex pheromones are the mostly studied semiochemicals in insects, especially in lepidopterans and coleopterans. They are with different functional groups, more number of carbon atoms ranging from 12 to 36 and with different structural configurations. These chemicals are often used for monitoring, mass trapping and mating disruption in different insects. These will be dealt appropriately in the forthcoming sections.

Chemical Ecology of Lepidopteran insects

Moths and butterflies use semiochemicals successfully in locating their host plants for feeding, mating places (even host plants), mates and ovipositional substrates. They also encounter various stimuli from non host plants, their natural enemies and repellent odours and act accordingly. Primarily the antennae are involved in the long range perception of volatiles, however, maxillary palp and labial pads are also involved in the perception of volatiles. Olfactory binding proteins or pheromone binding proteins, olfactory receptors and pheromone activating biopeptides have been greatly localized in various species of lepidopterous insects, involved in the variety of volatiles.

Helicoverpa armigera (Hubner)

The sex pheromone of *Helicoverpa armigera* and other species of *Helicoverpa* was identified as Z-11-hexadecenal and Z-9-hexadecenal, however the blend ratio varies between the species. Intra specific variations in the blend ratio was observed for different geographical populations (Bakthavatsalam un published data). The pheromone of *H. armigera* has been successfully exploited in pest management for monitoring the populations.

Besides the pheromone compounds, 1-hexanol, benzaldehyde, benzene acetaldehyde, and beta -phenylethyl alcohol from withered leaves of black poplar (Guo-XianRu *et al.,* 2001); 3-methyl-2-pentanal, myrcene, limonene and ocimene from carrot flowers (Ding-HongJian *et al.,* 1997); (2E)-hexenal, (3Z)-hexenylacetate and (3Z)-hexenyl-2-methylbutyrate, alpha -pinene, beta - myrcene, limonene, E- beta -ocimene and linalool (Rajapakse *et al.,* 2006); benzaldehyde, (+or-)-linalool, phenylacetaldehyde and (S)-(-)-limonene from marigold (Bruce *et al.,* 2002) were found to elicit beahvioural responses in males and female. The mixture of pheromone and plant volatiles phenylacetaldehyde and Z-3-hexenyl acetate also synergize the male trap catches of *H. armigera* (Kvedaras *et al.,* 2007; Ochieng *et al.,* 2002; Deng-JianYu *et al.* (2004). A synthetic plant based attracticide has been developed and field tested in by several workers (Gregg, *et al.,* 2010; Socorro *et al.,* 2010).

Chilo partellus (Swinhoe)

Chilo partellus is a serious graminovrous pest on maize, sorghum and rice. The pheromone of *C. partellus* was identified as : Z=-11-hexadecenal, and Z-11-hexadecenol (5-40: 1-3), although several minor compounds were identified (Nesbitt *et al.*, 1979; Hansson *et al.,* 1995).

Chilo partellus also responds to volatiles from plants such as 3E)-4,8-dimethylnona-1,3,7-triene from *Melinis minutiflora* (Kimani *et al.,* 2000); linalool, acetophenone, and 4-allylanisole from sorghum and maize (Birkett *et al.,* 2006); (E)-2-hexenal, (Z)-3-hexen-1-Ol, and (Z)-3-hexen-1-yl acetate from wild monocot host plants (Chamberlain *et al.,* 2006);

Leucinodes orbonalis (Guenee)

Leucinodes orbonalis, the brinjal shoot and fruit borer is a serious pest in Asian countries, particularly in India and Bangladesh. The females call the males during 22:00 hours to 24:00 hours and the sex pheromone was identified as a binary blend consisting of E-11-hexadecenyl acetate and E-11-hexadecen-1-0l as a minor compound (Gunawardena, 1992; Gunaverdena *et al.,* 1989; Attygalle *et al.,* 1988; Zhu *et al.,* 1987).

The pheromone components were successfully used in the mass trapping of *L. orbonalis* in India and Bangaladesh. Besides the direct benefit of reduction in the pest population, the number of pesticides were greatly reduced with resultant environmental benefits.

Besides the pheromone compounds, several plant volatiles do play a role in the searching, feeding and oviposition behaviour of several species of lepidopteran insects. Linalool, terpinene, ethyl acetate, octene, methyl salicylate, hexenol acetate, pinene, camphene, hexen-1-0l verbanool and methyl butenol are some of the compounds which act as attractant. Myrcene, methyl butanol, á-terpinene, camphor, hexanal, octanal, nonanal, 1-oct-3-0l, terpinene and verbenal were recorded as repellents for some species (Tasin *et al.,* 2012; Zhang *et al.,* 2013; Anfora *et al.,* 2014; Zhang *et al.,* 2015a).

Table 3 : Volatiles identified to elicit responses in lepidopteran insects

	Species name	Compound	Source	Function	Reference
1.	Several lepidopteran pollinators, including noctuid moths.	Linalool and its associated lilac aldehyde/ alcohol compounds (especially lilac aldehyde B) benzaldehyde and benzyl acetate	*Phlox divaricata*	Floral attractant	Majetic *et al.,* 2015
2.	*Ectropis obliqua* (Prout)	Myrcene, α-terpinene, β-terpinene, linalool, *cis*-verbenol, camphor, α-terpineol and verbenone,	*Rosmarinus officinalis* (Lamiaceae)	Repellent	Zhang *et al.*, 2015b
3.	*Kallima inachus* (Boisduval)	Ethyl acetate, butanone, α-pinene and ethanol	Rotting fruits (pear, apple, banana, watermelon, orange, and persimmon)	Foraging Attractant	Tang *et al.*, 2014
4.	*Holcocerus hippophaecolus* Hua, Chou, Fang and Chen	1-octene, methyl salicylate, and (*Z*)-3-hexen-1-ol acetate	The sea buckthorn, *Hippophae Hippophae*	Attractant	Wang *et al.*, 2014a
5.	European corn borer (ECB) (*Ostrinia nubilalis* Hüb.)	Monoterpènes (MT), homoterpenes and β-farnesene	Maize (*Zea mays*, L.)	Ovipositional Attractant	Leppik *et al.*, 2014
6.	*Dendrolimus superans* (Butler)	α-Pinene, myrcene and camphene	*Larix gmelinii* monoterpene volatiles	Ovipositional Attractant	Lin Jian *et al.,* 2014
7.	*Hypsipyla robusta* (Moore)	*Z*)-β-Ocimene, (*Z*)-3-hexen-1-yl acetate, hexan-1-ol, nonanal, (*Z*)-3-hexen-1-yl butanoate, 2-ethyl hexan-1-ol, decanal, β-	Mahogany trees	Ovipositional Attractant	Abraham *et al.,* 2014

Contd...

		caryophyllene, (*Z*)-3-hexen-1-yl hexanoate and germacrene D			
8.	*Phthorimaea operculella* (Zeller)	Hexanal, octanal, nonanal and 1-octen-3-ol	Potato	Repellent	Anfora *et al.,* 2014
9.	*Lobesia botrana* (Denis & Schiffermüller)	*S*)-(-)-Perillaldehyde and isoegomaketone	Non-host plant *Perilla frutescens*.	Attractant	Cattaneo *et al.,* 2014
10.	*Ectropis obliqua*. (Prout)	β-Myrcene, α-terpinene, γ-terpinene, linalool, *cis*-verbenol, camphor, α-terpineol, and verbenone	*Ocimum basilicum*, *Rosmarinus officinalis*, *Corymbia citriodora*, and *Ruta graveolens*,	Repellent	Zhang *et al.,* 2013
11.	*Sitotroga cerealella* (Olivier)	Nonanal, decanal, and geranyl acetone	Corn kernel	Attractant	Fouad *et al.,* 2013
12.	*Hyphantria cunea* (Drury)	β-Ocimene, hexanal, cis-3-hexenal, limonene, trans-2-hexenal, cyclohexanone, cis-2-penten-1-ol, 6-methyl-5-hepten-2-one, 4-hydroxy-4-methyl-2-pentanone, trans-3-hexen-1-ol, and 2,4-dimethyl-3-pentanol.	Mulberry (*Morus alba* L., Moraceae)	HIPV Attractant	Tang *et al.,* 2012
13.	*Tortrix viridana* L	α-Farnesene and germacrene D	*Quercus robur* L.	HIPV Repellent	Ghirardo *et al.,* 2012
14.	*Helicoverpa assulta* (Guenee)	(*E*)-β-Ocimene, octanal, (*Z*)-3-hexenyl acetate, (*Z*)-3-hexen-1-ol, nonanal, (*Z*)-3-hexenyl-2-methyl butyrate, decanal, linalool, and (*E*)-β-caryophyllene	Tobacco	Attractant	Sun *et al.,* 2012
15.	*Plutella xylostella* (L.)	(*E*)-2-Hexenal, (*Z*)-3-hexen-1-	Green leave volatile(GLV)	Attractant	Li PengYan *et al.,* 2012

Contd...

	(Lepidoptera: Plutellidae)	ol, and (*Z*)-3-hexenyl acetate	blend with synthetic pheromone		
16.	*Heortia vitessoides* Moore (Lepidoptera: Crambidae).	Hexanal, limonene, 2-hexanol, octanal, (*Z*)-3-hexenyl acetate, (*Z*)-3-hexen-1-ol, nonanal, decanal, and 2,6,10-trimethyl-dodecane	*Aquilaria sinensis* (Lour.) Gilg (Thymelaeaceae)	Attractant	Qiao HaiLi *et al.*, 2012
17.	*Lobesia botrana*	*E*)-β-Caryophyllene, (*Z*)-3-hexenyl acetate, 1-hexanol, or 1-octen-3-ol (*E*)-4,8-dimethyl-1,3,7-nonatriene, (*E*)-β-farnesene, (*Z*)-3-hexenol, or methyl salicylate	Grapevines VOC blend with female sex pheromone Attractant	Attractant	Arx *et al.*, 2012
18.	*Lobesia botrana*	3-Methyl-1-butanol	Fungus,*Botrytis cinerea* infected grapes *Vitis vinifera*	Repellent	Tasin *et al.*, 2012
19.	*Grapholitha molesta* Busek	(*Z*)-8-Dodecenyl acetate, (*E*)-8-dodecenyl acetate, (*Z*)-8-dodecenol and dodecanol. (*E*)-2-hexenal, (*Z*)-3-hexen-1-ol, (*Z*)-3-hexen-1-yl acetate, benzaldehyde and benzonitrile	Sex pheromone blend natural peach shoots volatiles	Attractant	Lu PengFei *et al.*, 2010
20.	*Cactoblastis cactorum* (Berg)	(*Z,E*)-9,12 Tetradecadien-1-ol acetate, 42% (*Z,E*)-9,12 tetradecadien-1-ol, and 4%	Sex pheromone, virgin female	Attractant	Heath *et al.*, 2006

Contd...

		(*Z*)-9-tetradecen-1-ol acetate			
21.	Guatemalan moth *Tecia solanivora*	Methyl phenylacetate β-caryophyllene, germacrene-D-4-ol, germacrene-D, kunzeaol, and (*E,E*)-α-farnesene.	Floral fragrance Tuber potato	Attractant	Karlsson *et al.,* 2009
22.	*Maruca vitrata* F.	A blend of limonene, 1,3-diethylbenzene, benzaldehyde, acetophenone, 4-ethylbenzaldehyde, 1-(4-ethylphenyl)-ethanone, 2-methyl-3-phenylpropanal, 1H-indol-4-ol, and 1,12 -(1,4-phenylene)bis-ethanone	Floral volatiles from *Lablab purpureus*	Attractant	Wang Pan *et al.*, 2014

Chemical Ecology of Coleopteran Insects

Among the orders that cause damage to crops, coleopterans contribute a major share. Considering their long life cycle and their cryptic behavior their management using the conventional approaches becomes difficult. This problems can be overcome by adopting the chemoecological measures

In coconut production the major pests that influence the crop yield are, rhinoceros beetle, *Oryctes rhinoceros*, red weevil, *Rhynchophorus ferrugineus*, black headed leaf eating caterpillar, *Opisina arenosella*, eriophyid mite, *Aceria guerreronis* and white grub, *Leucopholis coneophora.* Chemoecological approaches are well exploited for key pests of coconut.

Coconut red palm weevil, *Rhynchophorus ferrugineus* (Oliver)

The red palm weevil, *Rhynchophorus ferrugineus* (Oliver) is distributed in Asia and Europe. The host range of the weevil includes coconut, oil palm, date palm and sago (Wattanapongsiri, 1966). The adult female oviposit on young, damaged, stressed and healthy palms (Kalshoven, 1950). On hatching the larvae bore into the palm and develop into adult in two months (Giblin Davis *et al.*, 1996).

Trapping by baiting with coconut stem tissue was done to reduce the weevil populations in India (Abraham and Kurian, 1975). Evidence of male produced aggregation pheromone was established in laboratory bio assay by Abraham (1987).The aggregation pheromone, 4 methyl 5 nonanol (Ferrugineol) was synthesized by Hallet *et al.*, (1993). Coconut logs treated with toddy, yeast and acetic acid were effective in trapping *R. ferrugineus* (Kurian *et al.*, 1984).

Rhynchophorus ferrugineus weevils are opportunistic oligophages to early fermentation volatiles like ethanol emanating from wounded host (Gunatilake and Gunawardane,1986), on feeding the palm tissue they produce an aggregation pheromone (4 methyl 5 nonanol) that attract their conspecifics (Giblin-Davis *et al.*, 1996). Food volatiles strongly enhance the attraction of *Rhynchophorus* species to aggregation pheromone (Rochat *et al.*, 1993).

The palm esters, ethyl acetate, ethyl propionate, ethyl butyrate and ethyl isobutyrate were identified as kairomones for *R. phoenicis*, *R. palmarum*, *R.cruentatus*, *R. ferrugineus* and *R. vulnaratus* (Gries *et al.*, 1994). The American palm weevil, *R. palmarum* adults were attracted to odors of variety of plant tissues (pineapple, banana and coconut) that were used as baits in traps (Jaffe *et al.*, 1993; Oehlschlager *et al.*, 1993). Among the complex pattern of odorants emitted by host plants only a few key compounds were used to locate the host. The identification of natural volatiles emitted by host plants aided in development of synthetic blends (rhyncophorol + host volatiles) to trap *R. palmarum* (Rochat *et al.*, 2000).Synergy between pheromone and plant volatiles (PVs) was confirmed by analyzing the locomotory responses of *R. palmarum* in a four choice olfactometer (Saÿd *et al.*, 2003).

Racemic nonanoic lactone and 4 hydroxy 3 methoxy styrene from the steam volatiles of coconut bark caused electrophysiological response of *R. ferrugineus* antennae (Gunawardane *et al.*, 1998). Ethyl acetate, ethyl propionate, ethyl butyrate and ethyl isobutyrate synergized attraction of *R. cruentatus* to cruentol. None of the palm esters tested in combination with the pheromone were as attractive as palm or sugarcane tissue (Giblin-Davis *et al.*, 1994). Coconut petioles in traps attracted maximum weevils when they were 2-5 days old, the catch declined thereafter as volatile profile changed due to fermentation (Hallet *et al.*, 1993). The proportional changes in volatile from fermenting palm are attributed to abiotic conditions and microflora present (Nagnan *et al.*, 1992, Samarajeeva *et al.*, 1981). Fermented plant tissues produce a spectrum of odorants that are significantly different from those released by healthy plants (Giblin-Davis *et al.*, 1994; Rochat *et al.*, 2000). Fermented sap exuding from dead or wounded palms was highly attractive to *R. cruentatus* (Giblin-Davis *et al.*, 1996). Moist fermenting tissue from various palm species, fruits, sugarcane, pineapple and molasses are similarly attractive to palm weevils

(Giblin-Davis *et al.,* 1994). The efficiency of the kairomones in attracting insects depends on the odour quality and/or the amount released. Timing of volatile trapping to identify the compounds and their ratios in their matrix is essential to formulate an effective pherosynergestic blend.

Acetoin a volatile product of anaerobic fermentation with ethyl acetate was an effective synergist with aggregation pheromone of *R. palmarum.* Stimulation by plant volatile blends excited three olfactory receptor neurons. Pheromone baited traps containing plant volatile blends attracted twice as many *R. palmarum* as control (Said *et al.,* 2005).

Indigenously formulated CPCRI lure containing ferrugineol resulted in a weevil catch of 2– 6 weevils / month and it was 50 % efficient as compared to imported lures but was cost effective as compared to imported lures (Faleiro, 2003).

The electroantennogram assay on the adult *R. ferrugineus* showed higher antennal response to ethyl acetate (a major component in pineapple odor). A comparative study of the odorant binding protein genes and isolation of the partial OBP gene form *R. ferrugineus* was attempted. Amplification of putative OBP genes of *R. ferrugineus* yielded two genomic fragments. On sequencing they had homology with OBP of other insects (Rajesh *et al.*, 2008; Sree Smitha *et al.*, 2008).

Commercial lures are loaded in polymembrane dispensers and they have a release rate ranging from 2 – 20 mg / day. In an attempt to reduce the loss of the chemistry smart delivery devices were developed at CPCRI. The smart delivery aids in controlled release of pheromone for over six months as compared to commercial lures loaded in polymer membrane that are exhausted in 3 months.

Coconut rhinoceros beetle, *Oryctes rhinoceros.*

Thirty-nine species of *Oryctes* have been registered but only few of them have an impact on coconut production. They are distributed throughout South East Asia and South Pacific Islands. Apart from coconut, they also infest palmyra, date palm, wild date, areca, sago palm, pandanus, pineapple, colocasia, banana, oil palm and sugarcane. The pest occurs round the year with a spike in their population occurring during June to September, the period when the adults visit the crowns.

The black colored beetle bores holes and feeds on the unopened fronds and spathe. On opening, the damaged leaves show geometric cuts (V shaped) on leaf lets. If the damage is severe, several cuts can be seen one above the other. The beetles cause damage to seedlings, young and adult palms. Damage when done to the leaves, reduces the photosynthetic area and renders them unsuitable for thatching purpose, but when damage is done to spathe it causes direct crop loss upto 10 per cent (Nair, 1986). Ramachandran *et al.*, (1963) reported an yield loss of 5.5 – 9.1%. The beetles also cause death of the seedling/ young palms by destroying the growing points. Apart from feeding damage they serve as predisposers for red weevil attack and bud rot. Repeated damage done to the meristem may be lethal.

Male coconut rhinoceros beetles,*Oryctes rhinoceros* (L.), produce sex-specific compounds, , ethyl 4-methylheptanoate, and 4-methyloctanoic acid, the first of which is an aggregation pheromone (Hallet *et al.*, 1995). Ethyl 4-methyloctanoate was synthesized by Chem Tica and marketed as Oryctalure in the India. Buckets of 18 lit. capacity with black painted metal vanes were used as trap. The pheromone sachet was hung on the diamond shaped hole made in the vane. The trap was hung on a pole at 3-4 cm above the ground level. Coconut petioles were placed in the bucket trap for better attraction. A nonomatrix has been developed at CPCRI for the delivery of rhinoceros beetle pheromone.

Chemoreception in white grubs

Holotrichia is a genus of the melolonthine scarabs that occurs across the Indian subcontinent and through southeast and east Asia (Ward *et al.*, 2002). The chafer beetle, *Holotrichia serrata* F. (Coleoptera : Scarabaeidae) in its larval stage is a serious pest of coconut, (*Cocos nucifera* L.), sugarcane, (*Saccharum officinarum* L.), groundnut, (*Arachis hypogaea* L.) and vegetables in parts of Western and peninsular India (Ganeshaiah and Kumar,1993).The grubs cause damage by feeding on the roots and the adults emerge with the arrival of monsoon or heavy pre - monsoon showers (Yadav and Sharma, 1995). On emergence at dusk, they aggregate on plants like neem, (*Azadirachta indica* A. Juss), gulmohar, (*Delonix regia* L.), tamarind, (*Tamarindus indica* L.), mahagony, (*Swietenia mahagony* L.), drumstick, (*Moringa oleifera* Lam.) and subabul, (*Leucaena leucocephala* Lam.) for feeding and mating (A.R.V.Kumar Personal communication and Yadav and Sharma, 1995).

Attraction of adult males by females has been documented in *H. serrata* (Ganeshaiah and Kumar, 1993). The pheromone of *H. consanguinea* and *H. reynaudi*, has been isolated and identified as anisole (Leal *et al.*, 1996; Ward *et al.*, 2002). Chemical insecticides are widely used by farmers to manage the grubs of *Holotrichia* (Anitha *et al.*, 2006) and they result in varying degree of success in grub management. Indiscriminate use of insecticides results in buildup of residues and cause negative impact on non target organisms. Hence, it is imperative to search for alternative pest control methods. One such option is to exploit the behavioral features of the insect. Ethological control has been successfully applied in the management of scarabs (Leal *et al.,* 1996; Ruther *et al.*, 2000). Though *H. serrata* is an important pest, there are no reports on its olfactory response till date. As a primary step to identify physiologically active compounds causing antennal responses, the electroantennography was used.

The non aromatic esters, ethyl acetate and propyl acetate elicited a higher response in both male (-1.40 and – 1.35 mV respectively) and female (-.81 and -1.31 mV respectively) beetles while the hydrocarbon alkanes *viz.*, tridecane and nonane elicited the lowest responses.

Non aromatic esters – saturated open chain compounds were found to elicit significant EAG response from both males and females of *H. serrata*. The amplitude of the responses varied with the carbon chain length of the compound. Ethyl acetate a C4 elicited higher response in males. Female antenna was more responsive to propyl acetate. In either case the response decreased with increase in carbon chain length. In general the antennae of both male and female beetles were more responsive to esters followed by alcohol and hydrocarbon alkane.

Among the sex, the response of males to all the compounds tested was higher as compared to the females. Extract of pheromone glands in diethyl ether (-4.43 mV) when exposed to male antennae elicited a significantly higher response as compared to the extracts made in dichloromethane or hexane which were at par (-3.40 and -3.31 mV respectively).

Among the solvents used to obtain extracts from the pheromone gland, diethyl ether, dichloromethane and hexane caused maximum EAG amplitude in the male antennae of *H. serrata*. Extraction of abdominal glands of *H. consanguinea* beetles in dichloromethane and ether yielded a clean profile of anisole and indole; the compounds responsible for attraction of *H. consanguinea* males (Leal *et al.*, 1996).

Both adult male and female antennae responded to host volatiles. But there was a sexual dimorphism in the olfactory perception of the host volatiles by *H. serrata*. The male antenna was more sensitive to host extracts. Perhaps, this indicates the necessity of finding the host first for their mate location, apart from using the female produced pheromone as the cue. The use of green leaf volatiles as a sexual kairomone has been observed in *M. melolantha* and *H. consanguinea* males (Reinecke *et al.*, 2005; Yadav and Yadav, 2004). A combination of the synthetic sex attractant (R,Z)-5-(Idecenyl) dihydro-2(3H)-furanone with a 3:7 mixture of phenethyl propionate (PEP) and eugenol caught significantly more *Popillia japonica* (Klein *et al.*, 1981).

Semiochemicals are a vital tool for monitoring and mass trapping in plantation crops. Being perennial crops the pest occurs round the year. To depend on pesticides is a difficult task in terms of application in the target area and also the frequency at which they have to be taken up. Hence exploiting the ethology is an effective method to monitor and trap the pest population in plantation crops.

Sweet potato weevil

Sweetpotato weevils cause damage by feeding on storage roots or in the crown of the plant (Jansson *et al.*, 1990). The damage level can be determined only by destructive sampling of the storage roots (Sutherland, 1986). Hence monitoring of the adult population with pheromone traps is a vital component under the integrated pest management (IPM) program (Chiranjeevi and Reddy, 2003).

Heath *et al.* (1996) purified and synthesized the active component as (Z)-3-dodecen-1-ol (E)-2-butenoate. Subsequent studies have reported on new synthesis procedures or formulations that improve safety, reduce cost. Sugimoto *et al.* (1994) reported that the average attractive distance for a lure with 100 or 400g of pheromone was 55 and 64 m, respectively. Pheromone lures are used for a period of three months in the field (Pillai *et al.*, 1993).

Banana weevil

The banana root borer *Cosmopolites sordidus* (Germar) is a major pest of bananas throughout the world. The male-produced aggregation pheromone sordidin attracts both sexes, and mass trapping by using ground traps has the potential to replace inefficient insecticide treatments (Reddy *et al.,* 2009).

Cottonboll weevil

The cottonboll weevil *Anthonomus grandis* Boheman is a major pest of cotton in the Americas the males of which produce aggregation pheromone, grandlure (Tumlinson *et al.,* 1969). The grandlure traps when placed at a density of 14 traps per ha caused a strong reduction in weevil population.

Bark beetles

Bark beetles are area serious pest in coniferous forests worldwide, the discovery of the aggregation pheromones was used in area wide mass trapping campaigns. Placing pheromone blend of ipsdienol and E-myrcenol at the rate of one trap / ha. led to trapping of over 1.7 million beetles which led to reduction of tree damage (Schlyter *et al.,* 2003).

The chemicals identified for a selected colepteran insects is given in (Table-4).

Table 4 : Chemicals identified to elicit behavioural response in some coleopteran insects

Species	Compound	Function	Reference
Acanthocnemus nigricans (Hope)	5-methylfurfual, 4-methylguaiacol, 4-ethylguaiacol, eugenol and hydroxyacetone	Attractant	Paczkowski *et al.*, 2014
Aegorhinus nodipennis (Hope)	β-thujone, 1,8-cineole, p-cymene, and 4-terpineol,	Repellent	Tampe *et al.*, 2015
Anoplophora glabripennis Motschulsky	3E,6E-α-farnesene Mixture of 4-n-heptyloxy butenol and with (-) linalool, t-caryophyllene and Z-3-hexen-1-ol	Attractant Attractant	Crook *et al.,* 2014 Meng *et al.,* 2014
Anthonomus grandis (Boheman)	hexan-1-ol and hexanal	Aggregation pheromone	Dickens *et al.*, 1990
Apion miniatum Germ.	Blend of (Z)-3-hexenal+ (E)-2-hexenal + (Z)-3-hexenol + (E)-2-hexenol + (Z)-3-hexen-1-yl acetate	Attractant	Piesik *et al.,* 2014
Carpophilus antiquus (Melsheimer)	(3*E*,5*E*,7*E*,9*E*)-6,8-diethyl-4-methyl-3,5,7,9-dodecatetraene	Aggregation pheromone	Bartelt *et al.*, 1993a
Carpophilus freemani (Dobson)	(2*E*,4*E*,6*E*)-5-ethyl-3-methyl-2,4,6-nonatriene 2*E*, 4*E*,6*E*,8*E*)-7-ethyl-3,5-dimethyl-2,4,6,8-undecatetraene	Aggregation pheromone	Bartelt *et al*,, 1990b
Carpophilus hemipterus. (Linnaeus)	(2*E*,4*E*,6*E*,8*E*)-3,5,7-trimethyl-2,4,6,8-decatetraene 2*E*,4*E*,6*E*,8*E*)-3,5,7-trimethyl-2,4,6,8-undecatetraene	Aggregation pheromone	Bartelt *et.al*, 1990a
Callosobruchus maculatus (F.)	Blend of octanone, 3-octanol, linalool oxide, 1, octanol and nonanol	Attractant	Adhikary *et al.,* 2015
Ceutorhynchus assimilis (Payk)	3-butenyl and 4-pentenyl isothiocyanate (NCS) phenylacetonitrile, 4-pentenenitrile and 5-hexenenitrile	Attractant Attractant	Bartlet *et al.*, 1993b Bartlet *et al.*, 1997

Contd...

Conotrachelus psidii (Marshall)	(1R,2S,6R)-2-hydroxymethyl-2,6-dimethyl-3-oxabicyclo [4.2.0]octane (papayanol)	Attractant	Palacio-Cortés *et al.*, 2015
Cosmopolites sordidus (Germar)	α-pinene, β-pinene, β-myrcene, limonene, α-cubebene, α-copaene, β-caryophyllene and α-humulene	Attractant	Ndiege *et al.*, 1991
	1,8-Cineole	Attractant	Ndiege *et al.*, 1996
Cryptolestes ferrugineus (Stephens) and *C. pusillus* (Schonherr),	((3*Z*,11*S*)-3,11-dodecen-11-olide and (4*E*,8*E*)-4,8-dimethyl- decadienolide)	Aggregation pheromone	Chambers *et al.*, 1990
D. rufipennis (Kirby)	1-methyl-cyclohex-2-en-1-ol	Attractant	Gries, 1992
Dendroctonus terebrans (Olivier)	(-)-frontalin and (+)-*exo*-brevicomin	Attractant	Phillips *et al.*, 1990
Diabrotica virgifera virgifera (Chevrolat)	Linalool	Attractant	Hammack, 1997
Diabrotica spp.	2-phenyl-1-ethanol	Attractant	Hammack, 2003
Dryocoetes affaber (Mann.)	1:2 blend of (+)EXOB and (+)ENDOB	Aggregation pheromone	Camacho *et al.*, 1994
Gnatocerus cornutus (F.)	major (acoradiene) and the minor (α-cedren-14-al)	Attractant	Tebayashi *et al.*, 1998
Ips typographus (L.)	*trans*-conophthorin, (5*S*, 7*S*)-(-)-7-methyl-1, 6-dioxaspiro[4.5]decane	Attractant	Zhang QingHe *et al.*, 2000
Leptinotarsa decemlineata (Say)	Z)-3-Hexen-1-ol and 2-phenylethanol	Attractant	Weissbecker *et al.*, 1999
	(±)-linalool, methyl salicylate, and (*Z*)-3-hexenyl acetate.	Attractant	Dickens, 2002
Megaplatypus mutatus (Chapuis)	A-copaene	Attractant for males only	Lucia *et al.*, 2014
O. mercator (Fauvel) *O. surinamensis* (L.)	3-Methylbutanol	Attractant	Pierce *et al*, 1991
O. surinamensis	2-phenylethanol and ethanol	Attractant	Pierce *et.al*, 1991
Popillia japonica (Newman)	phenylacetonitrile and (*Z*)-jasmone	Attractant	Loughrin *et al.*, 1998

Contd...

Ips sexdentatus (Boerner)	ipsdienol and amitinol,	Attractant	Jactel *et al.*, 2001
Rhynchophorus cruentatus (F.)	5-Methyl-4-octanol	Aggregation pheromone	Weissling *et al.*, 1994
Tenebrio molitor (L.)	propionic acid	Repellent	Weaver *et al.*, 1990
Tomicus minor (L.)	α-pinene and β-phyllandrene	Attractant	Hua *et al.*, 2015

Chemical Ecology of Hemipteran insects

Several mirid bugs use sex pheromones for their communication (Laughlin, 1998), however aggregation pheromones by mirid bugs is not uncommon. Males of *Lygocoris pabulinus* are sensitive to butanoates and pentanoates while females are sentistive to hexan-1-ol, heptan-1-ol, 1-octen-3-Ool, 2-heptanone, hexanone, linalool, geraniol neol and methyl salicylate (Groot *et al.*, 1999). Several species of *Lygus* respond to hexyl butyrate, E-2 hexenyl butyrate and E-4-oxo-2-hexenal (Drijfhout *et al.*, 2002; Innocenzi *et al.*, 1998). Herivore induced plant volatiles such as 5-ethyl-5-2(5H) furanone, Z-3-hexenyl tiglate, E,E α-farnesen, β-caryophyllene, E,E-β-farnesene elicted good response in *Nezara virudula* L. and *Macrolohus caliginosus* (Wagner) (Moyari *et al.*, 2007; Willimas *et al.*, 2005). James (2003) observed that (E)-3 hexenyl aceteate attracted *Orius tristicolor* (White) and *Deraocoris brevis* (Uhler) while methyl salicylate attracted *Geocoris pallens* (Stal) and syrphids. E-hexenal released from the disturbed bugs often elicited good response in Lygus lineolaris *Lygus lineolaris* (Palisot de Beauvois) (Wardle *et al.*, 2003).

Volatiles from males acted as aggregation pheromone and female produced lipophilic fraction acted as contact sex pheromone for the pentatomids (Guarino *et al.*, 2007; Takita *et al.*, 2008). The pheromone compound E-2-hexenyl E-2 hexenoate of *Riptortus clavatus* (Thurnberg) attracted the pentatomid, *Piezodorus hybneri* (Gmelin) (Endo *et al.*, 2006). Sometimes stridulatory organ also produce volatiless such as (2E4Z)-decadienoate, methyl 2,6,10 trimethyl tridecanoate from *Euschstus heros* (F.) (Zhang *et al.*, 2003).

Table 5 : Semiochemicals identified for selected hemipteran insects

Species	Compound	Function	Reference
Aleurodicus dispersus (Russell)	(±)-2-hexanol	Attractant	Zheng *et al.*, 2014
Amrasca devastans (Dist.)	citral [3,7-dimethyl-2,6-octadienal] = carvacrol [2-methyl-5-(1-methylethyl)phenol] > citronellol [3,7-dimethyl-6-octen-1-ol] = farnesol [3,7,11-trimethyl-2,6,10-dodecatrien-1-ol] = geraniol [(E)-3,7-dimethyl-2,6-octadien-1-ol] = eucalyptus oil > neem oil = Cymbopogon oil.	antiovipositants	Saxena and Basit, 1982
	cineole [1,3,3-trimethyl-2-oxabicyclo[2.2.2]octane]	Attractant	Saxena, and Kumar, 1984
Batocera lineota Chev.	(*Z*)-3-Hexen-1-ol, Linalool	Attractant Repellent	Wang BaoXin *et al.*, 2014
Batocera horsfieldi (Hope)	(1*S*)-(")-β-pinene (*Z*)-3-hexen-1-ol	Repellent Attractant	Yang *et al.*, 2010
Bemisia tabaci (Gennadius)	α-pinene, p-cymene and β-phellandrene	Repellent	Saad *et al.*, 2014
Brevicoryne brassicae (Linnaeus)	4-pentenyl isothiocyanate	Repellent	Nottingham *et al.*, 1991
Bruchophagus gibbus (Boh.) or *Meligethes nigrescens* Steph.	hexyl acetate, 2-hexanone, 2-heptanone, acetophenone and acetoin	Attractant	Buttery *et al.*, 1984
Campylomma verbasci (Meyer)	(*E*)-crotyl butyrate	Attractant	Smith *et al.*, 1991
Coridius janus (Fabricius)	*E*-2-hexenal and N-tridecane	Repellent	Pathiratne *et al.*, 1991
Empoasca vitis (Göthe)	(*Z*)-3-hexenyl acetate	Attractant	Zhang Zheng Qun *et al.*, 2015
Euschistus spp.	methyl (2*E*,4*Z*)-2,4-decadienoate	Attractant	Aldrich *et al.*, 1995
Hyadaphis tataricae (Aizenberg)	(*E*)-β-farnesene	Attractant	Hedin *et al.*, 1991
Lonicera spp.			

Contd...

Leptoglossus australis (Fabricius)	Benzyl alcohol [benzenemethanol] (31%) and 2-phenylethanol [benzeneethanol] (42%) 2-octenyl propanoate (19%), (2*E*,4*E*)-2,4-hexadienyl acetate (6%) and (*E*)-2-hexen-1-ol (2%).	Attractant	Gough *et al.*, 1985
Lygus lineolaris (Palisot) de Beauvois	Hexyl butyrate and (*E*)-2-hexenyl butyrate	Attractant	Chinta *et al.*, 1994
Lygus rugulipennis (Nau, B. S)	hexyl butyrate, (*E*)-2-hexenyl butyrate and (*E*)-4-oxo-2-hexenal in 39:1:3 ratio.	Attractant	Innocenzi *et al.*, 1998
Megoura viciae (Buckton)	(*E*)-2-heptenal, 1-octenol-3, hexyl acetate, (*Z*)-3-hexenyl acetate, hexanol-1, hexanal, 2-heptanone and 3-octanone	Attractant	Visser and Piron, 1995
Musgraveia sulciventris (Stal)	trans-2-octenyl acetate	Attractant	Macleod *et al.*, 1975
Neotoxoptera formosana (Takahashi)	1,8-Cineole	Repellent	Hori and Komatsu, 1997
Nephotettix virescens (Distant)	β-Farnesene and trans-2-hexenol	Repellent	Calumpang *et al.*, 2014
Nezara viridula L.	(*Z*)-α-bisabolene (17%), *trans*- and *cis*-1,2-epoxides of (*Z*)-α-bisabolene (44 and 15%, respectively), (*E*)-nerolidol (1.4%) and nonadecane (7.4%)	Attractant	Aldrich *et al.*, 1987
	(*E*)-2-decenal	Attractant	Mattiacci *et al.*, 1991
	(*Z*)-α-bisabolene (1-methyl-4-(1,5-dimethyl-(*Z*)-1,4-hexadienyl)-cyclohexene), and *trans*- and *cis*-1,2-epoxides of (*Z*)-α-bisabolene. The*trans/cis*-1,2-epoxide	Attractant	Aldrich *et al.*, 1993

Contd...

Nilaparvata lugens (Stal)	α-pinene camphene linalool	Attractant Repellent Attractant	XianYing *et al.*, 2014; Zhang *et al.*, 2014
	3,7-dimethyl-2,6-octadienal	Repellent	Mahatheeranont, *et al.*, 1995
Oxycetonia jucunda (Falderman)	β-myrcene	Attractant	Ogihara, *et al.*, 1989
Pachylis laticornis (F.)	(E)-2-hexenyl tiglate [(E)-2-hexenyl (E)-2-methyl-2-butenoate] and (E)-2-hexenyl (E)-2-hexenoate,	Attractant	Aldrich *et al.*, 1982
Phorodon humuli (Schrank)	(*E*)-2-hexanal	Attractant	Campbell *et al.*, 1993
	70% 1*S* and 30% 1*R* nepetalactone	Attractant	Campbell *et al.*, 1990
Plautia stali Scott	methyl (*E,E,Z*)-2,4,6-decatrienoate	Attractant	Sugie *et al.*, 1996
Podisus maculiventris (Say),	*E*)-2-hexenal, (*E*)-2-octenal, (*E*)-2-hexenoic acid, benzaldehyde and nonanal	Attractant	Aldrich *et al.*, 1984
Pseudococcus comstocki (Kuw.)	2,6-dimethyl-1,5-heptadien-3-yl acetate	Attractant	Bierl-Leonhardt, *et al.*, 1982
Sitobion avenae (F.)	trans-2-hexenal and hexanal) benzaldehyde	Attractant	Yan, and Visser, 1982
Toumeyella parvicornis (Cockerell)	β-myrcene and β-phellandrene	Repellent	Green *et al.*, 2015
Toxoptera aurantii (Boyer de Fonscolombe)	(4a*S*,7*S*,7a*R*)-nepetalactone (I) and (1*R*,4a*S*,7*S*,7a*R*)-nepetalactol (II) in a ratio of 4.3-4.9:1	Attractant	Han BaoYu *et al.*, 2014
Trioza apicalis (Förster)	limonene	Attractant	Nehlin *et al.*, 1996

Chemical Ecology of Dipteran Insects

The dipteran insects, especially fruit flies respond to the semiochemicals like parapheromones and fruit volatiles. Parapheromones such as methyl eugenol, cuelure and trimedlure are known to attract fruit flies of a particular sub groups.

Besides, the odours from fruits do effectively elicit oviposition and repel some insects. Terpinolene, α-terpinene, linalool, β-ocimene and dimethyl trisulfide are a few compounds which modify the behaviour of *Lucilia sericata, Musca domestica* and *Aedes aegypti* (Zito *et al.*, 2014; Zito *et al.*, 2013; Dube *et al.*, 2011). Nevertheless the pheromone communication in dipteran insects poorly understood.

Table 6 : Chemicals which elicit beahvioural response in some dipteran insects

Species name	Compound	Source	Function	Reference
Anastrepha ludens (Loew) Mexican fruit fly.	Limonene, *trans*-ocimene. Linalool, β-pinene, and methyl salicylate, linalool and methyl salicylate	Bitter orange fruit, *Citrus aurantium* L.	Attractant	Rasgado *et al.*, 2009
	Ethyl butyrate, (*E*)-3-hexenol, (*Z*)-3-hexenol, hexanol, ethyl hexanoate, hexyl acetate, (*Z*)-3-hexenyl butyrate and ethyl octanoate.	Guava (*Psidium guajava* L.)	Attractant	Malo *et al.*, 2005
Anastrepha suspensa (Loew)	linalool [3,7-dimethyl-1,6-octadien-3-ol] in relation to limonene [1-methyl-4-(1-methylethenyl)cyclo hexene]	Citrus fruit	Repellent	Greany *et al.*, 1983
Anopheles arabiensis (Patton) and *Aedes aegypti* (L.)	β-Ocimene	Fresh and dried leaves and smoke from medicinal plants *Corymbia citriodora*, *Ocimum suave*, *cimum lamiifolium*, *Olea europaea* and *Ostostegia integrifolia*	Repellent	Dube *et al.*, 2011
Atherigona soccata (Rondanni) (sorghum shoot fly)	(*Z*)-3-hexen-1-yl acetate, (-)-α-pinene, (-)-(*E*)-caryophyllene, methyl salicylate,	Sorghum susceptible cultivar 'Swarna,'	Repellent	Padmaja *et al.*, 2010

Contd...

	octanal, decanal, 6-methyl-5-hepten-2-one and nonanal.			
Bactrocera dorsalis (Hendel).	Isobutyl acetate, decanal, β-caryophyllene,camphor, γ-terpinene, α-farnesene and 1,2-benzenedicarboxylic acid bis(2-methylpropyl) ester	*Syzygium samarangense* (Blume) fruit and wax-apple fruit	Attractant	Jin *et al.,* 2015.
Bactrocera oleae (Rossi)	Methyl thiolacetate, ammonia, 2-pentanone, 2-heptanone, ethyl tiglate and methyl thiocyanate.	Bacteria *Pseudomonas idor,* olive leaves and olives	Attractant	Liscia *et al.,* 2013
Bradysia sp	2-Methyl butyric acid methyl ester	*Rheum nobile*	Pollinator Attractant	Song Bo *et al.,* 2014
Ceratitis capitata (Wied.)	2-heptanone	Coffee fruit	Attractant	Prokopy *et al.,* 1998
Ceratitis capitata	Methyl (E)-6-nonenoate	Volatile extracts of adult males	Attractant	Ohinata *et al.,* 1979
Ceratitis capitata, Dacus oleae (Gmel.) and *Rhagoletis cerasi* (L.)	Heptanal,1-hexanol, octanal, nonanal and 1-nonanol	Fruit volatiles	Attractant	Guerin *et al.,* 1983
Ceratitis capitata	Hexenyl, hexyl, 3-methylbutyl, butyl and 2-methylpropyl esters.	Different peach cultivars	Repellent	Tabilio *et al.,* 2013.
Ceratitis capitata	Ammonium acetate	Addition of ammonium acetate to commercially available proteinaceous baits and to beer waste	Attractant	Piñero *et al.,* 2015.
Dasineura dielsi (Rubsaamen)	(Z)-3-hexen-1-ol acetate, 4-oxoisophorone, (Z)-β-ocimene, heptadecane, and nonadecane	Different flower stages and leaves of *Acacia cyclops*	Attractant	Kotze *et al.,* 2010

Delia antiqua (Mg.)	(-)-3-isothujone and (+)-3-thujone	Western red cedar (*Thuja plicata*)	Oviposition deterrent	Alfaro *et al.*, 1981
	n-Dipropyl disulfide (Pr_2S_2),	Onion	Ovipositional Attractant	Gouinguené *et al.*, 2005
Delia platura (Meigen)	4-hydroxy-4-methyl-2-pentanone, 1-hepten-3-one, 1-octen-3-ol, and 3-octanone	*Phaseolus vulgaris*	Ovipositional Attractant	Gouinguené *et al.*, 2006
Drosophila melanogaster (Meigen)	Acetic acid, 2-phenyl ethanol and acetoin	Vinegar	Attractant	
	Vinegar and geosmin	Microbiological off-flavor	Repellent	Becher *et al.*, 2010
Drosophila suzukii (Matsumura)	β-cyclocitral	Earlier ripening	Attractant fruit	Keesey *et al.*, 2015.
	Acetic acid, ethanol, acetoin and methionol	Fermented bait headspace	headspace	Cha *et al.*, 2014.
Drosophila sp.	(*z*)11-octadecenyl acetate (*cis*-vaccenyl acetate)	male cuticular hydrocarbon profile	Repellent/ suppressed court ship	Ejima *et al.*, 2007
Episyrphus balteatus (De Geer)	(*E*)-β-farnesene (EBF)	Aphid-infested potato plants	HIPV Ovipositional Attractant	Harmel *et al.*, 2007
Exorista japonica Townsend	Blend of host induced[(*E*)-4,8-dimethyl-1,3,7-nonatriene, indole, 3-hydroxy-2-butanone, and 2-methyl-1-propanol] and non specific [(*Z*)-3-hexen-1-yl acetate, (*E*)-2-hexenal, hexanal, (*Z*)-3-hexen-1-ol, and linalool)	Host infested and non specific blend (artificially damaged plants)	Attractant	Ichiki *et al.*, 2008
Liriomyza bryoniae (Kaltenbach)	Methyl salicylate (MeSa)	Synthetic methyl salicylate (MeSa)	Attractant	Buda *et al.*, 2008
Lucilia sericata Meigen, *L. caesar*, *Calliphora vicina*, *Protocalliphora*	Dimethyl trisulfide (DMTS).	Brood-site deceptive plants	Pollinator Attractant	Zito *et al.*, 2014

azurea and *Musca domestica*				
Lucilia sericata (Meigen)	Dimethyl trisulfide (DMTS).	Bimodal cue	Ovipositional Attractant	Brodie *et al.*, 2014.
Musca domestica	Terpinolene, α-terpinene and linalool	*Caralluma europaea* (Guss.)	Pollinator Attractant	Zito *et al.*, 2013.
Rhagoletis pomonella	3-Methylbutan-1-ol	Black hawthorn (*Crataegus. douglasii*) and the iornamental English hawthorn, *C. monogyna*	Attractant	Cha *et al.*, 2012.
Rhagoletis pomonella	Blend [Butyl butanoate (10%), propyl hexanoate (4%), butyl hexanoate (37%), hexyl butanoate (44%), and pentyl hexanoate (5%)]	Apple	Attractant	Zhang *et al.*, 1999
Anastrepha obliqua (Macquart)	(Z)-3-nonenol and (Z)-3-nonenol+β-farnesene	Live calling males and to Super Q extracts of calling males	Attractant	López *et al.*, 2011.

Chemical Ecology of Insects Belonging to Other Orders

Besides these major orders, chemical ecology of insects belonging to other orders such as Hymenoptera and Neuroptera, was studied to a lesser extent. The larvae of Chrysopids use chemical cues such as tricosane, pentacosane and hexacosane, emanating from the egg deposits or larval cuticle while the adults use the aminoacids such as L-tryptophan which were released from the honey dew secretions of the aphids (Bakthavatsalam and Singh, 1996; Bakthavatsalam *et al.,* 2007; Harrison and McEwen, 1998). Besides, pheromone of some chrysopid species was also studied (Zhang *et al.*, 2004).

Some limited studies have been done in insects belonging to other orders and pherobase (*www.pherobase.com*) is a good source for these references.

Application of Chemical Ecology in Pest Management

Compelling demand for the semiochemical for the pest management ushered the development of pheromones, parapheromones and plant volatiles for the pest management in addition to the use of semiochemicals for increasing the efficacy of natural enemies. Pheromones are mainly used for monitoring, mass trapping and mating disruption, while parapheromones are used for male annihilation techniques and plant volatiles basically for attracting the coleopteran insects.

Monitoring

Pheromones from insect pests are used for the quantitative and qualitative monitoring, while quantitative monitoring helps in studying the seasonal incidence, appearance of insects, trends in dispersal and estimating the population levels, qualitative monitoring enables us to document entry of invasive such as *Bactrocera invadens* Drew, Tsuruta & White. The use of appropriate blends with the quality dispensers (rubber speta, PVC tubes, vials, plywood *etc.*), with an exact quantity of pheromone with a release rate on par with a calling female sutiable traps (Isellve, bucket, fly trap, delta, sticky *etc.*) with the minimal numbers per unit area ensures the good captures for effective monitoring. Quite a good number of insects have been monitored for over a period of time. A basic correlation with the field level population of egg and larvae (of lepidopteran insects) enable us to easily develop simulation models for the prediction of the pest and to initiate suitable management options. Apart from lepidopteran pests, monitoring has been very effective in detection of mealybug infestation in grape wine orchards. Parapheromones like cuelure, methyl eugenol and trimedlure are commonly used for quarantine monitoring and biodiversity documentation studies.

However, the main problem faced by the scientists is that in several instances insignificant correlation exist between the field population of immature with the pheromone catches, mainly attributed to the movement of insects from other adjacent fields. However, monitoring for quarantine purpose and dispersal studies holds promise for several invasive pests.

Mass trapping

Mass trapping is the technique where the pheromone traps are used in sufficient numbers to effect reduction in the field population to such a level that the pest gradually declines. Though basically the pheromone septa, traps etc remain the same similar to monitoring, the number of traps used is increased manifold to effectively trap the insects. Most of lepidopterous insects have been effectively managed through the use of phreromones for mass trapping. *Cydia pomonella* L. and *Cydia molesta* (Busck) become classical examples for mass trapping in several European Countries through Area Wide Management Program. Integration of mass trapping as a component of IPM ensured the reduction in insecticide cost and usage, conservation of natural enemies and other environmental benefits.

Coleopteran insects such as coffee white stem borer, *Xylotrechus quadripes*, *Cylas formicarius* (Fab.), *Rhynchophorus ferrugineus* and *Oryctes rhinoceros*, pheromones were used for mass trapping (Soroker *et al.*, 2005; Jayarama *et al.,* 2007; Chiranjeevi and Reddy, 2003; Ponnamma *et al.*, 2002) and in several species of lepidopteran insects for *eg.*, *Chilo infuscatellus* Snell. and *C. sacchariphagous indicus* (Bojer) (David *et al.,* 1985; Chelvi *et al.,* 2010); *Pectnophora gossypiella* Saunders (Huber and Hoffmann, 1979, Patil *et al.*, 2008). However, for the successful implementation long term usage of mass trapping is essential for the total eradication of the pest as in the case of *Cydia pomonella* (Linnaeus), *Pectinophora gossypiella*, bark beetles, palm weevils, corn rootworm, *Anthonomus grandis* Boheman and *Lymantria dispar* (Linnaeus) (El Sayed *et al.,* 2006). Nevertheless in some insects for eg, *Anthonomus rubi*, the mass trapping technique was ineffective in bringing the down the population (Lethmayer *et al.,* 2004; Cross *et al.,* 2004) and several factors such as inefficient blend ratio, insufficient number of traps or quantum of pheromone and migration of insects from other regions contribute to the failure of mass trapping technology.

Aggregation pheromones of insects, especially coleopterans have been used extensively for mass trapping. Use of aggregation pheromones was practiced in the mass trapping of *Xylotrechus quadripes*, *R. ferrugineus* and while sex pheromone was used for *C. formicarius*.

Plant volatiles play an important role in mass trapping several coleopteran pests. Phenylacetaldehyde, ethyl acetate, ethanol, α-pinene, 1,8, cineole, anisaldehyde, β-caryophyllene, coumarin, and α-humulene are commonly used attractants for a variety of insects. Combination of plant volatiles with pheromone synergize catches of insects, and α-pinene is a common synergist.

Mating Disruption

Inundation of the field with pheromone in excessive quantities interrupts in the capacity of the males to locate their mates ultimately resulting in less fecundity and pest load in the field and this technology is known as mating disruption. To achieve mating disruption different types of dispensers (for *e.g.* RAK1+2, RAK1+2R. Isomate C) are used which are rather different from those used for monitoring. Integration of chemicals, with appropriate reduction in insecticide, and pheromone greatly reduced the incidence of pests such as *Lobeis botrana* (Denis and Schiffermuller), *Eupoecilia ambiguella* (Hubner) and *Cydia molesta* (Emery and Schmid, 2001; Nunez, 2000). Predators , parasitoids, and microbial pathogens have been quite successful for the management of several pests species when integrated with mating disruption. Reduction of chemicals in terms of number of rounds of insecticide sprays had been observed in several cases where mating disruption was used resulting in conservation of natural enemies and reduction in cost of cultivation.

However, in some instances, the mating disruption caused undesired effect such as appearence of new pest species and shifting of one host to other.

Male Annihilation Technique

Male Annihilation Technique (MAT) involves the use of a high density of bait stations consisting of a male lure combined with an insecticide to reduce the population of male fruit flies to such a low level that mating does not occur (www.spc.int/Ind/project/male-annihilation-technique). Methyl eugenol traps @ 6-10 per ac or cue lure @ 10-20 per acre were sufficient in male annihilation to substantially reduce the male ratios and the resultant reduction in damage. Mango is the crop where male annihilation technique is most successful in reducing infestation substantially (Verghese *et al.,* 2005; Verghese *et al.,* 2007; Reji *et al.,* 2012; Singh *et al.,* 2013; Shelly *et al.,* 2013).

Semiochemicals in Biological control

An entomophage effectively uses the semiochemical cues from the host plants (for habitat location), host insects (host location/oviposition), pheromone from conspecifics (for mating) besides cues from its enemies (alarm response) and accordingly their sensors are tuned for this. The semiochemicals, particularly the kairomones from host plants/host insects, synomones from infested plants and pheromones from conspecifics were exploited in biological control programs.

A profile of chemicals normally used by the parasitoids or predators include unsatured hydrocarbons like tricosane, triacontane, nonacosane, hexacosane, pentacosane etc. While some of these chemicals are attractants, many of them are considred as repellents.

Predators like larval chrysopids, utilize the scale components of larvae or egg deposits for locating their prey. Sometimes the honey dew secretions from aphids also are perceived a volatile cue by larval coccinellids. Plant volatiles like β-cayophyllene and volatile cues from aphids such as acid hydrolzysed L-tryptophan,and valine have been known arrestant and ovipositional attractant for adult chrysopids. Alarm pheromone of aphids, β-farnesene is used as lady bug attractant for commercial exploitation.

Synomones produced from host plants infested with herbivores alter the behaviour of parasitoids. The volatiles, both in quality and quantity, differ from the uninfested plants.

Conclusions

The chemical ecological studies on insects have gained significance due to their application in pest management programs. While in mega orders like Coleoptera, Diptera and Lepidoptera substantial contribution was made in chemical ecology, in several orders like Ephemeroptera, Odonata *etc.* the progress has been rather minimal. The chemical ecological studies in these orders are particularly significant in view of their contribution to scientific knowledge. Still several areas in chemical ecology such as chemicals involved in intra-specific recognition of members, pheromones/volatiles for coleopterans, ovipositional attractants for lepidoteran insects and chemical ecology of immature stages remain to be studied.

Molecular aspects in olfaction, particularly identification of olfactory genes for a few insects have been done. However, there is an urgent need to extent those studies to other insects belonging to Lepidotera, Diptera and Coleoptera for the sake of unraveling the mysteries of olfactory genes.

Creating farmers awareness, competitive pricing and marketing, policy and financial supports from Governments, and relaxation in regulatory procedures are the key factors to make pheromone technology a viable alternative.

Chemical ecology studies should be encouraged by universities by including a comprehensive curricula, financial support for students and creation of world class institutions in chemical ecology. Involvement of private partners with a 'win-win' situation with the Research organizations will surely benefit the farmers and end users with cheaper and quality semiochemical products.

Acknowledgement

The authors are grateful to Director, National Bureau of Agricultural Insect Resources, Bengaluru for the encouragement and support. The technical help received from Ms. Sowmya, Ms. Kushboo Sinha, and Ms. Thilagavathy, Senior Research Fellows at NBAIR is gratefully acknowledged.

References

Abraham, V.A. and Kurian,C. (1975). An integrated approach to the control *Rhychophorus ferrugineus* the red weevil of coconut palm. In Proc of 4 th session of the FAO technical work party on coconut production protection processing. Kingston, Jamaica, September 14 – 25.

Adhikary, P., Mukherjee, A. & Barik, A (2015). Attraction of *Callosobruchus maculatus* (F.) (Coleoptera: Bruchidae) to four varieties of *Lathyrus sativus* L. seed volatiles. *Bulletin of Entomological Research*,105(2): 187-201.

Akino, T and Yamoka, R. (1996). Purification of the trail pheromone of *Lasius fuliginosus* Latreille. *Japanese Journal of Applied Entomology and Zoology*, 40(3): 233-238.

Akino, T., Knapp, J. J., Thomas, J. A and Elmes, G. W. (1999). Chemical mimicry and host specificity in the butterfly *Maculinea rebeli*, a social parasite of Myrmica ant colonies. *Proceedings of the Royal Society of London Series B, Biological Sciences*, 266(1427): 1419-1426.

Aldrich, J. R., Kochansky, J. P., Lusby, W. R. and Dutky, S. R. (1982). Volatile male-specific natural products of a coreid bug (Hemiptera: Heteroptera). *Journal of Chemical Ecology*, 8 (11):1369-1376.

Aldrich, J. R., Numata, H., Borges, M., Bin, F., Waite, G. K. and Lusby, W. R. (1993). Artifacts and pheromone blends from *Nezara* spp. and other stink bugs (Heteroptera: Pentatomidae). *Zeitschrift für Naturforschung. Section C, Biosciences* , 48 (1-2) : 73-79.

Aldrich, J. R., Oliver, J. E., Lusby, W. R., Kochansky, J. P. and Lockwood, J. A. (1987). Pheromone strains of the cosmopolitan pest, *Nezara viridula* (Heteroptera: Pentatomidae). *Journal of Experimental Zoology,* 244(1) :171-175.

Aldrich, J. R., Rosi, M. C. & Bin, F.(1995). Behavioral correlates for minor volatile compounds from stink bugs (Heteroptera: Pentatomidae). *Journal of Chemical Ecology*, 21(12) :1907-1920.

Aldrich, J. R.; Lusby, W. R.; Kochansky, J. P.; Abrams, C. B. (1984). Volatile compounds from the predatory insect *Podisus maculiventris* (Hemiptera: Pentatomidae): male and female metathoracic scent gland and female dorsal abdominal gland secretions. *Journal of Chemical Ecology,* 10(4): 561-568.

Alfaro, R. I., Pierce, H. D., Jr., Borden, J. H. & Oehlschlager, A. C.(1981). Insect feeding and oviposition deterrents from western red cedar foliage. *Journal of Chemical Ecology*, 7:1, 39-48.

Allou, K., Morin, J. P., Kouassi, P., N'klo, F. H and Rochat, D. (2006). *Oryctes monoceros* trapping with synthetic pheromone and palm material in Ivory Coast. *Journal of Chemical Ecology*, 32(8): 1743-1754.

Aluja, M and Diaz Fleischer, F. (2006). Foraging behavior of *Anastrepha ludens*, *A. obliqua*, and *A. serpentina* in response to feces extracts containing host marking pheromone. *Journal of Chemical Ecology*, 32(2): 367-389.

Anfora, G., Vitagliano, S., Larsson, M. C., Witzgall, P., Tasin, M., Germinara, G. S. and Cristofaro, A. de. (2014). Disruption of *Phthorimaea operculella* (Lepidoptera: Gelechiidae) oviposition by the application of host plant volatiles. *Pest Management Science*,70 (4): 628-635.

Anitha, V., Rogers, D. J., Wightman, J. and Ward, A., (2006). Distribution and abundance of white grubs (Coleoptera : Scarabaeidae) on groundnut in Southern India. *Crop Protection,* 25 : 732 -740.

Arx, M. von., Schmidt-Büsser, D. and Guerin, P. M. (2012). Plant volatiles enhance behavioral responses of grapevine moth males, *Lobesia botrana* to sex pheromone. *Journal of Chemical Ecology*, 38(2): 222-225.

Attygalle, A. B., Schwarz, J. and Gunawardena, N. E. (1988). Sex pheromone of brinjal shoot and pod borer *Leuconodes orbonalis* Guenee (Lepidoptera: Pyralidae: Pyraustinae). *Zeitschrift fur Naturforschung, C Biosciences*, 43(9-10): 790-792.

Attygalle, A.B., Mutti, A., Rohe, W., Maschwitz, U., Garbe, W and Bestmann, H. J. (1998). Trail pheromone from the Pavan gland of the ant *Dolichoderus thoracicus* (Smith). *Naturwissenschaften*, 85(6): 275-277.

Ayasse, M. & Jarau S. (2014). Chemical Ecology of bumble bees. Annual Review of Entomology 59: 299-319.

Babu, B. N and Chauhan, K. R. (2009). Enantioselective synthesis of (S)-3,7-dimethyl-2-oxo-6-octene-1,3-diol: a Colorado potato beetle pheromone. *Tetrahedron Letters*, 50(1): 66-67.

Babuder, G.,Pohleven, F. and Brelih, S. (1996). Selectivity of synthetic aggregation pheromones LinopraxReg. and PheropraxReg. in the control of the bark beetles (Coleoptera, Scolytidae) in a timber storage yard. *Journal of Applied Entomology*, 120(3): 131-136.

Bakthavatsalam, N., Tandon, P. L. Patil, S. B., Bhemanna Hugar, and Hosamani, A. (2007). Kairomone formulations as reinforcing agents for increasing abundance of *Chrysoperla carnea* (Stephens) in cotton ecosystem. *Journal of Biological Control*, 21: 1-8.

Bakthavatsalam, N. and Singh, S. P. (1996). L –tryptophan as an ovipositional attractant for *Chrysoperla carnea* (Stephens) (Neuroptera: Crysopidae). *Journal of Biological Control*, 10 (1&2): 1-5.

Bartelt, R. J. and Hossain, M. S. (2006). Development of synthetic food-related attractant for *Carpophilus davidsoni* and its effectiveness in the stone fruit orchards in Southern Australia. *Journal of Chemical Ecology*, 32(10): 2145-2162.

Bartelt, R. J., Dowd, P. F., Plattner, R. D. and Weisleder, D. (1990a). Aggregation pheromone of dried fruit beetle, *Carpophilus hemipterus* wind-tunnel bioassay and identification of two novel tetraene hydrocarbons. *Journal of Chemical Ecology*, 16(4): 1015-1039.

Bartelt, R. J., Dowd, P. F., Shorey, H. H. and Weisleder, D. (1990b). Aggregation pheromone of *Carpophilus freemani* (Coleoptera: Nitidulidae): A blend of conjugated triene and tetraene hydrocarbons. *Chemoecology*, 1(3-4): 105-113.

Bartelt, R. J., Seaton, K. L. and Dowd, P. F. (1993a). Aggregation pheromone of *Carpophilus antiquus* (Coleoptera: Nitidulidae) and kairomonal use of *C. lugubris* pheromone by *C. antiquus*. *Journal of Chemical Ecology*, 19(10), 2203-2216.

Bartlet, E., Blight, M. M., Hick, A. J., and Williams, I. H. (1993b). The responses of the cabbage seed weevil (*Ceutorhynchus assimilis*) to the odour of oilseed rape (*Brassica napus*) and to some volatile isothiocyanates. *Entomologia experimentalis et applicata*, 68(3), 295-302.

Bartelt, R. J., Weaver, D. K. and Arbogast, R. T. (1995). Aggregation pheromone of *Carpophilus dimidiatus* (F.) (Coleoptera: Nitidulidae) and responses to *Carpophilus* pheromones in South Carolina. *Journal of Chemical Ecology*, 21(11): 1763-1779.

Bartlet, E., Blight, M. M., Lane, P. and Williams, I. H. (1997). The responses of the cabbage seed weevil *Ceutorhynchus assimilis* to volatile compounds from oilseed rape in a linear track olfactometer. *Entomologia experimentalis et applicata*, 85(3): 257-262.

Beale, M. H., Birkett, M. A; Bruce, T. J. A., Chamberlain, K., Field, L. M., Huttly, A. K., Martin, J. L., Parker, R., Phillips, A. L., Pickett, J. A., Prosser, I. M., Shewry, P. R., Smart, L. E., Wadhams, L. J., Woodcock, C. M. and Zhang YuHua. (2006). Aphid alarm pheromone produced by transgenic plants affects aphid and parasitoid behavior, *Proceedings of the National Academy of Sciences of the United States of America*, 103(27): 10509-10513.

Becher, P. G., Bengtsson, M., Hansson, B. S. and Witzgall, P. (2010). Flying the fly: long-range flight behavior of *Drosophila melanogaster* to attractive odors. *Journal of Chemical Ecology*, 36(6): 599-607.

Bell, W. J. and Carde R. T. (Eds) (1985). *Chemical Ecology of Insects*. Chapman & Hill, 524 pp.

Berens, U. and Scharf, H. D. (1995). The first stereoselective synthesis of racemic beta - multistriatin: a pheromone component of the European elm bark beetle *Scolytus multistriatus* (Marsh.). *Journal of Organic Chemistry*. 60(16): 5127-5134.

Bernal, J. S. and Luck, R. F. (2007). Mate finding via a trail sex pheromone by *Aphytis melinus* DeBach (Hymenoptera: Aphelinidae) males. *Journal of Insect Behavior*, 20(6): 515-525.

Bierl-Leonhardt, B. A., Moreno, D. S., Schwarz, M., Forster, H. S., Plimmer, J. R. and DeVilbiss, E. D. (1982). Isolation, identification, synthesis, and bioassay of the pheromone of the Comstock mealybug and some analogs. *Journal of Chemical Ecology*, 8(4): 689-699.

Birkett, M. A., Chamberlain, K., Khan, Z. R., Pickett, J. A., Toshova, T., Wadhams, L. J. and Woodcock, C. M. (2006). Electrophysiological responses of the lepidopterous stemborers *Chilo partellus* and *Busseola fusca* to volatiles from wild and cultivated host plants. *Journal of Chemical Ecology*, 32(11): 2475-2487.

Blatrix, R., Schulz, C., Jaisson, P., Francke, W. and Hefetz, A. (2002). Trail pheromone of ponerine ant *Gnamptogenys striatula*: 4-methylgeranyl esters from Dufour's gland. *Journal of Chemical Ecology*, 28(12): 2557-2567.

Borden, J. H., Pureswaran. D. S. and Lafontaine, J. P. (2008). Synergistic blends of monoterpenes for aggregation pheromones of the mountain pine beetle (Coleoptera: Curculionidae). *Journal of Economic Entomology*, 101(4): 1266-1275.

Breed, M. D., Deng, XiaoBao and Buchwald, R. (2007). Comparative nestmate recognition in Asian honey bees, *Apis florea, Apis andreniformis*, *Apis dorsata*, and *Apis cerana*. *Apidologie*, 38(5): 411-418.

Brodie, B., Gries, R., Martins, A., VanLaerhoven, S. and Gries, G. (2014). Bimodal cue complex signifies suitable oviposition sites to gravid females of the common green bottle fly. *Entomologia Experimentalis et Applicata*,153(2):114-127.

Brown, A. E., Riddick, E. W., Aldrich, J. R and Holmes, W. E. (2006). Identification of (-)- beta - caryophyllene as a gender-specific terpene produced by the multicolored Asian lady beetle. *Journal of Chemical Ecology*, 32(11): 2489-2499.

Bruce, T. J., Cork, A., Hall, D. R. and Dunkelblum, E. (2002). Laboratory and field evaluation of floral odours from African marigold, *Tagetes erecta,* and sweet pea, *Lathyrus odoratus*, as kairomones for the cotton bollworm *Helicoverpa armigera*. *Bulletin OILB/SROP*. 25(9): 315-322.

Buda, V. and Radziute, S. (2008). Kairomone attractant for the leafmining fly, *Liriomyza bryoniae* (Diptera, Agromyzidae). *Zeitschrift für Naturforschung. Section C, Biosciences*.63:7/ 8, 615-618.

Buttery, R. G., Kamm, J. A. and Ling, L. C. (1984). Volatile components of red clover leaves, flowers, and seed pods: possible insect attractants. *Journal of Agricultural and Food Chemistry,* 32(2) : 254-256.

Calumpang, S. M. F., Navasero, M. M., Burgonio, G. A. S. and Navasero, M. V. (2014). Repellency of volatile organic chemicals of kakawate, *Gliricidia sepium* (Jacq.) Walp., to rice green leafhopper, *Nephotettix virescens* (Distant) (Hemiptera: Cicadellidae). *Phillipine Agriculture Scientist*, 97(2): 148-154.

Camacho, A. D., Pierce Jr, H. D. and Borden, J. H. (1994). Aggregation pheromones in *Dryocoetes affaber* (Mann.) (Coleoptera: Scolytidae): stereoisomerism and species specificity. *Journal of Chemical Ecology*, 20(1), 111-124.

Campbell, C. A. M., Dawson, G. W., Griffiths, D. C., Pettersson, J., Pickett, J. A., Wadhams, L. J. and Woodcock, C. M. (1990). Sex attractant pheromone of damson-hop aphid *Phorodon humuli* (Homoptera, Aphididae). *Journal of Chemical Ecology*, 16(12): 3455-3465.

Campbell, C. A. M., Pettersson, J., Pickett, J. A., Wadhams, L. J. and Woodcock, C. M. (1993). Spring migration of damson-hop aphid, *Phorodon humuli* (Homoptera: Aphididae), and summer host plant-derived semiochemicals released on feeding. *Journal of Chemical Ecology*, 19 (7) : 1569-1576.

Carde, R. T. and Bell, W. J. (1995). *Chemical Ecology of Insects*. 2. Chapman & Hall, 433pp.

Carde, R. T. and Miller, (2004). *Advances in Insect Chemical Ecology*. Cambridge University press. 341 pp.

Cattaneo, A. M., Bengtsson, J. M., Borgonovo, G., Bassoli, A. and Anfora, G. (2014). Response of the European grapevine moth *Lobesia botrana* to somatosensory-active volatiles emitted by the non-host plant *Perilla frutescens. Physiological Entomology*, 39(3): 229-236.

Cha, D. H., Adams, T., Werle, C. T., Sampson, B. J., Adamczyk, J. J., Jr., Rogg, H. and Landolt, P. J. A. (2014). Four-component synthetic attractant for *Drosophila suzukii* (Diptera: Drosophilidae) isolated from fermented bait headspace. *Pest Management Science,* 70(2): 324-331.

Cha, D. H., Yee, W. L., Goughnour, R. B., Sim, S. B., Powell, T. H. Q., Feder, J. L. and Linn, C. E. (2012). Identification of host fruit volatiles from domestic apple (*Malus domestica*), native black hawthorn (*Crataegus douglasii*) and introduced ornamental hawthorn (*C. monogyna*) attractive to *Rhagoletis pomonella* flies from the western United States. *Journal of Chemical Ecology*, 38(3): 319-329.

Chamberlain, K., Khan, Z. R., Pickett, J. A., Toshova, T and Wadhams, L. J. (2006). Diel periodicity in the production of green leaf volatiles by wild and cultivated host plants of stemborer moths, *Chilo partellus* and *Busseola fusca. Journal of Chemical Ecology*, 32(3): 565-577.

Chambers, J., Morgan, C. P., White, P. R., Mori, K., Finnegan, D. E., and Pinniger, D. B. (1990). Rust-red grain beetle, *Cryptolestes ferrugineus*, and flat grain beetle, *Cryptolestes pusillus:* antennal and behavioral responses to synthetic components of their aggregation pheromones. *Journal of Chemical Ecology, 16*(12), 3353-3372.

Chambers, J., Wyk, C. B van., White, P. R., Gerrard, C. M. and Mori, K. (1996). Grain weevil, *Sitophilus granarius* (L.): antennal and behavioral responses to male-produced volatiles. *Journal of Chemical Ecology*, 22(9): 1639-1654.

Chelvi, C.T., Kandasamy, R. and Singh, J.P (2010). Case study on the use of pheromone technology for the control of shoot borer and internode borer in sugarcane. *Cooperative Sugar*, 41(7): 61-64.

Chen, L. and Fadamiro, H. Y. (2007). Behavioral and electroantennogram responses of phorid fly *Pseudacteon tricuspis* (Diptera: Phoridae) to red imported fire ant *Solenopsis invicta* odor and trail pheromone. *Journal of Insect Behavior*, 20(2): 267-287.

Chen GuoFa., Zhang QingHe., Wang YanJun., Liu GuangTian and Bai YongAn. (2009). Aggregation pheromone of bark beetle *Ips duplicatus*. *Journal of Northeast Forestry University*, 37(7): 96-98

Chinta, S., Dickens, J. C. and Aldrich, J. R. (1994). Olfactory reception of potential pheromones and plant odors by tarnished plant bug, *Lygus lineolaris* (Hemiptera: Miridae*). Journal of Chemical Ecology* , 20 (12) : 3251-3267.

Chiranjeevi, C. and Reddy, D. D. R. (2003). Evaluation of an integrated pest management practice for sweetpotato weevil *Cylas formicarius* (F) in Andhra Pradesh. *Journal of Research ANGRAU*, 31: 22–28.

Cosse, A. A., Bartelt, R. J., Zilkowski, B. W., Bean, D. W. and Petroski, R. J. (2005). The aggregation pheromone of *Diorhabda elongata*, a biological control agent of saltcedar (*Tamarix* spp.): identification of two behaviorally active components. *Journal of Chemical Ecology*, 31(3): 657-670.

Crook, D. J., Lance, D. R. and Mastro, V. C. (2014). Identification of a potential third component of the male-produced pheromone of *Anoplophora glabripennis* and its effect on behavior. *Journal of Chemical Ecology*, 40 (11/12):1241-1250.

Cross, J. V., Hall, D. R., Innocenzi, P. J. and & Burgress, C. M. (2004). Exploiting sex aggregation pheromone of strawberry blossom weevil *Anthonomu rubi*. *Bulletin OILB/SROP* 27(4): 125-132.

Cruz Lopez, L., Aguilar, S., Malo, E. A., Rincon, M., Guzman, M. and Rojas, J. C. (2007). Electroantennogram and behavioral responses of workers of the stingless bee *Oxytrigona mediorufa* to mandibular gland volatiles. *Entomologia Experimentalis et Applicata*, 123(1): 43-47.

Dagg, J. and Scheurer, S. (1998). Observations on some patterns of the males' sexual behaviour of certain aphid species indicate the existence of male marking pheromones. Aphids in natural and managed ecosystems. *Proceedings of the Fifth International Symposium on Aphids*, Leon, Spain, 15-19-September, 1997. 167-171.

Dallara, P. L., Seybold, S. J., Meyer, H., Tolasch, T., Francke, W. and Wood, D. L. (2000). Semiochemicals from three species of *Pityophthorus* (Coleoptera: Scolytidae): Identification and field response. *Canadian Entomologist*, 132(6): 889-906.

David, H., Nesbitt, B. F., Easwaramoorthy, S. and Nandagopal, V. (1985). Application of sex pheromones in sugarcane pest management. *Proceedings of the Indian Academy of Sciences, Animal Sciences*, 94(3): 333-339.

Deng JianYu., Huang YongPing., Wei HongYi. and Du JiaWei. (2004). EAG and behavioral responses of *Helicoverpa armigera* males to volatiles from poplar leaves and their combinations with sex pheromone. *Journal of Zhejiang University Science*, 5(12): 1577-1582.

Dicke, M. and Takken, W. (Eds) (2006). *Chemical ecology: From gene to ecosystem*. Springer Netherlands, 192pp.

Dickens, J. C. (2002). Behavioural responses of larvae of Colorado potato beetle, *Leptinotarsa decemlineata* (Coleoptera: Chrysomelidae), to host plant volatile blends attractive to adults. *Agricultural and Forest Entomology*, 4(4): 309-314.

Dickens, J. C., Jang, E. B., Light, D. M. and Alford, A. R. (1990). Enhancement of insect pheromone responses by green leaf volatiles. *Naturwissenschaften* , 77(1): 29-31.

Ding HongJian., Guo YuYuan and Wu CaiHong. (1997). Isolation and identification of semiochemicals from carrot flower and behavioral responses in cotton bollworm moths. *Acta Entomologica Sinica,* 40(Supp): 73-78.

Drijfhout, F. P., Groot, A .T., Posthumus, M.A., Beek, T. A. van and Groot, A .de. (2002). Coupled gas chromatographic-electroantennographic responses of *Lygocoris pabulinus* (L.) to female and male produced volatiles. *Chemoecology.* 12(2): 113-118.

Duarte, A. G., Lima, I. S. de., Navarro, D. M do. A. F and Sant'-ana, A. E. G. (2003). Capture of *Rhynchophorus palmarum* L. (Coleoptera: Curculionidae) in traps baited with aggregation pheromone and volatile compounds from pineapple. *Revista Brasileira de Fruticultura*, 25(1): 81-84.

Dube, F. F., Tadesse, K., Birgersson, G., Seyoum, E., Tekie, H., Ignell, R. and Hill, S. R. (2011). Fresh, dried or smoked? Repellent properties of volatiles emitted from ethnomedicinal plant leaves against malaria and yellow fever vectors in Ethiopia. *Malaria Journal*, 10: 375.

Dusenbury, D.B. (1992). *Sensory ecology. How Organisms Acquire and Respond to Information?* W.H.Freeman, New York.

Eisner, T. and Meinwald, J. (1995). *Chemical ecology: The chemistry of biotic interactions*. National Academy Press. Washington 221 pp.

Ejima, A., Smith, B. P. C., Lucas, C., Naters, W. van der G. van, Miller, C. J.,Carlson, J. R., Levine, J. D. and Griffith, L. C. (2007). Generalization of courtship learning in *Drosophila* is mediated by *cis*-vaccenyl acetate. *Current Biology*,17:7, 599-605.

Eller, F. J. and Bartelt, R. J. (1996). Grandisoic acid, a male-produced aggregation pheromone from the plum curculio, *Conotrachelus nenuphar*. *Journal of Natural Products*, 59(4): 451-453.

El Sayed, A. M., Suckling, D. M., Wearing, C. H. and Byers, J. A. (2006). Potential of mass trapping for longterm pest management and eradication of invasive species. *Journal of Economic Entomology*, 99(5): 1550-1564.

Emery, S. and Schmid, A. (2001). Control of grape moth in areas with a high initial population: mating disruption combined with a growth regulator treatment. *Revue Suisse de Viticulture, Arboriculture et Horticulture*, 33(2):101-105.

Endo, N., Wada, T., Mizutani, N., Moriya, S. and Sasaki, R. (2005). Ambiguous response of *Riptortus clavatus* (Heteroptera: Alydidae) to different blends of its aggregation pheromone components. *Applied Entomology and Zoology*, 40(1): 41-45.

Erbilgin, N., Gillette, N. E., Mori, S. R., Stein, J. D., Owen, D. R. and Wood, D. L. (2007). Acetophenone as an anti-attractant for the western pine beetle, *Dendroctonus brevicomis* LeConte (Coleoptera: Scolytidae). *Journal of Chemical Ecology,* 33(4):817-823.

Fierke, M. K., Skabeikis, D. D., Millar, J. G., Teale, S. A., McElfresh, J. S. and Hanks, L. M. (2012). Identification of a male-produced aggregation pheromone for *Monochamus scutellatus scutellatus* and an attractant for the congener *Monochamus notatus* (Coleoptera: Cerambycidae). *Journal of Economic Entomology*, 105(6): 2029-2034.

Fouad, H. A., Faroni, L. R. D., Vilela, E. F. and Lima, E. R. de. (2013). Flight responses of *Sitotroga cerealella* (Lepidoptera: Gelechiidae) to corn kernel volatiles in a wind tunnel. *Arthropod - Plant Interactions*, 7(6):651-658.

Fujiwara Tsujii, N., Yamagata, N., Takeda, T., Mizunami, M. and Yamaoka, R. (2006). Behavioral responses to the alarm pheromone of the ant *Camponotus obscuripes* (Hymenoptera: Formicidae). *Zoological Science*, 23(4): 353-358.

Gamboa, G. J., Grudzien, T. A., Espelie, K. E. and Bura, E.A. (1996). Kin recognition pheromones in social wasps: combining chemical and behavioural evidence. *Animal Behaviour*, 51(3): 625-629.

Ganeshaiah, K. N. and Kumar, A. R. V., (1993). Self organization and chemically mediated aggregation of adults in the whitegrub, *Holotichia serrata* . In T. N. Ananthakrishnan and A. Raman (Ed.) *Chemical Ecology of Phytophagous Insects*. New Delhi: Oxford & IBH Publishing Co. pp. 167 – 178.

Gblin Davis, R. M., Weissling, T. J., Oehlschlager, A. C., and Gonzales, L. M. (1994). Field response of *Rhynchophorus cruentatus* F. (Coleoptera: Curculionidae) to its aggregation pheromone and fermenting plant volatiles. *Florida Entomologist,* 77:164-177.

Ghirardo, A., Heller, W., Fladung, M., Schnitzler, J. P. and Schroeder, H. (2012). Function of defensive volatiles in pedunculate oak (*Quercus robur*) is tricked by the moth *Tortrix viridana*. *Plant, Cell and Environment*, 35(12): 2192-2207.

Giblin Davis, R. M., Pena, J. E., Oehlschlager, A. C. and Perez, A. L. (1996). Optimization of semiochemical-based trapping of *Metamasius hemipterus sericeus* (Olivier) (Coleoptera: Curculionidae). *Journal of Chemical Ecology*, 22(8): 1389-1410.

Giblin-Davis, R. M., Oehlschlager, A. C., Perez, A. L., Gries, G., Gries, R., Weissling, T. J., Chinchilla, C. M., Pen˜a, J. E., Hallett, R. H., Pierce Jr, H. D., Gonzalez, L. M., (1996). Chemical and behavioral ecology of palm weevils (Curculionidae: Rhynchophorinae). *Florida Entomologist,* 79: 153–167.

Gomez, J., Barrera, J. F., Rojas, J. C., Macias Samano, J., Liedo, J. P., Cruz Lopez, L and Badii, M. H. (2005). Volatile compounds released by disturbed females of *Cephalonomia stephanoderis* (Hymenoptera: Bethylidae): a parasitoid of the coffee berry borer *Hypothenemus hampei* (Coleoptera: Scolytidae). *Florida Entomologist*, 88(2): 180-187.

Gough, A. J. E., Games, D. E., Staddon, B. W. and Olagbemiro, T. O. (1985). Male produced volatiles from coreid bug *Leptoglossus australis* (Heteroptera). *Zeitschrift für Naturforschung*, 40(1/2) : 142-144.

Gouinguené, S. P. and Städler, E.(2006). Oviposition in *Delia platura* (Diptera, Anthomyiidae): The role of volatile and contact cues of bean. *Journal of Chemical Ecology*, 32(7): 1399-1413.

Gouinguené, S. P., Buser, H. R. and Städler, E. (2005). Host-plant leaf surface compounds influencing oviposition in *Delia antiqua*. *Chemoecology*, 15(4): 243-249.

Greany, P. D., Styer, S. C., Davis, P. L., Shaw, P. E. and Chambers, D. L. (1983). Biochemical resistance of citrus to fruit flies. Demonstration and elucidation of resistance to the Caribbean fruit fly, *Anastrepha suspensa*. *Entomologia Experimentalis et Applicata*, 34(1): 40-50.

Green, P. W. C., Hamilton, M. A., Sanchez, M. D., Corcoran, M. R., Manco, B. N. and Malumphy, C. P. (2015). The scope for using the volatile profiles of *Pinus caribaea* var. *bahamensis* as indicators of susceptibility to pine tortoise scale and as predictors of environmental stresses. *Verlag Helvetica Chimica Acta AG, Zürich, Switzerland, Chemistry & Biodiversity,* 12(4): 652-661.

Gregg, P. C., Socorro, A. P. del and Henderson, G. S. (2010). Development of a synthetic plant volatile-based attracticide for female noctuid moths. II. Bioassays of synthetic plant volatiles as attractants for the adults of the cotton bollworm, *Helicoverpa armigera* (Hubner) (Lepidoptera: Noctuidae). *Australian Journal of Entomology*, 49(1): 21-30.

Gries, G. (1992). Ratios of geometrical and optical isomers of pheromones: irrelevant or important in scolytids?. *Journal of Applied Entomology*, 114(1 5): 240-243.

Gries, G., Gries, R., Perez, A.L., Gonzales, L.M., Pierce, H.D. Jr, Oehlschlager, A.C., Rhainds, M., Zebeyou, M. and Kouame, B., (1994). Ethyl propionate: synergistic kairomone for African palm weevil, *Rhynchophorus phoenicis* L. (Coleoptera: Curculionidae). *Journal of Chemical Ecology,* 20: 889–897.

Groot, A. T., Drijfhout, F. P., Heijboer, A., Beek, T. A van and Visser, J. H. (2001). Disruption of sexual communication in the mirid bug *Lygocoris pabulinus* by hexyl butanoate. *Agricultural and Forest Entomology*, 3(1): 49-55.

Groot, A. T., Timmer, R., Gort, G., Lelyveld, G .P., Drijfhout, F. P., Beek, T. A. and Visser, J. H. (1999). Sex-related perception of insect and plant volatiles in *Lygocoris pabulinus*. *Journal of Chemical Ecology,* 25(10): 2357-2371.

Guerin, P. M., Remund, U., Boller, E. F., Katsoyannos, B., Delrio, G. and Cavalloro, R. (1983). Fruit fly electroantennogram and behaviour responses to some generally occurring fruit volatiles. Fruit flies of economic importance. *Proceedings of the CEC/IOBC International Symposium*, Athens, Greece, 16-19 November 1982. 248-251.

Gunatilake, R. and Gunawardena, N. E. (1986). Ethyl alcohol: A major attractant of red weevil *(Rhynchophorus ferrugineus). In : Proc Sri Lanka Association for the Advancement of Science. 42nd Annual Sessions*, p. 70.

Gunawardena, N. E. and Bandarage, U. K. (1995). 4-Methyl-5-nonanol (ferrugineol) as an aggregation pheromone of the coconut pest, *Rhynchophorus ferrugineus* F. (Coleoptera: Curculionidae): synthesis and use in a preliminary field assay. *Journal of the National Science Council of Sri Lanka*, 23(2): 71-79.

Gunawardena, N. E. and Herath, H. M. W. K. B. (1995). Enhancement of the activity of ferrugineol by N-pentanol in an attractant baited trap for the coconut pest, *Rhynchophorus ferrugineus* F. (Coleoptera: Curculionidae). *Journal of the National Science Council of Sri Lanka,* 23(2): 81-86.

Gunawardena, N. E. (1992). Convenient synthesis of (E)-11-hexadecenyl acetate, the female sex pheromone of the brinjal moth *Leucinodes orbonalis* Guenee. *Journal of the National Science Council of Sri Lanka,* 20(1): 71-80.

Gunawardena, N. E., Attygalle, A. B and Herath, H. M. W. K. B. (1989). The sex pheromone of the brinjal pest, *Leucinodes orbonalis* Guenee (Lepidoptera): problems and perspectives. *Journal of the National Science Council of Sri Lanka,* 17(2): 161-171.

Guo XianRu., Lu ShaoHui., Fan CaiLing., Yuan GuoHui and Ma JiSheng. (2001). Electrophysiological and behavioral response of *Helicoverpa armigera* (Lepidoptera: Noctuidae) to volatiles from withered leaves of black poplar. *Acta Agriculturae Boreali Sinica*, 16 (special issue): 22-27.

Haak, U., Holldobler, B., Bestmann, H. J. and Kern, F. (1996). Species-specificity in trail pheromones and Dufour's gland contents of *Camponotus atriceps* and C. *floridanus* (Hymenoptera: Formicidae). *Chemoecology*, 7(2): 85-93.

Hallet, R. H., Gries, G., Gries, R., Borden, J. H., Czyzewska, E., Oehlschlager, A. C.,Pierce, H. D., JR., Angerilli, N. P. D. and Rauf, A. 1993. Aggregation pheromones of two Asian palm weevils, *Rhynchophorus ferrugineus* and *R. vulneratus. Naturwissenschaften,* 80:328-331.

Hallet, R. H., Perez, A. L., Gries, G., Gries, R., Pierce, H. D. Jr. Yue-Junming, Oehlschlager, A. C., Gonzalez, L. M., Borden, J. H. and Yue, J. M. (1995). Aggregation pheromone on coconut rhinoceros beetle, *Oryctes rhinoceros* L. (Coleoptera : Scaraebidae). *Journal of Chemical Ecology,* 21 (10):1549-1570.

Hamilton, J. G. C., Hall, D. R and Kirk, W. D. J. (2005). Identification of a male-produced aggregation pheromone in the western flower thrips *Frankliniella occidentalis. Journal of Chemical Ecology*, 31(6): 1369-1379.

Hammack, L. (1997). Attractiveness of synthetic corn volatiles to feral northern and western corn rootworm beetles (Coleoptera: Chrysomelidae). *Environmental entomology*, 26(2): 311-317.

Hammack, L. (2003). Volatile semiochemical impact on trapping and distribution in maize of northern and western corn rootworm beetles (Coleoptera: Chrysomelidae). *Agricultural and Forest Entomology*, 5(2): 113-122.

Han BaoYu., Wang MengXin., Zheng YingCha., Niu YuQun., Pan Cheng., Cui Lin., Chauhan, K. R. and Zhang, Q. H. (2014). Sex pheromone of the tea aphid, *Toxoptera aurantii* (Boyer de Fonscolombe) (Hemiptera: Aphididae)., *Chemoecology*, 24 (5): 179-187.

Harmel, N., Almohamad, R., Fauconnier, M. L., Jardin, P. du., Verheggen, F., Marlier, M., Haubruge, E. and Francis, F.(2007). Role of terpenes from aphid-infested potato on searching and oviposition behavior of *Episyrphus balteatus. Insect Science*, 14(1): 57-63.

Harrison, S. J. and McEwen, P. K. (1998). Acid hydrolyzed L-tryptophan and its role in the attraction of green lace wing *Chrysoperla carnea* (Stephens) (Neuroptera: Chrysopidae). *Journal of Applied Entomology*, 122: 343-344.

Heath, R. R., Teal, P. E. A., Epsky, N. D., Dueben, B. D., Hight, S. D.; Bloem, S., Carpenter, J. E., Weissling, T. J., Kendra, P. E., Cibrian-Tovar, J. and Bloem, K. A. (2006). Pheromone-based attractant for males of *Cactoblastis cactorum* (Lepidoptera: Pyralidae). *Environmental Entomology*, 35(6): 1469-1476.

Hedin, P. A., Phillips, V. A. and Dysart, R. J. (1991). Volatile constituents from honeysuckle aphids, *Hyadaphis tataricae,* and the honeysuckle, *Lonicera* spp.: Search for assembling pheromones. *Journal of Agricultural and Food Chemistry,* 39 (7) : 1304-1306.

Hillery, A. E. and Fell, R. D. (2000). Chemistry and behavioral significance of rectal and accessory gland contents in *Camponotus pennsylvanicus* (Hymenoptera: Formicidae). *Annals of the Entomological Society of America*, 93(6): 1294-1299.

Hodges, R. J., Addo, S., Farman, D. I. and Hall, D. R. (2004). Optimising pheromone lures and trapping methodology for *Prostephanus truncatus* (Horn) (Coleoptera: Bostrichidae). *Journal of Stored Products Research*, 40(4): 439-449.

Holldobler, B., Oldham, N. J., Alpert, G. D. and Liebig, J. (2002). Predatory behavior and chemical communication in two *Metapone* species (Hymenoptera:Formicidae). *Chemoecology*, 12(3): 147-151.

Hori, M.; Komatsu, H. (1997). Repellency of rosemary oil and its components against the onion aphid, *Neotoxoptera formosana* (Takahashi) (Homoptera, Aphididae). *Applied Entomology and Zoology*, 32 (2) : 303-310.

Hua, Y., Hong-Wei, W., Wei, Y., and Chun-Ping, Y. (2015). Electrophysiological and behavioral responses of *Tomicus minor* (Coleoptera: Scolytidae) to host volatiles. *Entomologica Fennica*, 26(1), 15-24.

Huber,R.T. and Hoffmann, M. P. (1979). Development and evaluation of an oil trap for use in pink bollworm pheromone mass trapping and monitoring programs. *Journal of Economic Entomology,* 72(5): 695-697.

Hu-FuLiang and Xuan-HongZhuan. (2004). The recent advances in honeybee queen pheromone. *Entomological Knowledge*, 41(3): 208-211.

Hughes, W O. H., Howse, P. E., Vilela, E. F.. Knapp, J. J. and Goulson,D. (2002). Field evaluation of potential of alarm pheromone compounds to enhance baits for control of grass-cutting ants (Hymenoptera: Formicidae). *Journal of Economic Entomology*, 95(3): 537-543.

Hughes, W. O. H. and Goulson, D. (2001). Polyethism and the importance of context in the alarm reaction of the grass-cutting ant, *Atta capiguara*. *Behavioral Ecology and Sociobiology,* 49(6): 503-508.

Huh HyeSoon., Jang SinAe and Park ChungGyoo. (2009). Variation in aggregation pheromone secretion of bean bug, *Riptortus clavatus*. *Korean Journal of Applied Entomology*, 48(1): 73-79.

Huh Wan and Park ChungGyoo. (2005). Seasonal occurrence and attraction of egg parasitoid of bugs, *Ooencyrtus nezarae*, to aggregation pheromone of bean bug, *Riptortus clavatus*. *Korean Journal of Applied Entomology*. 44(2): 131-137.

Ichiki, R. T., Kainoh, Y., Kugimiya, S., Takabayashi, J. and Nakamura, S. (2008). Attraction to herbivore-induced plant volatiles by the host-foraging parasitoid fly *Exorista japonica*. *Journal of Chemical Ecology*, 34(5): 614-621.

Innocenzi, P. J., Hall, D. R., Cross, J. V., Masuh, H., Phythian, S. J., Chittamaru, S. and Guarino, S. (2004). Investigation of long-range female sex pheromone of the European tarnished plant bug, *Lygus rugulipennis*: Chemical , electrophysiological, and field studies. *Journal of Chemical Ecology,* 30(8): 1509-1529.

Innocenzi, P. J., Hall, D. R., Sumathi, C., Cross, J. V. and Jacobson, R. (1998). Studies of the sex pheromone of the European tarnished plant bug, *Lygus rugulipennis* (Heteroptera: Miridae). *Brighton Crop Protection Conference: Pests and Diseases 1998: Volume 3: Proceedings of an International Conference, Brighton, UK, 16-19-November-1998*. 829-832.

Innocenzi, P. J., Hall, D. R. and Cross, J. V. (2001). Components of male aggregation pheromone of strawberry blossom weevil, *Anthonomus rubi* Herbst. (Coleoptera: Curculionidae). *Journal of Chemical Ecology*, 27(6): 1203-1218.

Jactel, H., Van Halder, I., Menassieu, P., Zhang, Q. H. and Schlyter, F. (2001). Non-host volatiles disrupt the response of the stenographer bark beetle, *Ips sexdentatus* (Coleoptera: Scolytidae), to pheromone-baited traps and maritime pine logs. *Integrated Pest Management Reviews*, 6(3-4): 197-207.

Jaffe´, K., Sanchez, P., Cerda, H., Hernandez, J.V., Jaffe, R., Urdaneta, N., Guerra, G., Martinez, R. and Miras, B., (1993). Chemical ecology of the palm weevil *Rhynchophorus palmarum* (L.) (Coleoptera: Curculionidae): Attraction to host plants and a male produced aggregation pheromone. *Journal of Chemical Ecology,* 19: 1703– 1720.

James, D. G. (2003). Synthetic herbivore-induced plant volatiles as field attractants for beneficial insects. *Environmental Entomology*, 32(5): 977-982.

James, D. G., Bartelt, R. J and Moore, C. J. (1996). Mass-trapping of *Carpophilus* spp. (Coleoptera: Nitidulidae) in stone fruit orchards using synthetic aggregation pheromones and a coattractant: development of a strategy for population suppression. *Journal of Chemical Ecology*, 22(8): 1541-1556.

James, D. G., Mori, K., Aldrich, J. R and Oliver, J. E. (1994). Flight-mediated attraction of *Biprorulus bibax* Breddin (Hemiptera: Pentatomidae) to natural and synthetic aggregation pheromone. *Journal of Chemical Ecology*, 20(1): 71-80.

Janssen, E., Bestmann, H. J., Holldobler, B. and Kern, F. (1995). N,N-dimethyluracil and actinidine, two pheromones of the ponerine ant *Megaponera foetens* (Fab.) (Hymenoptera: Formicidae). *Journal of Chemical Ecology*, 21(12): 1947-1955.

Janssen, E., Holldobler, B. and Bestmann, H. J. (1999). A trail pheromone component of the African stink ant, *Pachycondyla* (*Paltothyreus*) *tarsata* Fabricius (Hymenoptera: Formicidae: Ponerinae). *Chemoecology*, 9(1): 9-11.

Jansson, R. K., Hunsberger, A. G. B., Lecrone, S. H. and O'Hair,S. K. (1990). Seasonal abundance, population growth, and within plant distribution of sweetpotato weevil (Coleoptera: Curculionidae) on sweetpotato in southern Florida. *Environmental Entomology*, 19: 313–321.

Jarau, S., Schulz, C. M., Hrncir, M., Francke, W., Zucchi, R., Barth, F. G. and Ayasse, M. (2006). Hexyl decanoate, the first trail pheromone compound identified in a stingless bee, *Trigona recursa*. *Journal of Chemical Ecology*, 32(7): 1555-1564.

Jayarama., D'Souza, M. V. and Hall D. R. (2007). (S)-2-hydroxy-3-decanone as sex pheromone for monitoring and control of coffee white stem borer, (*Xylotrechus quadripes*). *21st International Conference on coffee Science.* Montpellier, Frace, 11-15 September 2006, 1291-1300.

Jian, Chen., Henderson, G and Laine, R. A. (1998). Isolation and identification of 2-phenoxyethanol from a ballpoint pen ink as a trail-following substance of *Coptotermes formosanus* Shiraki and *Reticulitermes* sp. *Journal of Entomological Science*, 33(1): 97-105.

Jin Ju., Ruan ZanYu., Huang ZhenFu., Lai GuiYan., Huang SongSong and Fan XiaoLing. (2015). Attractions of volatiles from wax-apple fruit to the oriental fruit fly. *Journal of South China Agricultural University*. 36 (3): 71-77.

Jumean, Z., Rowland, E., Judd, G. J. R. and Gries, G. (2004). Male and female *Cydia pomonella* (Lepidoptera: Olethreutidae) larvae produce and respond to aggregation pheromone. *Canadian Entomologist*, 136(6): 871-873.

Kalshoven, L.G. E. (1950). *Pests of crops in Indonesia*. Translated by van der Laan (1981) 701.

Kamezawa, M., Tachibana, H., Ohtani, T. and Naoshima, Y. (1994). Biocatalytic synthesis of (S)-2-tridecanyl acetate and (S)-2-pentadecanyl acetate, aggregation pheromone components of *Drosophila mulleri* and *D. busckii*, by enantioselective hydrolysis with lipase. *Journal of Chemical Ecology*, 20(5): 1057-1061.

Karlsson, M. F., Birgersson, G., Cortes Prado, A. M., Bosa, F., Bengtsson, M. and Witzgall, P. (2009). Plant odor analysis of potato: response of Guatemalan moth to above- and below ground potato volatiles. *Journal of Agricultural and Food Chemistry*, 57 (13): 5903-5909.

Kashiwagi, T., Nakashima, T., Tebayashi, S. and Kim ChulSa. (2006). Determination of the absolute configuration of quercivorol, (1S,4R)-p-menth-2-en-1-ol, an aggregation pheromone of the ambrosia beetle *Platypus quercivorus* (Coleoptera: Platypodidae). *Bioscience, Biotechnology and Biochemistry*, 70(10): 2544-2546.

Keesey, I. W., Knaden, M. and Hansson, B. S. (2015). Olfactory specialization in *Drosophila suzukii* supports an ecological shift in host preference from rotten to fresh fruit. *Journal of Chemical Ecology*, 41(2): 121-128.

Kern, F., Klein, R. W., Janssen, E., Bestmann, H. J., Attygalle, A. B., Schafer, D. and Maschwitz, U. (1997). Mellein, a trail pheromone component of the ants *Lasius fuliginosus*. *Journal of Chemical Ecology*, 23(3): 779-792.

Khoo, C. C. H. and Keng HongTan. (2005). Rectal gland of *Bactrocera papayae*: ultrastructure, anatomy, and sequestration of autofluorescent compounds upon methyl eugenol consumption by the male fruit fly. *Microscopy Research and Technique*, 67(5): 219-226.

Khrimian, A., Shearer, P. W., Zhang, A. J., Hamilton, G. C. and Aldrich, J. R. (2008). Field trapping of the invasive brown marmorated stink bug, *Halyomorpha halys*, with geometric isomers of methyl 2,4,6-decatrienoate. *Journal of Agricultural and Food Chemistry*, 56(1): 197-203.

Kim JunHeon., Matsuyama, S. and Suzuki, T. (2005). 4,8-Dimethyldecanal, the aggregation pheromone of *Tribolium castaneum*, is biosynthesized through the fatty acid pathway. *Journal of Chemical Ecology*, 31(6): 1381-1400.

Kimani, S. M., Chhabra, S. C., Lwande, W., Khan, Z. R., Hassanali, A. and Pickett, J. A. (2000). Airborne volatiles from *Melinis minutiflora* P. Beauv., a non-host plant of the spotted stem borer. *Journal of Essential Oil Research*, 12(2): 221-224.

Kirk, W. D. J., MacDonald, K. M., Whittaker, M. S., Hamilton, J. G. C. and Jacobson, R. (1998). The oviposition behaviour of the western flower thrips, *Frankliniella occidentalis* (Pergande). *Proceedings: Sixth International Symposium on Thysanoptera,* Akdeniz University, Antalya, Turkey, 27 April - 1 May, 1998. 69-75.

Klein, M.G., Tumlinson, J.H.,Ladd,T.U., Doolittle, R.E. (1981). Japanese beetle (Coleoptera : Scarabaeidea) response to synthetic sex attractant plus phenethyl propionate: Eugenol. *Journal of Chemical Ecology*, 7: 1 –7.

Kleineidam, C. J., Obermayer, M., Halbich, W. and Rossler, W. (2005). A macroglomerulus in the antennal lobe of leaf-cutting ant workers and its possible functional significance. *Chemical Senses*, 30(5): 383-392.

Kohl, E., Holldobler, B. and Bestmann, H. J. (2001). Trail and recruitment pheromones in *Camponotus socius* (Hymenoptera: Formicidae). *Chemoecology*, 11(2): 67-73.

Kohl, E., Holldobler, B. and Bestmann, H. J. (2003). Trail pheromones and Dufour gland contents in three *Camponotus* species (*C. castaneus*, *C. balzani*, *C. sericeiventris*: Formicidae, Hymenoptera). *Chemoecology*, 13(3): 113-122.

Kotze, M. J., Jürgens, A., Johnson, S. D., Hoffmann, J. H., Jürgens, A. and Viljoen, A. M. (2010). Volatiles associated with different flower stages and leaves of *Acacia cyclops* and their potential role as host attractants for *Dasineura dielsi* (Diptera: Cecidomyiidae). *South African Journal of Botany*, 76 (4): 701-709.

Krupke, C. H. and Brunner, J. F. (2003). Parasitoids of the consperse stink bug (Hemiptera: Pentatomidae) in North Central Washington and attractiveness of a host-produced pheromone component. *Journal of Entomological Science*, 38(1): 84-92.

Kuhar, T. P., Mori, K. and Dickens, J. C. (2006). Potential of a synthetic aggregation pheromone for integrated pest management of Colorado potato beetle. *Agricultural and Forest Entomology,* 8(1): 77-81.

Kurian, C., Abraham,V. A. and Ponnama, K. A. (1984). Attractants – an aid in red palm weevil management. *In: Proc : PLACROSYM V,Dec. 15 -18, Kasaragod, India.*

Kvedaras, O. L., Socorro, A. P.del. and Gregg, P. C. (2007). Effects of phenylacetaldehyde and (Z)-3-hexenyl acetate on male response to synthetic sex pheromone in *Helicoverpa armigera* (Hubner) (Lepidoptera: Noctuidae). *Australian Journal of Entomology*, 46(3): 224-230.

Lacey, E. S., Ginzel, M. D., Millar, J. G. and Hanks, L. M. (2004). Male-produced aggregation pheromone of the cerambycid beetle *Neoclytus acuminatus acuminatus*. *Journal of Chemical Ecology*, 30(8): 1493-1507.

Lacey, E. S., Moreira, J. A., Millar, J. G., Ray, A. M and Hanks, L. M. (2007). Male-produced aggregation pheromone of the cerambycid beetle *Neoclytus mucronatus* mucronatus. *Entomologia Experimentalis et Applicata*, 122(2): 171-179.

Le Conte, Y. R., Mohammadi, A., and Robinson G. E. (2001). Primer effect of brood pheromone on honey bee behavioural development. *Proceedings of Royal Society : Biological Sciences* 268 (1463): 163-168.

Leal, W. S., Yadava, C. P. S. and Vijayvergia, J. N., (1996). Aggregation of scarab beetle *Holotrichia consanguinea* in response to female released pheromone suggests secondary function hypothesis for semiochemical. *Journal of Chemical Ecology,* 22**:** 1557 -1566.

Lebedeva, K. V., Vendilo, N. V., Mitroshin, D. B., Pletnev, V. A., Maslov, A. D., Komarova, I. A., Matusevich, L. S., Zharkov,.D.G., Kobel' kov, M. E., Kuchin, A. V., Khairetdinov, R. R. and Kukharkin, D. V. (2003). Use of Verbenol, a pheromone of *Ips typographus*, for spruce protection in the Moscow region. *Lesnoe Khozyaistvo*, (1): 33-35.

Lee KyuChul., Kang ChangHeon., Lee DongWoon., Lee SangMyeong., Park ChungGyoo and Choo HoYul. (2002). Seasonal occurrence trends of hemipteran bug pests monitored by mercury light and aggregation pheromone traps in sweet persimmon orchards. *Korean Journal of Applied Entomology*, 41(4): 233-238.

Leppik, E. and Frérot, B. (2014). Maize field odorscape during the oviposition flight of the European corn borer. *Chemoecology*, 24 (6): 221-228.

Leskey, T. C. and Hogmire, H. W. (2005). Monitoring stink bugs (Hemiptera: Pentatomidae) in mid-Atlantic apple and peach orchards. *Journal of Economic Entomology*. 98(1): 143-153.

Lethmayer, C., Hausdorf, H. and Blumel, S (2004). The first field experiences with sex aggregation pheromones of the strawberry blossom weevil, *Anthonomus rubi*, in Austria. *Bulletin OILB/ SROP*, 27(4): 133-139.

Li PengYan., Zhu JunWei and Qin YuChuan. (2012). Enhanced attraction of *Plutella xylostella* (Lepidoptera: Plutellidae) to pheromone-baited traps with the addition of green leaf volatiles. *Journal of Economic Entomology*,105(4): 1149-1156.

Lin Jian, Liu WenBo, Meng ZhaoJun. and Yan ShanChun . (2014). *Larix gmelinii* monoterpene volatiles from slow-release micro-capsules affecting host selection behavior of *Dendrolimus superans*. *Journal of Beijing Forestry University*, 36(3): 42-47.

Liscia, A., Angioni, P., Sacchetti, P., Poddighe, S., Granchietti, A., Setzu, M. D. and Belcari, A. (2013). Characterization of olfactory sensilla of the olive fly: behavioral and electrophysiological responses to volatile organic compounds from the host plant and bacterial filtrate. *Journal of Insect Physiology*, 59(7): 705-716.

López-Guillén, G., Cruz López, L., Malo, E. A. and Rojas, J. C. (2011). Olfactory responses of *Anastrepha obliqua* (Diptera: Tephritidae) to volatiles emitted by calling males. *Florida Entomologist*, 94(4): 874-881.

Loughrin, J. H., Potter, D. A., and Hamilton-Kemp, T. R. (1998). Attraction of Japanese beetles (Coleoptera: Scarabaeidae) to host plant volatiles in field trapping experiments. *Environmental Entomology*, 27(2), 395-400.

Lu PengFei., Huang LingQiao. and Wang ChenZhu.(2010).Semiochemicals used in chemical communication in the oriental fruit moth,*Grapholitha molesta* Busck (Lepidoptera: Tortricidae). *Acta Entomologica Sinica*, 53(12):1390-1403.

Lucia, A., González-Audino, P., Masuh, H. and Hindawi (2014). Volatile organic compounds from the clone Populus × canadensis "Conti" associated with *Megaplatypus mutatus* attack. *Psyche: A Journal of Entomology,* pp 6.

Luxova, A., Urbanova, K., Valterova, I., Terzo, M. B. and Karlson, A. K. (2004). Absolute configuration of chiral terpenes in marking pheromones of bumble bees and cuckoo bumble bees. *Chirality*, 16(4): 228-233.

Majetic, C. J., Wiggam, S. D., Ferguson, C. J. and Raguso, R. A. (2015). Timing is everything: temporal variation in floral scent, and its connections to pollinator behavior and female reproductive success in *Phlox divaricata*. *American Midland Naturalist*, 173(2): 191-207.

Malo, E. A., Cruz-López, L., Toledo, J., Mazo, A. del, Virgen, A. and Rojas, J. C. (2005).Behavioral and electrophysiological responses of the Mexican fruit fly (Diptera: Tephritidae) to guava volatiles. *Florida Entomologist*, 88(4): 364-371.

Mattiacci, L., Vinson, S. B., Williams, H. J., Aldrich, J. R., Colazza, S. and Bin, F. (1991). Kairomones for the egg parasitoid *Trissolcus basalis* (Woll.): isolation and identification of a compound from the metathoracic glands of *Nezara viridula* L. *Redia*, 74 (3): 167-168.

McLaughlin, J .R. (1998). The status of *Lygus* pheromone research. (1998). Proceedings Beltwide Cotton Conferences, San Diego, California, USA, 5-9-January-1998-Volume-2. 1998; 938-940.

Meng, P. S., Trotter, R. T., Keena, M. A., Baker, T. C., Yan, S., Schwartzberg, E. G. and Hoover, K. (2014). Effects of pheromone and plant volatile release rates and ratios on trapping *Anoplophora glabripennis* (Coleoptera: Cerambycidae) in China. *Environmental entomology*, 43(5), 1379-1388.

Messer, C., Dettner, K ., Schulz, S. and Francke, W. (1999). Phenolic compounds in *Neanura muscorum* (Collembola, Neanuridae) and the role of 1,3-dimethoxybenzene as an alarm substance. *Pedobiologia*, 43(2): 174-182.

Micha, S. G and Wyss, U. (1996). Aphid alarm pheromone (E)- beta -farnesene: a host finding kairomone for the aphid primary parasitoid *Aphidius uzbekistanicus* (Hymenoptera: Aphidiinae). *Chemoecology*, 7(3): 132-139.

Miller, J. and Haynes, K. (1998). *Methods in Chemical Ecology Vol. 1. Chemical Methods*. Springer US 390 pp.

Ming QingLei and Lewis, S. M. (2010). Mate recognition and sex differences in cuticular hydrocarbons of the diurnal firefly *Ellychnia corrusca* (Coleoptera: Lampyridae). *Annals of the Entomological Society of America*, 103(1): 128-133.

Mizutani, N. (2006). Pheromones of male stink bugs and their attractiveness to their parasitoids. *Japanese Journal of Applied Entomology and Zoology*, 50(2): 87-99.

Mizutani, N., Wada, T., Higuchi, H., Ono, M. and Leal, W. S. (1999). Effect of the synthetic aggregation pheromone of *Riptortus clavatus* on the density of and parasitism by the egg parasitoid *Ooencyrtus nezarae* Ishii (Hymenoptera: Encyrtidae) in soyabean fields. *Japanese Journal of Applied Entomology and Zoology*, 43(4): 195-202.

Moayeri, H .R .S., Ashouri, A., Poll, L. and Enkegaard, A. (2007). Olfactory response of a predatory mirid to herbivore induced plant volatiles: multiple herbivory vs. single herbivory. *Journal of Applied Entomology.* 131(5): 326-332.

Mohammedi, A., Paris, A., Crauser, D. and Le Conte, Y. (1998). Primer and releaser brood pheromone in honeybees: New evidence on worker ovary development. *Naturwissenschaften*, 85: 455-458.

Mondor, E. B., Baird, D. S., Slessor, K. N. and Roitberg, B. D. (2000). Ontogeny of alarm pheromone secretion in pea aphid, *Acyrthosiphon pisum. Journal of Chemical Ecology*, 26(12): 2875-2882.

Morgan, E. D., Brand, J. M., Mori, K. and Keegans, S. J. (2004). The trail pheromone of the ant *Crematogaster castanea. Chemoecology*, 14(2): 119-120.

Mori, K. (2007). Synthetic studies aimed at the elucidation of the stereostructure of the aggregation pheromone, 2-methyl-6-(4'-methylenebicyclo[3.1.0]hexyl)hept-2-en-1-ol, produced by the male stink bug *Erysarcoris lewisi. Tetrahedron Asymmetry*, 18(7): 838-846.

Morin, J. P., Rochat, D., Malosse, C., Lettere, M., Chenon, R. D. de., Wibwo, H. and Descoins, C. (1996). Ethyl 4-methyloctanoate, major component of *Oryctes rhinoceros* (L.) (Coleoptera, Dynastidae) male pheromone. *Comptes Rendus de l' Academie des Sciences Serie III, Sciences de la Vie,* 319(7): 595-602.

Morris, B. D., Smyth, R. R., Foster, S. P., Hoffmann, M. P., Roelofs, W. L., Franke, S. and Francke, W. (2005). Vittatalactone, a beta -lactone from the striped cucumber beetle, *Acalymma vittatum. Journal of Natural Products*, 68(1): 26-30.

Mu, LiLi and Li, GuoQing. (2007). Recent advances in marking pheromones in predatory insects and mites. *Acta Phytophylacica Sinica*, 34(1): 96-102.

Müller-Schwarze, (2001). *Hands on Chemical Ecology. Simple field and laboratory exercises.* Springer Science & Business Media. 170pp..

Nagnan, P., Cain, A. H. and Rochat, D. (1992). Extraction and identification of volatile compounds of fermented oil palm sap (palm wine), candidate attractants for the palm weevil. *Oléagineaux* 47: 135-142.

Nair, M.R.G.K. (1986). *Insects and mites on crops of India.* ICAR, New Delhi, India 83p.

Nakashima, T., Saito, S., Kobayashi, M., Kinuura, H and Tokoro, M. (2005). A challenge to Japanese oak wilt - structure and attractivity of aggregation pheromone in *Platypus quercivorus. Aroma Research*, 6(4): 348-351.

Ndiege, I. O., Budenberg, W. J., Lwande, W., and Hassanali, A. (1991). Volatile components of banana pseudostem of a cultivar susceptible to the banana weevil. *Phytochemistry*, 30(12), 3929-3930.

Ndiege, I. O., Budenberg, W. J., Otieno, D. O., and Hassanali, A. (1996). 1, 8-cineole: An attractant for the banana weevil, *Cosmopolites sordidus. .Phytochemistry*, 42(2), 369-371.

Nehlin, G., Valterová, I. and Borg-Karlson, A. K. (1996). Monoterpenes released from Apiaceae and the egg-laying preferences of the carrot psyllid, *Trioza apicalis. Entomologia Experimentalis et Applicata*, 80 (1): 83-86.

Niemeyer, H. and Watzek, G. (1996). Experiments on monoterpenes in combination with PheropraxReg. and ChalcopraxReg. in pheromone traps for catching the bark beetles *Ips typographus* L. and *Pityogenes chalcographus* L. (Col., Scolytidae). *Anzeiger fur Schadlingskunde, Pflanzenschutz Umweltschutz*, 69 (5): 109-110.

Nihei, K., Shibata, K. and Kubo, I. (2002). (+)-2,3-dehydro-10-oxo-alpha -isosparteine in *Uresiphita reversalis* larvae fed on *Cytisus monspessulanus* leaves. *Phytochemistry*, 61(8): 987-990.

Nishida R. (2014). Chemical ecology of insect plant interactions: Ecological significance of plant secondary metabolites. *Bioscience, Biotechnology and Biochemistry* 78(1): 1-13.

Nottingham, S. F., Hardie, J., Dawson, G. W., Hick, A. J., Pickett, J. A., Wadhams, L. J. and Woodcock, C. M. (1991). Behavioural and electrophysiological responses of aphids to host and nonhost plant volatiles. *Journal of Chemical Ecology,* 17 (6) : 1231-1242.

Nufio, C. R. and Papaj, D. R. (2001). Host marking behavior in phytophagous insects and parasitoids. *Entomologia Experimentalis et Applicata,* 99(3): 273-293.

Nunez, S. (2000). Entomological outlook for IFP implementation in Uruguay. *Acta Horticulturae*, (525): 363-365.

Ochieng, S. A., Park, K. C. and Baker, T. C. (2002). Host plant volatiles synergize responses of sex pheromone-specific olfactory receptor neurons in male *Helicoverpa zea. Journal of Comparative Physiology A, Sensory, Neural and Behavioral Physiology,* 188(4): 325.

Oehlschlager, A. C., Chinchilla, C. M., Gonzalez, L. M., Jiron, L. F., Mexzon, R. G. and Morgan, B. (1993). Development of a pheromone-based trapping system for *Rhynchophorus palmarum* (Coleoptera: Curculionidae). *Journal of Economic Entomol*ogy, 86: 1381-1392.

Oehlschlager, A. C., Prior, R. N. B., Perez, A. L., Gries, R., Gries, G., Pierce, H. D. Jr. and Laup, S. (1995). Structure, chirality, and field testing of a male-produced aggregation pheromone of Asian palm weevil *Rhynchophorus bilineatus* (Montr.) (Coleoptera: Curculionidae). *Journal of Chemical Ecology*, 21(10): 1619-1629.

Ogihara, K., Munesada, K., Yamamitsu, T. and Suga, T. (1989). Fragrant and biologically active constituents of the citrus cultivar Jyabon. *Phytochemistry,* 28(4) : 1061-1067.

Ohinata, K., Jacobson, M., Nakagawa, S., Urago, T., Fujimoto, M. and Higa, H.(1979). Methyl (E)-6-nonenoate: a new Mediterranean fruit fly male attractant. *Journal of Economic Entomology*, 72(4): 648-650.

Paczkowski, S., Paczkowska, M., Dippel, S., Flematti, G. and Schütz, S. (2014). Volatile combustion products of wood attract *Acanthocnemus nigricans* (Coleoptera: Acanthocnemidae). *Journal of Insect Behavior*, 27(2): 228-238.

Padmaja, P. G., Woodcock, C. M. and Bruce, T. J. (2010). A electrophysiological and behavioral responses of sorghum shoot fly, *Atherigona soccata*, to sorghum volatiles. *Journal of Chemical Ecology*, 36(12): 1346-1353.

Palacio-Cortés, A. M., Valente, F., Saad, E. B., Tröger, A., Francke, W. and Zarbin, P. H. G. (2015). (1*R*,2*S*,6*R*)-papayanol, aggregation pheromone of the guava weevil, *Conotrachelus psidii*. *Journal of the Brazilian Chemical Society*, 26(4):784-789.

Pathiratne, A., Gunawardena, N. E. and Liyanage, S. K. J. (1991). Investigations on toxic, antifeedant and repellent properties of the defensive secretion of *Coridius janus* (Hemiptera: Pentatomidae) and a synthetic mixture of its major volatile constituents. *Journal of the National Science Council of Sri Lanka* , 19 (2): 77-90.

Patil, S. B., Udikeri, S. S., Hirekurubar, R. B. and Guruprasad, G. S. (2008). Management of pink bollworm, *Pectinophora gossypiella* (Saunders) through mass trapping in cotton. *Pesticide Research Journal,* 20(2): 210-213.

Peppuy, A., Robert, A., Semon, E., Bonnard, O., Ngo Truong Son and Bordereau, C (2001). Species specificity of trail pheromones of fungus-growing termites from northern Vietnam. *Insectes Sociaux*, 48(3): 245-250.

Phillips, T. W., Nation, J. L., Wilkinson, R. C., Foltz, J. L., Pierce, H. D. and Oehlschlager, A. C. (1990). Response specificity of *Dendroctonus terebrans* (Coleoptera: Scolytidae) to enantiomers of its sex pheromones. *Annals of the Entomological Society of America*, 83(2), 251-257.

Pierce, A. M., Pierce Jr, H. D., Borden, J. H., and Oehlschlager, A. C. (1991). Fungal volatiles: semiochemicals for stored-product beetles (Coleoptera: Cucujidae). *Journal of Chemical Ecology*, 17(3), 581-597.

Piesik, D., Kalka, I., Wenda-Piesik, A. and Bocianowski, J. (2014). *Apion miniatum* Germ. herbivory on the mossy sorrel, *Rumex confertus* Willd.: induced plant volatiles and weevil orientation responses. *Journal of Environmental Studies*, 23(6): 2149-2156.

Pillai, K. S., Rajamma, P. and Palaniswami, P. S. (1993). New technique in the control of sweetpotato weevil using synthetic sex pheromone in India. *International Journal of Pest Management,* 39: 84–89.

Piñero, J. C., Souder, S. K., Smith, T. R., Fox, A. J. and Vargas, R. I. (2015). Ammonium acetate enhances the attractiveness of a variety of protein-based baits to female *Ceratitis capitata* (Diptera: Tephritidae). *Journal of Economic Entomology,* 108(2): 694-700.

Ponnamma, K.N., Lalitha, N. and Khan, A.S (2002). Bioefficacy of the pheromone rhinolure in the integrated pest management for rhinoceros beetle, *Oryctes rhinoceros* L. a major pest of oil palm. *Proceedings of the 15th Plantation Crops Symposium Placrosym XV, Mysore, India,10-13December,*2002, 525-530.

Prokopy, R. J., Hu XingPing., Jang, E. B., Vargas, R. I. and Warthen, J. D. (1998). Attraction of mature *Ceratitis capitata* females to 2-heptanone, a component of coffee fruit odor. *Journal of Chemical Ecology*, 24 (8): 1293-1304.

Qiao HaiLi., Lu PengFei.,Chen Jun.,Ma WeiSi.,Qin RongMin. and Li XiangMing.(2012). Antennal and behavioural responses of *Heortia vitessoides* females to host plant volatiles of *Aquilaria sinensis*. *Entomologia Experimentalis et Applicata*. 143 (3): 269-279.

Ragoussis, V., Giannikopoulos, A., Skoka, E and Grivas, P. (2007). Efficient synthesis of (+or-)-4-methyloctanoic acid, aggregation pheromone of rhinoceros beetles of the genus *Oryctes* (Coleoptera: Dynastidae, Scarabaeidae). *Journal of Agricultural and Food Chemistry*, 55(13): 5050-5052.

Rajapakse, C. N. K., Walter, G. H., Moore, C. J., Hull, C. D. and Cribb, B. W. (2006). Host recognition by a polyphagous lepidopteran (*Helicoverpa armigera*): primary host plants, host produced volatiles and neurosensory stimulation. *Physiological Entomology*. 31(3): 270-277.

Ramachandran, C.P., Kurien, C., and Mathew, J. (1963). Assessment of damage to coconut due to *Oryctes rhinoceros*. Nature and damage caused by beetle and factor involved in the estimation of loss. *Indian Coconut Journal,* 17: 3-12.

Ramirez Lucas, P., Rochat, D. and Zagatti, P. (1996). Field trapping of *Metamasius hemipterus* with synthetic aggregation pheromone. *Entomologia Experimentalis et Applicata*, 80(3): 453-460.

Rao, S., Cosse, A. A., Zilkowski, B. W. and Bartelt, R. J. (2003). Aggregation pheromone of the cereal leaf beetle: field evaluation and emission from males in the laboratory. *Journal of Chemical Ecology*, 29(9): 2165-2175.

Rasgado, M. A., Malo, E. A., Cruz-López, L., Rojas, J. C. and Toledo, J. (2009). Olfactory response of the Mexican fruit fly (Diptera: Tephritidae) to *Citrus aurantium* volatiles. *Journal of Economic Entomology.* 102(2): 585-594.

Ray, A. M., Millar, J. G., McElfresh, J. S., Swift, I. P., Barbour, J. D. and Hanks, L. M. (2009). Male-produced aggregation pheromone of the cerambycid beetle *Rosalia funebris*. *Journal of Chemical Ecology*, 35(1): 96-103.

Razkin, J., Gil, P and Gonzalez, A. (1996). Stereoselective synthesis of dominicalure 1 and 2: components of aggregation pheromone from male lesser grain borer *Rhyzopertha dominica* (F.). *Journal of Chemical Ecology*, 22(4): 673-680.

Reddy, G. V. P., Cruz. T. and Guerrero, A. (2009). Development of an efficient pheromone-based trapping method for the banana root borer *Cosmopolites sordidus*. *Journal of Chemical Ecology*, 35:111–117.

Reinecke, A.; Ruther, J., Hilker, M., (2005). Electrophysiological and behavioural responses of *Melolantha melolantha* to saturated and unsaturated aliphatic alcohols. *Entomol Experimentalis et Applicata,* 115: 33 -40.

Reji, R. O. P., Paul, A. and Jiji, T. (2012). Field evaluation of methyl eugenol trap for the management of Oriental fruit fly, *Bactrocera dorsalis* Hendel (Diptera: Tephritidae) infesting mango. *Pest Management in Horticultural Ecosystems*, 18(1): 19-23.

Rochat, D. and Avand-Faghih, A., (2000). Trapping of red Palm weevil (*Rhynchophorus ferrugineus)* in Iran with selective attractants. In: Kleeberg, H., Zebitz, C.P.W. (Eds.), *Practice oriented results on use and production of neem-ingredients and pheromones* VI. Germany, pp. 219–224.

Rochat, D., Ramirez-Lucas, P., Malosse, C., Zagatti, P. and Mori, K. (1993). Pheromones and pheromone-related volatiles of four Rhynchophorinae weevils (Coleoptera: Curculionidae). *Bulletin of the International Organization for Biological Control/West Palearctic Regional Section,* 16: 178–184.

Rochat, D., Kazem Mohammadpoor., Malosse, C., Avand Faghih, A., Lettere, M., Beauhaire, J., Morin, J. P., Pezier, A., Renou, M. and Abdollahi, G. A. (2004). Male aggregation pheromone of date palm fruit stalk borer *Oryctes elegans*. *Journal of Chemical Ecology*, 30(2): 387-407.

Romeo, J. (Ed). (2005). *Chemical eclogy and phytochemistry of forest ecosystems. Proceedings of Phyotochemical Society of North America*. Elsevier.

Ruiz Montiel, C., Gonzalez Hernandez, H., Leyva, J., Llanderal Cazares, C., Cruz Lopez, L. and Rojas, J. C. (2003). Evidence for a male-produced aggregation pheromone in *Scyphophorus acupunctatus* Gyllenhal (Coleoptera: Curculionidae). *Journal of Economic Entomology*, 96(4): 1126-1131.

Ruther, J., Reinecke, A., Tolasch,T. and Hilker, M. (2000). Mate finding in the forest cocokchafer, *Melolantha hippocastani* mediated by volatiles from plants and females. *Physiological Entomology,* 25:172 -179.

Saad, K. A., Mohamad Roff, M. N., Mohd Shukri, M. A., Razali Mirad., Mansour, S. A. A., Ismail Abuzid., Hanifah, Y. M. and Idris, A. B. (2014). Artificial damage induction in the leaves of chilli plants leads to the release of volatiles that alter the host plant selection behaviour of *Bemisia tabaci* (Hemiptera: Aleyrodidae). *Journal of Entomology*, 11 (5): 273-282.

Said, I., Renou, M., Morin, J. P., Ferreira, J. M. S and Rochat, D. (2005). Interactions between acetoin, a plant volatile, and pheromone in *Rhynchophorus palmarum*: behavioral and olfactory neuron responses. *Journal of Chemical Ecology*, 31(8): 1789-1805.

Samarajeewa, U., M. R. Adams, M. R. and Robinson, J. M. (1981). Major volatiles in Sri Lanka arrack, a palm wine distillate. *Journal of Food Techno*logy 16: 437-444.

Sandstrom, P., Ginzel, M. D., Bearfield, J. C., Welch, W. H., Blomquist, G. J. and Tittiger, C. (2008). Myrcene hydroxylases do not determine enantiomeric composition of pheromonal ipsdienol in *Ips* spp. *Journal of Chemical Ecology*, 34(12): 1584-1592.

Santangelo, E. M., Correa, A. G. and Zarbin, P. H. G. (2006). Synthesis of (4R,8R)- and (4S,8R)-4,8-dimethyldecanal: the common aggregation pheromone of flour beetles. *Tetrahedron Letters*, 47(29): 5135-5137.

Saran, R. K., Millar, J. G. and Rust, M. K. (2007). Role of (3Z,6Z,8E)-dodecatrien-1-ol in trail following, feeding, and mating behavior of *Reticulitermes hesperus*. *Journal of Chemical Ecology*, 33(2): 369-389.

Saxena, K. N. and Kumar, H. (1984). Interference of sonic communication and mating in leafhopper *Amrasca devastans* (Distant) by certain volatiles. *Journal of Chemical Ecology,* 10(10):1521-1531.

Saxena, K. N. and Basit, A. (1982). Inhibition of oviposition by volatiles of certain plants and chemicals in the leafhopper *Amrasca devastans* (Distant). *Journal of Chemical Ecology*, 8(2): 329-338.

Schllyter F., Zhang, Q. H., Liu, G. T. and JI, L. Z. (2003). A successful case of pheromone mass trapping of the bark beetle *Ips duplicatus* in a forest island, analysed by 20-year time-series data. *Integrated Pest Managment Review,* 6:185–196.

Schorkopf, D. L. P., Jarau, S., Francke, W., Twele, R., Zucchi, R., Hrncir, M., Schmidt, V. M., Ayasse, M. and Barth, F. G. (2007). Spitting out information: Trigona bees deposit saliva to signal resource locations. *Proceedings of the Royal Society of London Series B, Biological Sciences*, 274(1611): 895-898.

Schulz, C. M., Lehmann, L., Blatrix, R., Jaisson, P., Hefetz, A. and Francke, W. (2002). Identification of new homoterpene esters from Dufour's gland of the ponerine ant *Gnamptogenys striatula*. *Journal of Chemical Ecology*, 28(12): 2541-2555.

Schulz, S. (2005). *The chemistry of pheromones and other semiochemicals. II.* Topics in Current Chemistry, 240, Springer, Heidelberg, 341pp.

Seidelmann, K. (2006). The courtship-inhibiting pheromone is ignored by female-deprived gregarious desert locust males. *Biology Letters*, 2(4): 525-527.

Selitskaya, O. G. and Shamshev, I. V. (1994). Response of the lesser grain borer (Coleoptera, Bostrichidae) to components of the aggregation pheromone. *Zoologicheskii Zhurnal,* 73(12): 43-48.

Shelly, T., Renshaw, J., Dunivin, R., Morris, T., Giles, T., Andress, E., Diaz, A., War, M., Nishimoto, J. and Kurashima, R. (2013). Release-recapture of *Bactrocera* fruit flies (Diptera: Tephritidae): comparing the efficacy of liquid and solid formulations of male lures in Florida, California and Hawaii. *Florida Entomologist*, 96(2): 305-317.

Silk, P. J., Sweeney, J., Wu, J., Sopow, S., Mayo, P. D. and Magee, D. (2011). Contact sex pheromones identified for two species of long horned beetles (Coloeptera: Cerambycidae) *Tetropium fusum* and *T. cinnamopterum* in the subfamily Spondylidinae. *Environmental Entomology*, 40 (3): 714-726.

Sillam Dusses, D., Semon, E., Lacey, M. J., Robert, A., Lenz, M. and Bordereau, C. (2007). Trail-following pheromones in basal termites, with special reference to *Mastotermes darwiniensis*. *Journal of Chemical Ecology*, 33(10): 1960-1977.

Sillam Dusses, D., Semon, E., Robert, A. and Bordereau, C. (2009). (Z)-Dodec-3-en-1-ol, a common major component of the trail-following pheromone in the termites Kalotermitidae. *Chemoecology*, 19(2): 103-108.

Sillam Dusses,D., Semon, E., Moreau, C., Valterova, I., Sobotnik, J., Robert, A. and Bordereau, C. (2005). Neocembrene - A, a major component of the trail-following pheromone in the genus *Prorhinotermes* (Insecta, Isoptera, Rhinotermitidae). *Chemoecology*, 15(1): 1-6.

Singh, M., Gupta, D. and Gupta, P. R. (2013). Population suppression of fruit flies (*Bactrocera* spp.) in mango (*Mangifera indica*) orchards. *Indian Journal of Agricultural Sciences*, 83(10): 1064-1068.

Smith, R. F., Pierce, H. D., Jr. and Borden, J. H. (1991). Sex pheromone of the mullein bug, *Campylomma verbasci* (Meyer) (Heteroptera: Miridae). *Journal of Chemical Ecology*, 17 (7) : 1437-1447.

Sobotnik, J., Hanus, R., Kalinova, B., Piskorski, R., Cvacka, J., Bourguignon, T. and Roisin, Y. (2008). (E,E)- alpha -farnesene, an alarm pheromone of the termite *Prorhinotermes canalifrons*. *Journal of Chemical Ecology*, 34(4): 478-486.

Socorro, A. P. del., Gregg, P. C., Alter, D. and Moore, C. J. (2010). Development of a synthetic plant volatile-based attracticide for female noctuid moths. I. Potential sources of volatiles attractive to *Helicoverpa armigera* (Hubner) (Lepidoptera: Noctuidae). *Australian Journal of Entomology*, 49(1): 10-20.

Song Bo., Chen Gao., Stöcklin, J., Peng DeLi., Niu Yang., Li ZhiMin, and Sun Hang. (2014). A new pollinating seed-consuming mutualism between *Rheum nobile* and a fly fungus gnat, *Bradysia* sp., involving pollinator attraction by a specific floral compound, *New Phytologist*, 203(4): 1109-1118.

Soroker, V., Blumberg, D., Haberman, A., Hamburger Rishard, M., Reneh, S., Talebaev, S., Anshelevich, L. and Harari, A. R. (2005).Current status of red palm weevil infestation in date palm plantations in Israel. *Phytoparasitica*, 33(1): 97-106.

Stoeffler, M., Maier, T. S., Tolasch, T and Steidle, J. L. M . (2007). Foreign language skills in rove-beetles? Evidence for chemical mimicry of ant alarm pheromones in *Myrmecophilous pella* beetles (Coleoptera: Staphylinidae). *Journal of Chemical Ecology*, 33(7): 1382-192.

Suckling, D. M., Peck, R. W., Stringer, L. D., Snook, K. and Banko, P. C. (2010). Trail pheromone disruption of Argentine ant trail formation and foraging. *Journal of Chemical Ecology,* 36(1): 122-128.

Suckling, D. M., Stringer, L. D., Corn, J. E., Bunn, B., El-Sayed, A. M. and Meer, R. K vander. (2012). Aerosol delivery of trail pheromone disrupts the foraging of the red imported fire ant, *Solenopsis invicta*. *Pest Management Science*, 68(12): 1572-1578.

Sugeno, W., Hori, M. and Matsuda, K. (2006). Identification of the contact sex pheromone of *Gastrophysa atrocyanea* (Coleoptera: Chrysomelidae). *Applied Entomology and Zoology*, 41(2): 269-276.

Sugie, H., Yoshida, M., Kawasaki, K., Noguchi, H., Moriya, S., Takagi, K., Fukuda, H., Fujiie, A., Yamanaka, M., Ohira, Y., Tsutsumi, T., Tsuda, K., Fukumoto, K., Yamashita, M. and Suzuki, H. (1996). Identification of the aggregation pheromone of the brown-winged green bug, *Plautia stali* Scott (Heteroptera: Pentatomidae). *Applied Entomology and Zoology*, 31(3) : 427-431.

Sugimoto, T., Sakuratani, Y., Setokuchi, O., Kamikado, T., Kiritani, K. and Okada, T. (1994). Estimations of the attractive area of pheromone traps and dispersal distance, of male adults of sweetpotato weevil *Cylas formicarius* (Fabricius) (Coleoptera) *Applied Entomology Zoology*, 29: 349–358

Sun JiuGuang., Huang LingQiao. and Wang ChenZhu. (2012). Electrophysiological and behavioral responses of *Helicoverpa assulta* (Lepidoptera: Noctuidae) to tobacco volatiles. *Arthropod-Plant Interactions*. 6(3): 375-384.

Sun XiaoLing., Gao ChangQi., Cheng Bin., Wang WenYi., Zhou GuangPing., Lu YongBin and Wang-Bin. (2006). Adult emergence rule of *Ips typographus* Linnaeus monitored by pheromone. *Journal of Northeast Forestry University*, 34(3): 7-8.

Sutherland, J. A. (1986). Damage by *Cylas formicarius* Fab. to sweetpotato vines and tubers, and the effect of infestations on total yield in Papau, New Guinea. *Tropical Pest Management*, 32: 316–323.

Tabilio, M. R., Fiorini, D., Marcantoni, E., Materazzi, S., Delfini, M., Salvador, F. R. de. and Musmeci, S. (2013). Impact of the Mediterranean fruit fly (Medfly) *Ceratitis capitata* on different peach cultivars: the possible role of peach volatile compounds..*Food Chemistry*, 140(1/2): 375-381.

Takita, M., Sugie, H., Tabata, J., Ishii, S. and Hiradate, S. (2008). Isolation and estimation of the aggregation pheromone from *Eysarcoris lewisi* (Distant) (Heteroptera: Pentatomidae). *Applied Entomology and Zoology*, 43(1): 11-17.

Tampe, J., Parra, L.., Huaiquil, K., Mutis, A. and Quiroz, A. (2015). Repellent effect and metabolite volatile profile of the essential oil of *Achillea millefolium* against *Aegorhinus nodipennis* (Hope) (Coleoptera: Curculionidae). *Neotropical Entomology*, 44(3):279-285.

Tang Rui, Zhang JinPing and Zhang ZhongNing. (2012). Electrophysiological and behavioral responses of male fall webworm moths (*Hyphantria cunea*) to herbivory-induced mulberry (*Morus alba*) leaf volatiles.*PLoS* ONE.7, (11): e49256.

Tang YuChong., Chen XiaoMing and Zhou ChengLi. (2014). The attractiveness of volatile compounds to foraging *Kallima inachus* (Boisduval). *Chinese Journal of Applied Entomology*, 51(5): 1327-1335.

Tashiro, T., Kurosawa, S. and Mori, K. (2004). Revision of the structure of the major aggregation pheromone of the broad-horned flour beetle (*Gnatocerus cornutus*) to (1S,4R,5R)- alpha - acoradiene by its synthesis. *Bioscience, Biotechnology and Biochemistry*. 68(3): 663-670.

Tasin, M., Knudsen, G. K. and Pertot, I. (2012). Smelling a diseased host: grapevine moth responses to healthy and fungus-infected grapes.*Animal Behaviour*, . 83(2): 555-562.

Tebayashi, S. I., Hirai, N., Suzuki, T., Matsuyama, S., Nakakita, H., Nemoto, T. and Nakanishi, H. (1998). α-Cedren-14-al: minor aggregation pheromone component of *Gnatocerus cornutus* (F.) (Coleoptera: Tenebrionidae). *Journal of Pesticide Science*, 23(4), 402-406.

Tokoro, M., Takahashi, M. and Yamaoka, R. (1994). (Z,E,E)-dodecatrien-1-ol: a minor component of trail pheromone of termite, *Coptotermes formosanus* Shiraki. *Journal of Chemical Ecology*, 20(1): 199-215.

Torres Contreras, H., Olivares Donoso, R. and Niemeyer, H. M. (2007). Solitary foraging in the ancestral South American ant, *Pogonomyrmex vermiculatus*. Is it due to constraints in the production or perception of trail pheromones? *Journal of Chemical Ecology*, 33(2): 435-440.

Tumlinson J. H., Guelder, R. C., Hardee, D. D., Thompson, A. C., Hedin, P. A., and Minyard, J. P. (1969). Sex pheromones produced by male boll weevil isolation, identification and synthesis. *Science,* 166:1010–1012.

Ubler, E., Kern, F., Bestmann, H. J., Holldobler, B and Attygalle, A. B. (1995). Trail pheromone of two formicine ants, *Camponotus silvicola* and *C. rufipes* (Hymenoptera: Formicidae). *Naturwissenschaften*, 82(11): 523-525.

Unelius, C. R., Schiebe, C., Bohman, B., Andersson, M. N., and Schlyter, F. (2014). Non-host volatile blend optimization for forest protection against the european spruce bark beetle, *Ips typographus*. *PloS one*, 9(1): e85381.

Valterova, I., Kunze,,J., Gumbert,,A., Luxova,,A., Liblikas, I., Kalinova, B. and Borg Karlson, A. K. (2007). Male bumble bee pheromonal components in the scent of deceit pollinated orchids; unrecognized pollinator cues? *Arthropod Plant Interactions*, 1(3): 137-145.

Veit, U. (2001). Influence of different environmental factors on the activity of ant trail pheromones. *Mitteilungen der Deutschen Gesellschaft fur allgemeine und angewandte Entomologie*, 13(1-6): 381-384.

Verghese, A., Sreedevi, K., and Kalleshwaraswamy, C. M. (2005). *Pest Managment in Horticultural ecosystems- Special issue on fruit fly* 11(2)::81-180.

Verghese, A., Sreedevi. K. and Nagaraju, D. K. (2007). Pre and post harvest IPM for the mango fruitf fly *Bactrocera dorsalis* (Hendel)- Fruti flies of Economic Importance.: from Bais to applied Knowledge. *Proc 7th International Symposium of fruit flies of Eonomic Importance, 10-15 September 2006, Salvadar, Brazil* 179-182.

Visser, J. H. and Piron, P. G. M. (1995). Olfactory antennal responses to plant volatiles in apterous virginoparae of the vetch aphid *Megoura viciae*. *Entomologia Experimentalis et Applicata*, 77 (1) : 37-46.

Wajnberg, E. and Colazza, S. (2013). *Chemical eology of insect parasitoids*. John Wiley & Sons, 312 pp.

Wang Rong., Zong ShiXiang., Yu LinFeng., Lu PengFei. and Luo YouQing. (2014c) . Rhythms of volatile release from female and male sea buckthorn plants and electrophysiological response of sea buckthorn carpenter moths. *Journal of Plant Interactions*. 9(1): 763-774.

Ward, A., Moore, C., Anitha, V., Wightman, J. and Rogers, D. J., (2002). Identification of sex pheromone of *Holotrichia reynaudi*. *Journal of Chemical Ecology,* 28 (3) : 515 -522.

Wardle, A .R., Borden, J .H., Pierce, H .D, Jr. and Gries, R. (2003). Volatile compounds released by disturbed and calm adults of the tarnished plant bug, *Lygus lineolaris*. *Journal of Chemical Ecology*. 29(4): 931-944.

Wattanapongsiri.A. (1966). *A revision of the genera Rhynchophorus and Dynamis (Coleoptera: Curculionidae)*. Department of Agriculture, Science Bulletin, Bangkok, 1: 1-328.

Weaver, D. K., McFarlane, J. E. and Alli, I. (1990). Repellency of volatile fatty acids present in frass of larval yellow mealworms, *Tenebrio molitor* L. (Coleoptera: Tenebrionidae), to larval conspecifics. *Journal of Chemical Ecology*, 16 (2): 585-593.

Weissbecker, B., Van Loon, J. J., and Dicke, M. (1999). Electroantennogram responses of a predator, *Perillus bioculatus*, and its prey, *Leptinotarsa decemlineata,* to plant volatiles. *Journal of Chemical Ecology*, 25(10): 2313-2325.

Weissling, T. J., Giblin-Davis, R. M., Gries, G., Gries, R., Perez, A. L., Pierce Jr, H. D., and Oehlschlager, A. C. (1994). Aggregation pheromone of palmetto weevil, *Rhynchophorus cruentatus* (F.) (Coleoptera: Curculionidae). *Journal of Chemical Ecology*, 20(3): 505-515.

Williams, L, III., Rodriguez Saona, C., Pare, P .W. and Crafts Brandner, S. J. (2005). The piercing-sucking herbivores *Lygus hesperus* and *Nezara viridula* induce volatile emissions in plants. *Archives of Insect Biochemistry and Physiology*, 58(2): 84-96.

Williams, N. L. J. and Borden, J. H. (2004). Response of *Dryocoetes confusus* and *D. autographus* (Coleoptera: Scolytidae) to enantiospecific pheromone baits. *Canadian Entomologist*, 136(3): 419-425.

Wood, W. F., Palmer, T. M. and Stanton, M. L. (2002). A comparison of volatiles in mandibular glands from three *Crematogaster* ant symbionts of the whistling thorn acacia. *Biochemical Systematics and Ecology*, 30(3): 217-222.

XianYing, Z., ZhiGuo, H., ChangYan, Y., and Fei, H. (2014). Effects of volatiles in twenty non-host plants on the repellent and attractive behaviors of brown planthopper, *Nilaparvata lugens*. *Journal of South China Agricultural University*,35(3): 63-68.

Yadav, C. P. S. and Sharma, G. K., (1995). *Indian white grubs and their management*. All India Coordinated Research Project on White grubs, Technical Bulletin No.2. Indian Council of Agricultural Research, New Delhi.

Yadav, V. K. and Yadav, N., (2004). Identification of the chemical mediating attraction of *Holotrichia consanguinea* beetles to its most preferred host tree. *Paper presented in IV International Crop Science Congress,* Brisbane, 2004.

Yan, F. S., Visser, J. H., Visser, J.H. and Minks, A. K. (1982). Electroantennogram responses of the cereal aphid *Sitobion avenae* to plant volatile components. *Proceedings of the 5th International Symposium on Insect-Plant Relationships,* Wageningen, the Netherlands, 1-4 March 1982, pp 387-388.

Yang, H., Yang, W., Yang, M. F., Yang, C. P., Zhu, T. H. and Huang, Q. (2010). Effects of plant volatiles on the EAG and behavioral responses of *Batocera horsfieldi* Hope (Coleoptera: Cerambycidae). *Journal of Agricultural and Urban Entomology*, 27(1) : 20-32.

Yasuda, T., Mizutani, N., Honda, Y., Endo, N., Yamaguchi, T., Moriya, S., Fukuda, T. and Sasaki, R. (2007). A supplemental component of aggregation attractant pheromone in the bean bug *Riptortus clavatus* (Thunberg) (Heteroptera: Alydidae), related to food exploitation. *Applied Entomology and Zoology,* 42(1): 161-166.

Yun Feng, Z., Zeng ZhiJiang., Yan WeiYu and Wu XiaoBo. (2010). Effects of three aliphatic esters of brood pheromone on worker feeding and capping behavior and queen development of *Apis cerana cerana* and *A. mellifera ligustica*. *Acta Entomologica Sinica*, 53(2): 154-159.

Zada, A., Ben Yehuda, S., Dunkelblum,, E., Harel, M., Assael, F. and Mendel, Z. (2004). Synthesis and biological activity of the four stereoisomers of 4-methyl-3-heptanol: main component of the aggregation pheromone of *Scolytus amygdali*. *Journal of Chemical Ecology,* 30(3): 631-641.

Zahn, D. K., Moreira, J. A. and Millar, J. G. (2008). Identification, synthesis, and bioassay of a male-specific aggregation pheromone from the harlequin bug, *Murgantia histrionica*. *Journal of Chemical Ecology*, 34(2): 238-251.

Zarbin, P. H. G., Arrigoni, E de B., Reckziegel, A., Moreira, J. A., Baraldi, P. T. and Vieira, P. C. (2003). Identification of male-specific chiral compound from the sugarcane weevil *Sphenophorus levis*. *Journal of Chemical Ecology*, 29(2): 377-386.

Zhang AiJun., Linn, C.Jr., Wright, S., Prokopy, R., Reissig, W. and Roelofs, W. (1999). Identification of a new blend of apple volatiles attractive to the apple maggot, *Rhagoletis pomonella*. *Journal of Chemical Ecology*, 25(6): 1221-1232.

Zhang, Q. H., Chauhan, K. R., Erbe, I T\F., Vellore, A. R., and Alrdirch. J. R. (2004). Semiochemistry of the golden eyed lacewing *Chrysopa oculata*: Attraction of males to a male produced pheromone. *Journal of Chemical Ecology*, 30(9): 1849-1870.

Zhang ZhengQun. and Chen ZongMao (2015a). Non-host plant essential oil volatiles with potential for a 'push-pull' strategy to control the tea green leafhopper, *Empoasca vitis*. *Experimentalis et Applicata*, 156 (1) :77-87.

Zhang ZhengQun., Bian Lei., Sun XiaoLing., Luo ZongXiu., Xin ZhaoJun., Luo FengJian. and Chen ZongMao. (2015b). Electrophysiological and behavioural responses of the tea geometrid *Ectropis obliqua* (Lepidoptera: Geometridae) to volatiles from a non-host plant, rosemary,*Rosmarinus officinalis* (Lamiaceae). *Pest Management Science*, 71(1): 96-104.

Zhang ZhengQun., Sun XiaoLing., Xin ZhaoJun., Luo ZongXiu., Gao Yu., Bian Lei and Chen ZongMao. (2013). Identification and field evaluation of non-host volatiles disturbing host location by the tea geometrid, *Ectropis obliqua*. *Journal of Chemical Ecology*. 39(10): 1284-1296.

Zhang, Q. H. and Aldrich, J. R. (2003). Pheromones of milkweed bugs (Heteroptera: Lygaeidae) attract wayward plant bugs: *Phytocoris* mirid sex pheromone. *Journal of Chemical Ecology*, 29(8): 1835-1851.

Zhang, Q. H., Ma JianHai., Zhao FengYu., Song LiWen and Sun JiangHua. (2009). Aggregation pheromone of the Qinghai spruce bark beetle, *Ips nitidus* Eich. *Journal of Chemical Ecology*, 35(5): 610-617.

Zhang, Q. H., Schlyter, F., and Birgersson, G. (2000). Bark volatiles from nonhost angiosperm trees of spruce bark beetle, *Ips typographus* (L.) (Coleoptera: Scolytidae): Chemical and electrophysiological analysis. *Chemoecology*, 10(2): 69-80.

Zheng, L. X., Wu, W. J. and Fu, Y. G. (2014). (±)-2-Hexanol from *Pterocarpus indicus* leaves as attractant for female *Aleurodicus dispersus* (Hemiptera: Aleyrodidae) Pretoria, South Africa. *African Entomology*, 22 (2):267-272.

Zhu JunWei., Unelius, R. C., Park KyeChung., Ochieng, S. A., Obrycki, J. J .and Baker, T. C. (2000). Identification of (Z)-4-tridecene from defensive secretion of green lacewing, *Chrysoperla carnea*. *Journal of Chemical Ecology*, 26(10): 2421-2434.

Zhu, P. C., Kong, F.L., Yu, S.D., Yu, Y.Q., Jin, S.P., Hu, X. H. and Xu, J. W. (1987). Identification of the sex pheromone of eggplant borer *Leucinodes orbonalis* Guenee (Lepidoptera: Pyralidae). *Zeitschrift fur Naturforschung, C Biosciences*. 42(11-12): 1343-1344.

Zito, P., Guarino, S., Peri, E., Sajeva, M. and Colazza, S.(2013). Electrophysiological and behavioural responses of the housefly to "sweet" volatiles of the flowers of *Caralluma europaea* (Guss.) N.E. Br. *Arthropod - Plant Interactions*, 7(5): 485-489.

Zito, P., Sajeva, M., Raspi, A. and Dötterl, S. (2014). Dimethyl disulfide and dimethyl trisulfide: so similar yet so different in evoking biological responses in saprophilous flies.*Chemoecology*, 24(6): 261-267.

Integrated Pest Management in the Tropics; pp. 687-701

New India Publishing Agency, New Delhi (India)

CHAPTER - 21

Plant Chemicals in Insect Plant Interactions

Usha Rani Pathipati
Biology and Biotechnology Division
Indian Institute of Chemical Technology, Hyderabad-500007, Telangana
email: usharani65@gmail, usharani65@yahoo.com

21.1 Introduction

Plants and insects share a wonderful relationship. Both are dependent on each other at different levels and chemicals emitted by plants connect them to a large extent. "Plants" contribute to a major share of occupancy on earth and makes several organisms such as insects & microbes and also humans to depend on them for their survival. This dependency leading to several interesting interactions between plants & insects which are often utilized by humans for their own benefit. Plants maintain incredible communication with their associated organisms. Since plants are rich banks of chemicals and constantly emit them in to its immediate environment which contributes in communication between insects and plants. Insects are the major organisms that have a long relation with the plants and there are several incredible ways how plants can interact with insect species. There is constant war between insects & plants for their successful survival. Chemicals contribute a major bridge/ barrier for this war. Insect in general use them for their advantage either to get attracted towards to feed on them (synomones) to get repelled (repellents) away to save themselves from ingesting toxic chemicals. Recent advances and innovations of different techniques, instruments and equipment in analytical sciences made the identification of these chemicals easier, leading the subject of insect–plant interactions to higher avenues. A multidisciplinary approach is needed to study these complex insect-plant interactions which are important for understanding ecosystems and needed to be investigated thoroughly. In recent times these studies are gaining tremendous importance.

Insects are the prominent pests of agriculture, horticulture and stored products and also include domestic pests, vectors etc. They all depend on plants and plant emitted chemicals for their and their offspring's development. After successful mating, a gravid female's prime task is to find a suitable host plant where she can lay her precious eggs, so that, her off springs can grow with minimum obstacles. Insect's decisions on host plant choice are determined majorly by plant chemistry. Plants being a reservoir of chemicals and comprise multitude of quantitatively and qualitatively different chemicals which even in normal stage without any stimulation also release them into their surrounding atmosphere. Since most of these chemicals are chiefly volatiles in nature, play an important role in insect behavior and plant–insect interactions and have wider importance in plant biology (Holopainen *et al.*, 2004). In a word, insect's survival and reproduction depend on plant volatile perception and responding to them in appropriate manner.

The plant volatiles emitted signify the plant's identity (through species specific volatiles) and even suggest the status (infested by other pests or pathogens). The insects, their natural enemies and even neighboring plants respond to these volatiles emitted into plant's surrounding atmosphere. Plant constitutive volatiles emitted through their vegetative parts may attract the insect larvae for feeding purpose and gravid females for oviposition, but volatiles from feeding damaged plants may act as repellents to both (Fig. 1). Feeding or oviposition occurs when herbivore finds an appropriate plant that is accepted as a suitable host (Miller and Strickler 1984). The release of volatiles from an individual plant might vary with the environmental conditions that influence the plant's physiology. Several crop plants like corn, lima bean and cotton emit reduced quantities of volatiles due to low light intensity (Pare and Tumlinson, 1999). Analysis of volatiles emitted by different plant communities and species due to the feeding by different insect species suggests that activation of a common set of biosynthetic pathways are responsible for these emissions which are distinguished by broad spectrum of insect natural enemies like parasitoids and predators (Pare and Tumlinson, 1999).

Another interesting aspect of insect and plant connection is their relation with the metal ions. Some plant species growing in certain soils are able to hyper accumulate metals such as Zn, Cu, Co, Ni, Mg *etc.* and when their concentration exceeds certain limits inside the plant, that can influence the herbivore choices and act as feeding deterrents. This phenomena can be considered as plant's elemental defense and the reason for this accumulation of metals in some plants is to facilitate plant growth and reduce herbivory.

HEALTHY PLANT
Caterpillar approaching the healthy/ Uninfested Plant

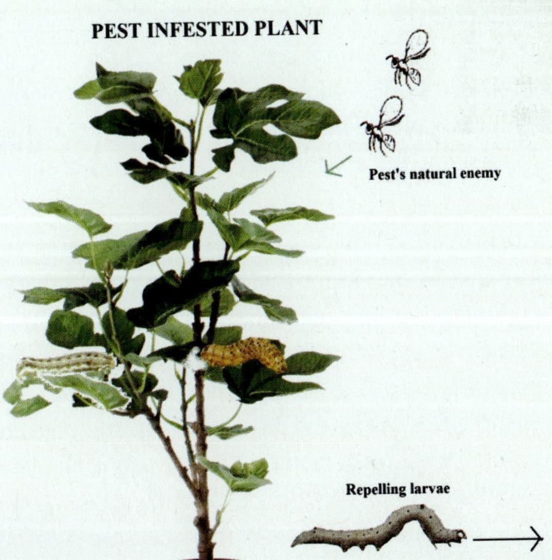
PEST INFESTED PLANT
Pest's natural enemy
Repelling larvae
Insect feeding induced plant chemicals repelling the approaching herbivore (Direct defense) and attracting the pest's natural enemy (Indirect defense)

Fig. 1

21.2 Role of Plant Chemicals in Insect Host Plant Location

The plant produced chemicals influence of the insect behavior, feeding behavior and sexual or reproductive behavior. The ultimate consumption of plant material by phytophagous insects is preceded by a complicated series of behavioral interactions between the food plant and the insect. The herbivore insect first and foremost has to reach the plant to utilize it for its own benefit, and after reaching it must then decide whether or not the plant is suitable as food, and has to finally settle on the plant for further action. Insect while searching for a suitable host plant either to feed upon or lay her eggs, locates the plant by recognizing its characteristic odour and taste. One major hindrance is that a host seeking female herbivore has to distinguish its correct and suitable host plant from the enormous number and varieties of plants releasing wide varieties of chemical cues such as species –specific or specific ratios of ubiquitous compounds and differentiate them from back ground multiple odors (Bruce and Pickett, 2011). This makes the task difficult for a host searching herbivore, as vast majority of plant species having varying compositions and ratios of emissions that often exists at a close distance and the chemicals emitted from a diverse group of plants stimulate insects to different extents. These bouquets of compounds mediating the behavior of herbivores render leaves attractive or disagreeable to herbivores which seek the plants for their nutritional as well as oviposition requirements. The chemical identity of the volatile chemicals released varies with the plant species. The behavioral elements which make up this intricate sequence depend upon the insect receiving various sensory stimuli from the plant.

Apart from providing host location cues to the approaching insect herbivore, in some cases it was also evident that the feeding induced volatiles attract the conspecifics and increase their incidence on the plant. Sun *et al.* (2010) through behavioral bioassays proved that volatile compounds from plants infested by tea weevil, *Myllocerinus aurolineatus* (Voss) (Coleoptera: Curculionidae) were attractive to conspecific weevils and the weevil infestations dramatically increased the emission of volatiles (Z)-3-hexenal, (Z)-3-hexenol, (E)-beta-ocimene, linalool, phenyl ethyl alcohol *etc.* A beetle pest *Meligethes aeneus* F. (Coleoptera: Nitidulidae), is a pollen-feeding insect on oilseed rape, *Brassica napus* L. Both male and female *M. aeneus* were attracted to higher concentrations of VOCs, particularly to the three common voaltiles (β-caryophyllene, (*E*)-β-farnesene and (*Z*)-β-ocimene) emitted by these plants 72 h after floral bud injury and oviposition by their conspecifics but avoided much higher doses (Dariusz *et al.,* 2013).

Insects have well developed sensory system which is responsible for perception of the plant emitted chemicals and the importance of insect antennae in plant volatile chemical recognition is demonstrated by several authors

(Usha Rani and Nakamuta, 2001, Lei Guo and Guo Qing Li, 2009). They studied the plant chemical perception by Asian corn borer moths, *Ostrinia furnacalis* (Guenée) (Lepidoptera: Crambidae) in a Electroantennogram towards the common plant chemicals, myristic, palmitic, stearic and oleic acids and their corresponding methyl esters and found that mated females with both antennae amputated, in contrast to intact females and females with one antenna removed, could not discriminate between controls and the treatments with a blend of oviposition deterring fatty acids. The insect oral cavity also possess gustatory sense organs that are assumed to aid insects detect and reject potentially toxic foods. Ablation of these gustatory sensilla from the larval *Manduca sexta* (Lepidoptera: Spingidae) prevented them from showing aversive response to plant chemicals (Glendinning, 2002). Scolytid bark beetles while tunneling in conifers cause the release of volatile oleoresins from the tree. These volatiles, together with the beetle pheromones which are released at the same time, attract flying dipterous predators to beetle-infested trees.

Since plant chemicals play a vital role in host location by insects, it is even possible to mask the attractive chemicals to prevent the insect from locating the correct host plant. The weeds allowed to grow on the background of cruciferous plants reduce the number of pest insects on the crop when it is intercropped with another plant species, or when the crop is under sown with living mulch (Finch and Collins, 2001). Obviously, if placing non-host plants in the vicinity of host plants reduces the numbers of insects that actually find their host plants, then this could provide a clue as to how insects use chemicals for locating their hosts. It has been suggested that when the background of crop plants growing in bare soil is made more diverse by allowing other non-host plant species to grow in the inter-row spaces, that the additional diversity "disrupts" insects from selecting otherwise acceptable host plants. The allyl sulphide (CH_2=$CHCH_2SCH_3$), n-propyl sulphide ($C_3H_7SC_3H_7$), methyl disulphide (CH_3SSCH_3), and propyl and iso-propyl mercaptan (C_3H_7SH) and ($(CH_3)_2$ CHSH) are the Sulphur containing compounds which are characteristic of the odour of onions are used by the female onion maggot *Hylemya antiqua* to locate the plant and are attractive to her larvae (Matsumoto, 1970). The volatile chemistry and the role of phytochemicals in host plant resistance of Sweet potato weevil, *Cylas formicarius* (Fabricius) (Insecta: Coleoptera: Curculionidae) was studied by Wang and Kays (2002) and they found that volatile extracts from aerial parts as well as the storage roots were attractive to female weevils. Several terpene compounds were identified from the head space volatiles through Gas chromatography and Gas Chromatography-mass spectroscopy.

Plant chemicals are vital not only to the actively feeding stages of the herbivore insects, but also influence the egg-laying behaviour of many phytophagous insects. However, here their role is less predominant and they take their place beside other factors such as texture, surface configuration, and colour. In the case of the codling moth, gravid females will lay some eggs in the absence of any olfactory stimulant at all, but the presence of the apple volatile α-farnesene stimulates a much heavier egg-lay (Wearing and Hutchins, 1973). A recent report from Max Planck Institute for Chemical Ecology, Jena, Germany states the ability of *Manduca sexta* moths to recognize changes in the volatile chemical profile released by the tobacco plants that are damaged by the feeding of the Manduca larvae, make them immediately to shift to another plants that are less likely to be attacked by insects to avoid competition from the larvae of the same species for resources (Allmann *et al.,* 2013). Adult female *M. sexta* moths are also able to detect the changes in the green volatile profile emitted by Sacred Datura plants that have been damaged by *M. sexta* caterpillars.

Normally in the plant-insect relationship main beneficiary is the insect, which derives food and shelter from the plant. But there are cases where the plant appears to derive all the benefit. Many insects are attracted to flowers and feed on the nectar. While doing this, they transfer pollen from one flower to another. Some species of orchids, however, produce no nectar and have evolved a special mechanism for ensuring pollination. The flowers of *Ophrys* spp. produce chemicals which are identical with the chemicals used as a sex attractant by females of some wasp species (Scoliidae, Sphecidae). Males of these species are attracted to the flowers by the odour and attempt to copulate with them. The flowers themselves are constructed in such a way as to encourage mating activity and during this the male wasps cover their head with pollen, which is then carried on to the next flower (Kullenberg, 1961).

21.3 Plants and Insect Feeding Stress

Herbivory leads to the loss of biomass. While feeding, the phytophagous insects not only remove leaf tissue as such and sometimes even parasitize the plant's vascular transport system, some tunnel through the stems and roots, while a few bore and mine. All these patterns of feeding generally induce the plants to react or respond to the attack. The plant responses are shown usually in the form of production of volatile as well as contact chemicals which are characteristic of the plant that are emitted through the plant leaf surface. These volatile organic compounds (VOCs) which are different than the undamaged or mechanically damaged plants and play a major conspicuous role in plant insect interactions. All these VOCs provide cues to the host searching parasitoids, predators or other members of the third trophic levels towards plant that harbor their host (Turlings *et al.,* 1995; DeMoraes *et al.*, 1998). These responses mediate interactions within organisms on multiple trophic levels leading to complex ecological consequences (Kessler and Halitschke, 2007).

Insect feeding induced stress is the subject of interest in recent years and is considered important in the studies on insect–plant interactions and in understanding the plant defense strategies. These induced plant chemicals may directly affect the feeding herbivore performance by reducing its feeding intensity, quantity of leaf damage or indirectly influence the herbivore natural enemies by attracting them towards the plant providing the information about the attacker (herbivore). (Z)-3- hexenol, a C_6-alcohol synthesized in the lipoxygenase pathway was proved to be the most important info-chemical for the herbivore repellence or attractance and natural enemy attraction (Wei and Kang, 2011). The plant volatiles emitted by diamondback moth *Plutella xylostella* was found to attract the parasitoids, *Trichogramma chilonis* and *Cotesia plutellae* and the predator *Chrysoperla carnea* towards the pest ((Reddy *et al.*, 2002). Certain Green leaf volatiles (GLVs) such as C_6 alcohols and aldehydes found in green leaves and shoots (Visser *et al.*, 1979) and methyl salicylate, methyl jasmonate and ethylene are involved in stress signaling within and between individual plants (Farmer and Ryan, 1990). Though the emission of herbivore induced volatiles occur in majority of the plants till now they are reported to be caused by feeding of Lepidoptera, Diptera, Thysanoptera, Coleoptera, Hemiptera and mites (Mumm and Dicke, 2010). What makes the plant recognize the pest attack/feeding? It is found that the chemicals present in the oral secretions and regurgitates of the feeding herbivores, such as Fatty acid–amino acid conjugates (FAC) mediate these insect specific plant responses.

Plant tissue damage not only due to the herbivore feeding, (Dicke *et al.*, 1990a, b) but also herbivore oviposition (Hilker, 2006) can modify the plants' volatile profiles by the induction of chemical blends. Previous studies mostly focused on interactions between a plant, pest and its parasite interactions on the above ground. However, it was found recently that roots can also emit volatiles that attract enemies of below ground feeding herbivore (Rasmann *et al.*, 2005; Ali *et al.*, 2010).

Majority of the plants contain species specific chemicals since the time of their birth, but the feeding of a phytophagous insect induces the production of additional toxic secondary metabolite chemicals that hamper with the further feeding, insect growth, development, hormonal system and reproduction. The important phenomena that occur due to pest feeding is the production of protease inhibitors which harm the digestive system of the plant chewing insects. Certain chemicals that act upon the herbivore can discourage its feeding. The herbivore feeding often leads to the expression of defense genes in plants as a direct defense response that stimulate the production of secondary metabolites. These induced secondary metabolites usually indicate the presence of the feeding herbivore to the approaching and prospective feeder of the same plant. Many moths avoid laying eggs on the plants which are already having insects feeding on them may be to prevent competition for their offspring. So the pest feeding induced plant secondary metabolites has the effect of deterring other herbivores from laying eggs (Kessler and Baldwin, 2001).

Several studies lead to the isolations, identification of the induced plant chemicals (Torto,2004; Dicke *et al.*, 2009). However, it is a difficult task to identify the chemicals involved in the blend which are complex mixtures often containing up to 200 components, particularly those which attract the herbivore natural enemies. These behaviorally active components are totally identified in only a very a few systems, *i:e* Lima bean, cucumber, potato, tobacco, brinjal, sweet pepper, rose, apple *etc.* (Mumm and Dicke, 2010). The relation between plants and insects is not only a complex one but very fascinating and scientists have been unraveling these complex interactions since 1980s when Marcel Dicke, professor of insect-plant interactions at Wageningen University in the Netherlands, says he was "the first to show that plants communicate with the enemies of their enemies (Dicke and Sabelis, 1988)."

Plant chemicals show profound influence on insect senses, participate widely in host location by the approaching pests, and stimulate oviposition, thus increasing the pest populations on plants. However, this concept took a great turn after 1980's after it has become clear that plants respond to insect herbivory with the induced production of chemicals to attract herbivore natural enemies(Pare and Tumlinson, 1999; De Moraes *et al.*, 1998). All these herbivory-induced changes are mediated by elaborate signaling networks, which include receptors/ sensors, Ca^{2+} influxes, kinase cascades, reactive oxygen species, and phytohormone signaling pathways.

Recent studies confirm that these chemicals generated by pest fed plants could mediate interactions between damaged and undamaged plants (Dicke and Bruin, 2001). In general, these induced volatile chemicals have differential impact on various members of the community that interacts with the emitting plants.

Herbivorous predator or parasite usually of very small size has to search for their pray or host which is again of a very smaller size, among a complex vegetation that consists of various chemicals emitted through their surfaces. In such circumstances, any type of information that can guide the natural enemy to its prey/host will reduce its searching time and enhance the chance of finding its food or oviposition sight, thus contributing to the success of its newer generation. Pest feeding induced volatiles provides reliable information about the pest location and even identity. These volatile chemicals produced as a response to herbivory often comprise a complex mixture of tens up to more than 200 compounds, the composition of which may vary with the herbivore species, herbivore developmental stage Renwick and Vite, (1969), Hunt *et al.* (1986) *etc.* Thus, these induced plant chemicals combine high detestability with considerable availability (Abdelganwad *et al.*, 2014) and constitute a valuable information to herbivore's natural enemies. Predators need to know the type of plants eaten by their prey and this is achieved only by recognizing the chemicals emanating from host plants of their prey.

Oxidative components of several plants are altered due to the feeding by phytophagous insects. All these changes in enzyme activities further responsible for the production of the VOCs. The enzymes like (polyphenol oxidase), POD (Peroxidase), and phenyl ammonia lyase were increased due to infestation of the insect, *Retithrips syriacus* in all the three host plants, *Ricinus communis*, *Eucalyptus globulus*, *Manihot utilissima* and further lead to the enhancement of phenols in the plant (Ananthakrishnan *et al.*, 1992). The phloem- feeding aphids, *Brevicoryne brassicae,* caused increased activities of the oxidative enzymes such as PPO and POD by feeding on cabbage plants and a decrease in SOD (Superoxide dismutase) and catalase and also altered the phytochemicals as well as biochemical in the plant. A decline in total soluble protein content occurred in aphid infested cabbage leaves. Levels of free amino acids, total carbohydrates were lower in the phloem-feeding aphid infested plants compared to normal plants (Khattab, 2007).

It is also interesting to know that herbivore feeding modify the primary metabolites of the plants. There are many studies reporting the biochemical and physiological changes in the plant due to herbivory. Laboratory studies conducted by Prasannalaxmi and Usha Rani, (Unpublished) further supported the role of primary metabolites in the defense mechanism of the plant against the insect attack. Brinjal, *Solanum melongena* infested by *Leucinodes orbonalis* caused an enhanced biochemicals including amino acids, carbohydrates, phenols and proteins. *Spodoptera litura* feeding on sweet potato plants also exhibited changes in the primary metabolites. The carbohydrates quantity was reduced with enhanced amino acids, phenols, and proteins. Whereas, cotton plants attacked by *S. litura* resulted in reduced amino acid and carbohydrate contents and increased total phenols and proteins ((Usha Rani and Pratyusha, 2013). A study was conducted to demonstrate the induced defense responses of chilli plant against herbivory by different larval instars of the tobacco cutworm, *Spodoptera litura* (F) at different time intervals (0, 24, 48 and 72 h) after feeding. Change in primary metabolite content was observed in the plants damaged with different larval stages. Second instar feeding caused gradual decrease in the carbohydrate content till 72h after damage. Decrease in carbohydrate content was recorded in chilli plants till 48h post feeding by third instar *S. litura*. Herbivory by later instars i:e fourth and fifth stage larvae failed to induce any change in carbohydrate content initially at 0h but a slight decrease in the quantity occurred after 24h

(Vijaya and Usha Rani, 2015, Unpublished). Feeding by sucking pests on plants too evoke changes in primary metabolites. Changes occurred in different biochemical and nutritional profiles of the castor, *Ricinus communis* plant, which depended on the mode of feeding of the phytophagous insect i.e., leaf surface or the internal tissue of the capsule. There was a significant decrease in carbohydrates due to sucking pest, borer insect attack and mechanical damage. The damage caused by the leaf chewing insects- *Achaea janata* and *S. litura* resulted in the increased carbohydrate and phenol contents on *R. communis* leaves. Although proteins are not secondary metabolites, certain types of proteins have functions similar to secondary metabolites, including protection of the plant against insects. There was an enhancement in protein content of wounded plants. However, these changes were not found in sucking pest, *Empoasca flavescens* infested leaves and mechanically damaged castor plants (Usha Rani *et al.*, 2009). These changes in "primary metabolites" may change the efficacy of secondary metabolites. It is interesting to note that, though both *A. janata* and *S. litura* has similar feeding modes and even the same feeding extent the feeding induced changes were of different nature.

Impact of different pests feeding on rice plants was studied by Usha Rani and Jyothsna (2010). The borer pest—yellow stem borer (YSB), *Scirpophagaincertulas* (W); surface feeder, leaf roller (LR), *Cnaphalocrosismedinalis* (G) and a sucking pest—brown plant hopper (BPH), *Nilaparvatalugens* (S) fed rice plants showed varied biochemical parameters compared to normal rice plants. A significant increase was recorded in the carbohydrate content and quantity of phenols of the rice plants infested with all the above three insect species in comparison with normal healthy rice plants. Feeding of *Amsacta albistriga*, *Aproaerema modicella* and *Spilosoma obliqua* on groundnut plants resulted in decreased total soluble sugars with enhanced total phenol content and amino acids compared to healthy groundnut plants. Total protein content showed minor differences between the normal and infested plants (Pratyusha and Usha Rani, 2013).

Castor plants exhibited clear and obvious changes in primary and secondary metabolite profile due to damage caused by *Trialeurodes ricini*. All the four biochemical components (total carbohydrates, proteins, amino acids and phenols) and five enzymes (PAL, PPO, CAT, SOD and POD) showed variation in the infested castor leaves when compared to the healthy ones (Sandhyarani and Usha Rani, 2015).

21.4 Insect Pheromones and Plant Chemicals

Another interesting relation between plants and insects is the plant's role as substrates for herbivore reproduction. Many plant chemicals are also used as precursors for pheromone production. Pheromones are the key common system among the insect species that are necessary for successful mating. Some host plant chemicals influence the pheromone in different ways. The interaction between species is manifested by host plant produced chemicals. Several beetles feed and mate on host plants as a result of attraction to insect sex pheromone. Since some chemicals from the beetle's host plants are used in making of pheromone and pheromone precursors by the beetles, the beetles will feed on the host plant and the same chemicals are sequestered on making of their own pheromone.

Among the plant - insect interactions the effects of host plant on pheromone behavior of herbivore adults is particularly important. The synergism between insect pheromone and plant odors can increase the attraction of insect's natural enemies which offers a new strategy for biological control. Biosynthesis and release of mating signals as well as production of eggs may be influenced or even determined by plant chemicals (Hilker and Meiners, 2011). Several herbivorous insects use host plants for the production and use of sex pheromones through larval or adult sequestration of chemically active compounds and pheromone precursors (Nishida, 2002). One good example of sequestration of plant chemicals by larvae and their subsequent use by adult males in sex attraction is shown in *Utetheisa ornatrix* (Arctiidae). The courtship pheromone of this insect derives from pyrrolizidine alkaloids (PAs) *e.g.* monocrotaline obtained by feeding and ingesting these chemicals from the host plant *Crotalaria spectabilis* during their larval stage (Conner *et al.,* 1990). The larvae of *U. ornatrix* have the capacity to retain the ingested alkaloids through metamorphosis till the adult stage. A few other studies also emphasized the association between feeding and sex attraction in certain insect species. Host plants of many *Dendroctonus* beetles serve as pheromone or pheromone precursors to the beetles apart from catering to their nutritonal needs. A very good example is the exposure of male *Ips paraconfusus* beetles to myrcene volatiles which increased the amounts of ipsenol and ipsdienol in hindgut tissues. Application of 2H-labeled myrcene and (2) - and (þ)-a-pinene resulted in production of 2H-ipsenol and 2H ipsdienol and cis- and trans-verbenol (Renwick *et al.,* 1976). The African palm weevil *Rynchophorus phoenicis*(Curculionidae), utilizes a mixture of volatile esters produced by its host plant, *Elaeisquineensis* (the oil palm) which induces male beetles to release the pheromone (E)-6-methyl-2-hepten-4-ol (rhyncophorol) (Jaffe *et al.*, 1993).

Males of certain species of orchid bees and tephritid fruit flies collect terpenoid compounds from orchid plants and use them in the production of aggregation or sex pheromone (Dressler, 1982). A very good example of host plant's involvement in pheromone production can be seen in Helicoverpa spp, one of the major pests of tomato and corn. The (VOCs) emitted by tomato or corn plants initiate sex pheromone production in female abdominal glands and the female moths away from their host plants cannot initiate the production of the pheromone. In another similar case involving European cockchafer, *Melolontha melolontha* the male insects were not attractive to the pheromone alone in traps and the addition of plant volatiles, mimicking the bouquet of mechanically damaged leaves the pheromone activity was enhanced (Reinecke *et al.*, 2002). In Lepidoptera too similar phenomena of synergism was reported. The pheromone [a mixture of (Z)-11-hexadecenal, (Z)-11-hexadecenyl acetate, (Z)-11-hexadecenol] could induce a higher attractant activity in unmated males of the diamondback moth *P. xylostella* only after the addition of green leaf volatiles from their host plant i:e cabbage (*Brassica oleracea*) (Reddy *et al.*, 2002).

Understanding the communication between insects and plants mediated by biologically active phytochemicals has several advantages. A major increase in research on plant secondary metabolites impact on the third tropic level is also important to come to some conclusions and to understand thoroughly of indirect plant defense mechanisms. It is necessary to provide understanding of the herbivore natural enemies response to elevated levels of defense chemicals in an injured plant. This kind of studies will provide deep insight into plant insect interactions and the role of chemicals in plants defense strategy will be well defined. It is expected that elevated levels of plant defense compounds such as phenols will lower the host plant quality and often this will induce both lengthened larval developmental times and greater mortality. All these may challenge the population dynamics of insect natural enemies, particularly the egg parasitoids like Trichogramma species. Plant influences on the effectiveness of natural enemies have important evolutionary consequences for the herbivores involved (Price *et al.,* 1980). At the same time some secondary metabolite compounds when differed in their quantities in the plant as a result of defensive reaction to herbivore attack might also act as kairomonal cues to wandering parasitoids thus enabling easier and faster host location. The ovipositing Trichgramma are sensitive to variations in host quality. Plants under herbivore attack are known to have changes in their secondary metabolite chemicals and these quantitative or qualitative changes will affect the utilization of the semio chemical-associated cues such as synomones and kairomones by the herbivore natural enemies, thereby disrupting or even enhancing the ability of these insects to successfully locate their hosts or sometimes enhancing their utilization. Therefore, ultimately both the parasitoid's reproductive success and its effectiveness as biological control agent depending on the changes in the secondary metabolite compounds.

According to Price *et al.* (1980) modified secondary metabolism in their host plants may affect the suitability of herbivores to their parasitoids and the variation in the quality and composition of food consumed also may result in differences in the chemical composition of the herbivore, that can affect its susceptibility to predator attack.

Acknowledgements

I gratefully acknowledge the cooperation rendered by my dear students, Pratyusha Sambagi and Jyothsna Yasur during the preparation of the manuscript. Thanks are due to Director, Indian Institute of Chemical Technology for the encouragement.

References

Abdelgawad H., Peshev D., Zinta G., Van den Ende W., Janssens I. A., Asard H. (2014). Climate extreme effects on the chemical composition of temperate grassland species under ambient and elevated CO_2: a comparison of fructan and non-fructan accumulators. PLoS ONE 9:e92044. 10.1371/journal.pone.0092044.

Ali JG, Alborn HT, Stelinski LL (2010). Subterranean herbivore-induced volatiles released by citrus roots upon feeding by *Diaprepes abbreviatus* recruit entomopathogenic nematodes. *J. Chem. Ecol.* 36, 361-368.

Allmann,S, Anna Späthe, Sonja Bisch- Knaden, Mario Kallenbach, Andreas Reinecke, Silke Sachse, Ian T Baldwin and Bill S Hansson, (2013); eLife 2:e00421. DOI: 10.7554/eLife.00421 pp1-23.

Ananthakrishnan, T.N., Gopichandran, R., Gurusubramanian, G., (1992). Influence of chemical profiles of host plants on the infestation diversity of *Retithrips syriacus*. *J. Biosci.* 17, 483–489.

Bruce TJ, Pickett JA (2011). Perception of plant volatile blends by herbivorous insects–finding the right mix. *Phytochem*, 72:1605–1611.

Conner, W.F. et al. (1990). Courtship pheromone production and body size as correlates of larval diet in males of the arctiid moth, *Utethisa ornatrix*. *Journal of Chemical Ecology,* 16: 543–552.

Dariusz P, Delaney KJ, Wenda-Piesik A, Sendel S, Tabaka P, Buszewski B (2013). *Meligethes aeneus* pollen-feeding suppresses, and oviposition induces, *Brassica napus* volatiles: beetle attraction/repellence to lilac aldehydes and veratrole. *Chemoecology*, 23:241–250.

De Moraes CM, Lewis WJ, Paré PW, Alborn HT, Tumlinson JH (1998). Herbivore-infested plants selectively attract parasitoids. *Nature*, 393: 570–573.

Dicke M, Bruin J (2001). Chemical information transfer between plans: back to the future. *Biochem Syst Ecol* 29: 981–994.

Dicke M, Van Beek TA, Posthumus MA, Ben Dom N, Van Bokhoven H, De Groot AE (1990). Isolation and identification of volatile kairomone that affects acarine predator-prey interactions – involvement of host plant in its production. *Journal of Chemical Ecology,* 16: 381–396.

Dicke M, Van Loon JJA, Soler R (2009). Chemical complexity of volatiles from plants induced by multiple attack. *Nature Chemical Biology*, 5, 317–324.

Dicke, M and Sabelis, MW. (1988). How plants obtain predatory mites as bodyguards. *Netherlands J Zool.*, 38: 148–163.

Dressler, R.L. (1982). Biology of the orchid bees (Euglossini). Annu.Rev. Ecol. Syst. 13, 373–394

Farmer EE, Ryan CA (1990). Interplant communication: airborne methyl jasmonate induces synthesis of proteinase inhibitors in plant leaves. *Proc Natl Acad Sci USA* 87: 7713–7716.

Finch S, Collier RH (2001). Host plant selection by insects. Horticulture Research International. Wellesbourne, Warwick, United Kingdom.

Glendinning JI (2002). How do herbivorous insects cope with noxious secondary plant compounds in their diet? *Entmol. Exp. Appl.*, 104: 15–25.

Guo L, Li GQ (2009). Olfactory perception of oviposition-deterring fatty acids and their methyl esters by the Asian corn borer, *Ostrinia furnacalis. J. Insect Sci.,* 9: 67.

Hilker M, Meiners T (2006). Early herbivore alert: insect eggs induce plant defense. *J Chem Ecol* 32 1379–1397.

Hunt, D.W.A. et al. (1986). Sex-specific production of ipsdienol and myrcenol by *Dendroctonus ponderosae* (Coleoptera: Scolytidae) exposed to myrcene vapors. *J. Chem. Ecol.* 12: 1579–1586.

Jaffe, K. *et al.* (1993). Chemical ecology of the palm weevil *Rhynchophorus palmarum* (L.) (Coleoptera: Curculionidae): attraction to host plants and to a male-produced aggregation pheromone. *J. Chem. Ecol.* 19: 1703–1720.

Kessler A, Halitschke R (2007). Specificity and complexity: the impact of herbivore-induced plant responses on arthropod community structure. *Curr. Opi. Plant Biol.*, 10: 409–414.

Khattab H, Khattab I (2005). Responses of eucalypt trees to the insect feeding (gall forming psyllid). *Int J Agric Biol* 7:979–984.

Kullenberg B (1961). Studies in Ophrys pollination. Zool. Bidrag. Uppsala 34: 1-340.

Matsumoto, Y. (1970). Volatile organic sulfur compounds as insect attractants with special referenceto host selection. In Control of Insect Behavior by Natural Products, ed. D. L. Wood, R. M. Silverstein, M. Nakajima, pp. 1 33-60. New York: Academic. 345 pp.

Mumm R, Dicke M. (2010). Variation in natural plant products and the attraction of bodyguards involved in indirect plant defense. *Can. J. Zool.,* 88: 628–667.

Nishida, R. (2002). Sequestration of defensive substances from plants by Lepidoptera. *Annu. Rev. Entomol.* 47: 57–92.

Pare PW, Tumlinson JH (1999). Plant Volatiles as a Defense against Insect Herbivores. *Plant Physiol*, 121: 325-332.

Prasannalaxmi K and Usha Rani P (2015). Interactions between Herbivore *Leucinodes orbonalis* G. and its Host plant *Solanum melongena* L.: A Study on Insect Induced Direct Plant Responses. (Communicated).

Pratyusha S, Usha Rani P (2013). Induction of phenolic acids and metals in *Arachis hypogaea* L. plants due to feeding of three lepidopteran pests. Arthropod-Plant Interactions, 7: 517-525.

Rasmann S, Köllner TG, Degenhardt J, Hiltpold I, Toepfer S, Kuhlmann U, Gershenzon J, Turlings TCJ (2005). Recruitment of entomopathogenic nematodes by insect-damaged maize roots. *Nature* 434: 732–737.

Reddy, G.V.P. *et al.,* (2002). Olfactory responses of *Plutella xylostella* natural enemies to host pheromone, larval frass, and green leaf volatiles. *J. Chem. Ecol.* 28, 131–143.

Reinecke, A. *et al.,* (2002). The scent of food and defense: green leaf volatiles and toluquinone as sex attractant mediate mate finding in the European cockchafer *Melolontha melolontha. Ecol. Lett.* 5: 257–263.

Renwick, J.A.A. and Vite, J.P. (1969). Bark beetle attractants: mechanism of colonization by Dendroctonus frontalis. *Nature* 224: 1222–1223.

Renwick, J.A.A. et al. (1976). Selective production of cis and transverbenol from (2)- and (þ)-a-pinene by a bark beetle. Science 191: 199–201.

Sandhyarani Kurra, Usha Rani Pathipati (2015). Whitefly, *Trialeurodes ricini* (Genn) feeding stress induced defense responses in castor, *Ricinus communis* L. plants. *Journal of Asia-Pac. Entomol.,* 18: 425–431.

Sun XL, Wang GC, Cai XM, Jin S, Gao Y, Chen ZM (2010). The tea weevil, Myllocerinu *saurolineatus*, is attracted to volatiles induced by conspecifics. *J. Chem. Ecol.* 36(4): 388-395.

Torto B (2004). Chemical signals as attractants, repellents and aggregation stimulants. In: Hardege JD (ed) Encyclopedia of Life Support Systems (EOLSS), Developed under the Auspices of the UNESCO. Eolss Publishers, Oxford.

Turlings TCJ, Loughrin JH, McCall PJ, Röse USR, Lewis WJ, Tumlinson JH (1995). How caterpillar-damaged plants protect themselves by attracting parasitic wasps. *Proc. Natl. Acad. Sci. USA*, 92: 4169–4174.

Usha Rani P and Nakamuta K. (2001). Morphology of Antennal Sensilla and Sexual Dimorphism of *Trogossita japonica* (Coleoptera: Trogossidae). *Ann. Entomol. Soci. Am.,* 94 (6): 917-927.

Usha Rani P and Pratyusha S (2013). Defensive role of *Gossypium hirusutum* L. Antioxidative enzymes and phenolic acids in response to *Spodoptera litura* F. feeding. *J. Asia Pac. Entomol.*, 16: 131-136.

Usha Rani, P., Jyothsna, Y. (2010). Biochemical and enzymatic changes in rice plants as a mechanism of defense. *Acta Physiol. Plant.* 32: 695–701.

Visser JH, Van Straten S, Maarse H (1979). Isolation and identification of volatiles in the foliage of potato, *Solanurn tuberosum*, a host plant of the colorado beetle, *Leptinotarsa decernlineata. J Chem Ecol* 5:13-25.

Vuorinen T, Reddy GVP, Nerg A-M, Holopainen JK (2004). Monoterpene and herbivore-induced emissions from cabbage plants grown at elevated atmospheric CO_2 concentration. *Atmos Environ*, 38**:** 675–682.

Wang Y and Kays S (2002). Volatile chemistry in relation to SweetpotatoWeevil, *Cylas formicarious* behavior, *J. Amer. Soc. Hort. Sci.* 127 (4) 656-662.

Wearing CH, Hutchins RFN (1973). Alpha-farnesene, a naturally occurring oviposition stimulant for the codting moth, *Laspeyresia pomonella. J. Insect Physiol.*, 19: 125 1 - 1256.

Wei J, Kang L (2011). Roles of (*Z*)–3–hexenol in plant–insect interactions. *Plant Signal. Behav.* 6, 369–371.

Integrated Pest Management in the Tropics; pp. 703-715

New India Publishing Agency, New Delhi (India)

Chapter - 22

Ecology of Pest Management in Tropics

Dharam P. Abrol
Division of Entomology, Faculty of Agriculture, Main Campus Chatha, Sher-E-Kashmir, University of Agricultural Sciences and Technolog, Jammu, 180009 (J&K)
e-mail:dharam_abrol@rediffmail.com,, honeybee_abrol@yahoo.com

22.1 Introduction

Plant pests have been part of man's environment for a long time. Traditional agriculture, which included diverse crops and indigenous crop management practices, provided little opportunity for pests to build up and spread rapidly (Smith and Allen, 1954). Crop production is constantly threatened by increasing difficulties in controlling insect pests, which are a major impediment to improved livelihoods of poor farmers. However, in the quest to increase food production for the growing population, use of pesticides became a norm in crop management. Pesticides became an integral part of the process owing to losses from weeds, diseases and pests reducing the harvestable produce. Use of pesticides along with high yielding varieties, irrigation and improved seed did result in increased productivity. But the rampant use of these chemicals, under the adage, 'if little is good, a lot more will be better' has played havoc with human and other life forms.

During the 1980s, the negative effects of pesticides were already being seen – many pest species were becoming resistant (thus needing heavier dosages and more frequent applications), secondary pests were emerging, the number of acute poisonings was very high, and the overall impact of pesticide use in the ecosystem was becoming apparent. This clearly showed the need for 'alternative methods' and for a 'different pest-management system' for worldwide agriculture. Farmers needed to look at pests and diseases as part of the ecosystem, and focus on prevention through the diversification of production systems. A major component of managing pests and diseases was the relationship between pest incidence, soil fertility and soil organic matter content. It was increasingly realized that poor soils not only result in poor plant growth but also increase plant vulnerability to pest attacks.

During late 1990s, IPM developed into a comprehensive methodology, based on farmers' better understanding of their own agro-ecosystems. IPM first considered the integration of control methods and target pests as a combination of biological-control methods, host-plant resistance, cultural control and selective chemical control (Knipling, 1979). When it was realized that many agricultural practices influence pest incidence, farm management was also added to the equation as an important part of an IPM approach, together with management of natural resources (Altieri, 1994; Bartlett, 1956, 1964; van den Bosch and Stern, 1962; Metcalf and Luckmann, 1994; Bottrell, 1979). Gradually, social aspects were also included. These included pest and disease management, the role of local organizations and the importance of indigenous knowledge in the process of developing the necessary skills and confidence to make ecologically sound and cost-effective decisions on crop healthIn the last 20 years, farmers have been able to stabilize and increase their yields and incomes. Many examples have shown a reduced reliance on pesticides and their use. Lobbying and advocacy actions have led to new rules and regulations in many countries, forbidding the commercialization and use of the most toxic products, and at the same time greatly increasing the awareness of farmers and consumers. However, the problems still persist. In addition, these are not limited to large-scale farmers or to intensive agriculture situations. Majority of farming communities are still grappling with chemical options.

Although reliable statistics are difficult to find, the impact of pests and diseases in worldwide agriculture is considerable, lowering yields and overall production, resulting in losses that are equivalent to millions of dollars (Dahlberg 1979). Agricultural production in India is limited by many factors such as soil, water, genetic potential of the crop and the organisms that feed on or compete with food plants. Food crops the world over are damaged by more than 10,000 species of insects and the overall estimated yield losses from different pests are reported to be US$500 billion globally and Rs 100,000 crores in India alone. About 42.1% of attainable production is lost due to pest attack; however, if no control measures are adopted, the figure would be 69.8%. These losses are not only unpredictable but are also greater in fragile ecosystems on which the majority of small and marginal farmers are dependent.

Furthermore, the most common and easy access to chemical pesticides has only made problems worse. Farmer suicides driven by indiscriminate investments and use of chemical pesticides have become common in countries as diverse as India and Ecuador. Pesticide poisoning is a common story, having reached epidemics proportions in many countries. The first report of poisoning owing to pesticides in India came from Kerala in 1958, where over 100 people died after consuming wheat flour contaminated with parathion. Farmers and labourers who are regularly in contact with pesticides suffer from severe health problems, seriously affecting them and their families. Indiscriminate use of pesticides has also resulted in the contamination of soils and groundwater, leading to the disappearance of fish and birds. Because of the lack of regulations, or of the difficulties in effectively enforcing them, these problems are greater in rural areas of the non-industrialized world.

Insect pest control continues to be a challenge for agricultural producers and researchers. Insect resistance to commonly used pesticides and the removal of toxic pesticides from the market have taken their toll on the ability of agricultural producers to produce high-quality, pest-free crops within economical means. In addition to this, they must not endanger their workers or the environment. We depend on agriculture for food, feed, and fibre, making it an essential part of our economy. Many people take agriculture for granted while voicing concerns over the adverse effects of agricultural production practices on the environment. Insect Pest Management presents a balanced overview of environmentally safe and ecologically sound practices for managing insects.

In the history of the world, no century can match the population growth of the one that has just come to a close. We entered the 20th century with fewer than 2 billion people and left it with 6 billion. At the current pace, when the population increases by about 78 billion per year, the global population is projected to rise to over 8 billion by 2025 (Hinrichsen and Robey 2000) and is heading for 9.2 billion by 2050 (Dempewolf *et al.,* 2010), thus the population may grow even more in the 21st century, but in a very different way. The new century's growth will occur almost exclusively in the developing countries – among people with limited financial resources. The developed countries which almost doubled their population in the 20th century will grow slowly or not at all. Over half of the world population growth will occur in Asia and one-third will be in Africa. The world farmers will thus need to produce enough food for the expanding population (Dhaliwal and Arora 2001; Nwilene *et al.,* 2008).

The future of the world food supply is closely linked to the pattern of population growth. In many countries over the past two decades, growth in food supply has lagged beyond population growth. Worldwide, the grain harvest increased by 1% annually between 1990 and 1997, a rate of growth substantially slower than the average population growth of 1.6% in the developing world (Hinrichsen and Robey, 2000). Since humans started practicing agriculture about

10,000 years ago, the increased need for food that resulted from increasing population or rising living standards was met by bringing a greater area of land under cultivation. However, during the last century, large needs for food were mostly met by increasing productivity per unit of land. This trend is likely to continue as the populations increase and the land is required for purposes other than agriculture (FAO 2001).

Insect pest management is an apparently complex subject which is often too difficult to comprehend fully. Even separate components of the subject present diverse and complicated interactions and unless the system is formalized in some way, it is very difficult to maintain a balanced and holistic perspective. This can lead to biases in emphasis, a narrowing of approach and adopted options, poor decision-making and communication.

IPM is today a common approach to dealing with insect pests in agriculture, although reliance on synthetic organic pesticides remains high. As an applied science, IPM has a structure that incorporates knowledge and information from several sub-disciplines and technologies, a central one of which is insect ecology. At least, insect ecology is widely acknowledged to have such a role, and most of the entomological inputs into IPM are comprehensively covered in a range of textbooks (Dent, 1997; Kennedy and Sutton; 2000; Norton and Mumford, 1993; Pedigo, 1999; Pimentel, 1997; Ruberson, 1999; Speight *et al.*, 1999), at least one of which is undergoing progressive development on the World Wide Web (Radcliffe and Hutchison, 2002). However, the actual relationship of insect ecology to IPM remains somewhat abstract for it has never really been specified in concrete terms. Even those ecological aspects that relate directly to IPM are not given good coverage in IPM texts. Neither do insect ecology texts spell out the way in which the ecological principles they cover are significant to IPM, or at least their claims in this regard are not particularly convincing (Walter, 1995). How one sets out to acquire an appropriate understanding of the biology and ecology of pests and beneficial organisms for IPM purposes is usually neglected in ecology texts. This occurs despite claims by Perkins (1982) and others that IPM provides an intellectual framework not only for practice, but also for research. The story of IPM is therefore incomplete.

Working from a partially developed framework is undesirable; it reduces the chances of successful outcomes and that is in turn economically unsatisfactory. The relationship that IPM shares with ecological considerations is perhaps omitted because the term ‘applied’ has two distinct meanings in relation to IPM. The first deals with the ‘application of understanding’. The development of each control method now available was contingent on at least a certain amount of understanding of the pest’s life cycle, physiology, behaviour or ecology, and this knowledge was then applied to develop a practical solution to a problem. Most of the studies never deal explicitly with this meaning, but tend to focus on the second meaning, which deals with the 'application of techniques'. Control techniques (or management practices) that are already available, and which were perhaps developed independently of the problem situation at hand, are recommended to growers and applied by them to suppress insect numbers. Most of the textbooks tend to provide instruction on application of the second type perhaps because the information is available and the methods have usually worked. The original problem and the difficulties encountered and lessons learned in solving that problem (the development of understanding) are rarely considered. This leaves an educational gap that is eventually likely to carry over into attempts to solve new IPM problems, with probable negative effects.

22.2 What is IPM?

The origin of IPM can be dated back to the response of governments, extension systems, and farmers to the emergence of problems associated with the reliance on chemical controls for insect pests that started after the Second World War and intensified during the Green Revolution. The search for solutions to these problems led to the development of a more holistic view of what constituted an agro-ecosystem and how human interventions could either enhance or disrupt a given agro-ecosystem. With the introduction of chemical insecticides following World War II, plant protection specialists were able, almost without restriction, to make use of powerful poisons. The goal of plant protection specialists in the early 1950s was 100% control (eradication). Entomologists such as Smith and Michelbacher (1943) sounded warnings against this unrestricted use and proposed an approach that tolerated low levels of pest populations to allow natural enemy populations an opportunity to suppress pest populations. They proposed an approach to crop protection that was based on the analysis of the ecology of a given agro-ecosystem and included biological controls. Their work resulted in the development of what was known as Integrated Pest Control (IPC). They proposed that the goal of pest-control strategies was to tolerate pest populations that were below specified threshold levels (Litsinger *et al.,* 1982). IPC eventually became IPM. The stress on decision-making, integration of tactics, and the allowance of tolerable thresholds was seen as a move away from control and towards management 1.3.1 Initial 'Principles of IPC'.

1. The use of chemical pesticides, without regard to the complexities of the agro-ecosystems in which they are used has been a major cause of disruption and undesirable side effects. Undesirable side effects include: target pest resistance and/or resurgence, secondary pest outbreaks, residue problems, and environmental pollution.

2. The agro-ecosystem is a unit composed of the total complex of organisms in the crop area together with the overall conditioning environment. There must be an analysis of the agro-ecosystem to determine population dynamics and mortality factors operating on pest populations.
3. The kinds of crops, agronomic practises, patterns of land use, weather, total complexity, and self-sufficiency of the agro-ecosystem affect the stability of an agro-ecosystem. As complexity increases, particularly among trophic interactions, there is usually an increase in the stability of the agro-ecosystem. IPC should seek to preserve or improve this complexity.
4. Levels or limits of tolerable damage are more important than pest population levels. Tolerable levels of damage vary with market conditions, stage of the crop, local conditions or grower economics, and the personal values of the people concerned. These levels will vary widely. The presence of pests is not an indication of a threat of economic damage to the crop.
5. In more sophisticated programmes, individual fields are surveyed for populations of pests, parasites, predators, and pathogens. On the basis of this information and a consideration of the time of year, stage of growth of the crop, and weather conditions, a prediction can be made of population trends and potential damage. This type of sampling and prediction requires a solid base of fundamental biological and ecological data.
6. All but the most sterile of manmade environments have some biotic agents influencing pest populations. Appropriate consideration must be given to biotic control agents. In some fortunate situations, the biotic agents are all that is necessary to have satisfactory economic control. The failure of natural enemies to keep a given pest under control should not cause us to invoke control practises that disrupt the controlling action of natural enemies of other species in the same agro-ecosystem (Smith and Reynolds, 1966).

22.2.1 Approaches to plant protection

The increase in pesticide application coincided with the introduction of new, higher yielding varieties during the Green Revolution, which were assumed to be irresistible to pests; cost of pesticide was seen as an acceptable insurance 'premium'. Because farmers were seen as being unable to deal with the complexities of pest control, prophylactic use of insecticides was promoted.

This control tactic relied on 'calendar-based' applications of pesticides. Farmers were told to apply insecticides based on the number of days after transplanting or the stage of plant development. In Indonesia, the calendar-based approach in rice cultivation has resulted in farmers applying broad-spectrum systemic insecticides to the nursery bed and at transplanting. Farmers then made additional applications of non-systemic broad-spectrum insecticides around 40 days after transplanting and at or around the milky stage of seed development.Tactical IPM first arrived in tropical rice production systems with the introduction of economic threshold levels (ETLs). This tactic seemed to be a rational way to reduce the amount of insecticides that were being applied under the calendar-based system. ETLs failed to take into account the role of natural enemies in suppressing pest populations. General ETLs were developed for insecticide use throughout a given country. Farmers were told to count the number of a pest, for example, brown planthopper (BPH), and spray when there was a certain amount of BPH per hill of rice. National forecasting systems were another approach taken to control pests. These systems took the responsibility for decision-making away from the farmer and allowed professionals to take charge. Forecasting systems based their control tactics on ETLs. The greater the fear of a particular pest, the lower the ETL and the more pressure put on farmers to apply insecticides. Surveillance systems sometimes made 'recommendations' to farmers with regard to insecticide applications. The systems also included pest eradication brigades to handle 'outbreaks'.

Prophylactic spraying, the use of ETLs, and forecasting systems, all placed limitations on the management role of rice farmers. Crop protection experts assumed that 'professionals' could take better decisions than could farmers. It was not until 1990, when the Indonesian government instituted a farmer IPM education programme, that a truly farmer-based IPM 'system' was promoted. The process of getting to this point required several discoveries on the part of researchers and the realization on the part of both scientists and government that it was appropriate for farmers to be the ultimate decision makers in an IPM crop protection system.

22.2.2 Ecological principles of IPM

Scientists began working out some of the basics of rice field ecology in the late 1970s and early 1980s. Their attention was generally focused on BPH, which at the time was ravaging rice fields across Southeast Asia. They discovered that BPH outbreaks were insecticide induced (Kenmore, 1980) and that breeding rice varieties resistant to BPH under continued pressure of insecticide use was futile (Gallagher *et al.,* 1994). Researchers soon worked out an ecologically-based means of BPH control (Ooi, 1988), yet it was to take some time before the broader ecological approach became part of the mainstream effort.

Whereas the specific conditions critical for management decisions with regard to inputs vary over a small spatial scale, agro-ecosystems do have a general structure and dynamics that are reasonably consistent for the entire system. In essence, it is possible for us to think in terms of a general theory for the structure and dynamics of specific agricultural ecosystems. IPM is not a 'theory' in a strict scientific sense, rather, it is merely a set of practical guidelines for how to best manage a specific crop.

The existence of diverse populations of natural enemies, supported by abundant alternative food species, assures that populations of pests are consistently maintained at low levels. In effect, the structured biodiversity of arthropods in tropical irrigated rice functions to consistently suppress pest populations by denying pests refuge in time or space. All of the key variables can be found in any rice ecosystem – only when the process is disrupted do pest populations explode, causing serious damage. Given this set of implications for rice, IPM practice can be determined. The use of insecticides disrupts and destabilizes natural enemy populations. The use of insecticides is by far the most common cause of pest outbreaks, especially for pests such as the rice BPH. These kinds of pest outbreaks are generally referred to as 'pesticide-induced resurgence'. Several factors combine to enable resurgence to occur:

1. Eggs of many pests, such as BPH, are not susceptible to chemical sprays.
2. Insecticides create a refuge for the development of pest populations by reducing the abundance of natural enemies.
3. Migratory abilities of pests are generally better and their generation many times faster than those of natural enemies. Certain landscape designs can cause delays in the arrival of natural enemies after long dry fallow periods. In many tropical areas, the potential exists for year-round cultivation of rice. However, some areas, both by natural constraints or government design, are planted synchronously over thousands of hectares, and have long (3–4 months) dry fallow periods. These are the areas in which natural enemy populations are weakest and where pest outbreaks are most frequent (Settle *et al.*, 1996).

After such a long dry spell, when the next season of rice is planted, it takes up to half a season to build up predator populations that would otherwise have been there from the beginning. Again, as pests are better recolonizers than natural enemies, the potential for outbreak is much higher in these synchronous, large-scale areas. Much attention has been given to the ideas of 'synchronous planting' and 'breaking the pest cycle with long dry fallow periods' to the point that these ideas have become ingrained almost as 'fundamental principles' of IPM. However, a close look at the empirical data and experiments that purport to demonstrate these principles will show that the support is very weak. Soils, high in organic matter, are the foundation for a 'healthy ecosystem'. Soil organic matter is not emphasized as much for irrigated rice as for dryland cropping systems because of the lesser need for good soil texture. However, as we have seen, soil organic matter is the foundation for energy cycles that ultimately support high populations of natural enemies.

Pest-management strategies can be viewed in the context of whole-farming systems. In whole-farming systems, pest-management methods are integrated into other management components of agronomic systems such as crop fertilization, cultivation, cropping patterns, and farm economics. Such alternative farm management strategies that promote soil and plant health, and water quality have been recommended by the National Research Council in their report *Alternative Agriculture* (National Research Council, 1989).

When natural ecosystems are left undisturbed they generally exist in a healthy balance. For example, the numbers of pest and predator insect species balance one another, so plants can grow without facing a high pest challenge. Some traditional agricultural systems maintain that balance, through planting of diverse crops and indigenous crop management practices that give little opportunity for pest numbers to build up. But efforts to increase food production to support a growing population have promoted different farming methods that have brought with them increased losses to pests and diseases. As a result, use of chemical pesticides to minimize those losses has become widespread.

In the 1980s, it became apparent that overuse of chemical pesticides was having some damaging side effects. For example, pest species developed resistance to the chemicals, demanding stronger or more frequent applications. People applying the pesticides were also experiencing health problems, for example, from inhalation of pesticide fumes or contamination on their skin. There were also instances of consumers being poisoned, when pesticide guidelines were not correctly followed. Beyond this, chemical pesticides were also causing damage to ecosystems, killing plants, animals, insects and thereby resulting in the thinking of ecological pest management.

22.3 Ecological Pest Management

IPM has had many successes, but the magnitude of the problem which farmers currently face forces us to look once again at pests and diseases and at their management. Pests and diseases are not an isolated part of agriculture, but rather a symptom of a broader problem, and need thus to be seen, and managed, accordingly (Levins and Wilson, 1980). There is a need for an ecologically-based pest management when addressing the presence of pests and diseases in agriculture (Turner, 1992; Higley and Pedigo,1996). Building on the wide range of experiences gained with IPM during the last two decades, an Ecological Pest Management (EPM) approach is one which focuses on managing pests as one component of a larger ecosystem (National Research Council, 1996). As such, EPM is based on the following three points.

(a) Minimizing the disturbances which are caused by agriculture

A successful pest-management strategy is based on mimicking nature, redesigning a farm so that it resembles a complex ecosystem. This will mean maximizing a farm's positive ecological processes (such as nitrogen fixation, nutrient mineralization), while at the same time, minimizing undesirable processes such as nutrient loss or erosion. In many cases, it may be convenient to reduce tillage and thus achieve minimal soil disturbance; in many others, it will be necessary to include perennial species and enhance a farm's overall diversity. The interaction of different species can have interesting results, contributing to the system's overall resilience. Undisturbed ecosystems include a host of plants, birds, animals – ensuring a balance in the crop–pest–predator relationships. Abundance of natural predators minimizes the need for specific plant protection measures.

(b) Decreasing plants' vulnerability

Field observations show that pests prefer to attack plants under stress. Current studies such as those falling under the theory known as trophobiosis, talk of a plant's 'internal balance', directly related to its nutritional state. The best way to prevent the attack of pests and diseases is thus by providing a healthy and balanced environment and food supply. There are many factors which can affect a plant's internal balance and thus lessen or increase its susceptibility to pest and disease attacks. These are related to the plant itself (such as adaptation to the local climate or its age) or to the environment (climate, light, temperature, humidity, wind). Plant vulnerability is also related to the different management practices which regularly take place in a farm, such as spacing, tilling, pruning, or the time of planting.

(c) Understanding pests and acting accordingly

In pest and disease management, one of the main considerations is the way the pest species behave – some show abilities to reproduce often and disperse widely, others are able to withstand competition or adverse conditions. To 'know your enemy' is thus a key strategy in every pest-management approach.

This knowledge then needs to be translated into action, considering, for example, the release of beneficial insects at a particular moment, adding bird nesting sites to a farm, or changing the sowing time of certain crops.

These principles are clearly visible in many traditional low input agricultural systems, where ecological principles form the basis for all pest-management strategies. Some of the examples to keep pests away are: raising companion crops emitting strong odour and use of botanical insect repellents. In brief, these refer to working with nature, and not against it. In agricultural terms, this means growing plants in the right soil, at the right time; nourishing the soil and relying on a system's biodiversity as a natural means to safeguard the whole system's health. Traditional wisdom is being maintained by many societies, while it is also being recreated in many 'modern' farms. Managing a farm, however, and relying on its ecological processes, requires a thorough understanding of how these work. Therefore, EPM is, above all, based on farmers' skills, abilities and knowledge.

22.4 Conclusions

The adoption of a more ecological approach, in which the actions and interactions of the component technologies are fully understood within the context of the local agro-ecosystem, could lead to the development of more effective and sustainable 'truly integrated' pest management. The need for this approach derives from the fact that the single solution, 'magic bullet' approach has undermined our basic understanding of how individual components actually work and has oriented IPM towards quick fixes, often underpinned by commercial incentives. Developing truly integrated pest management that addresses the problems of sustainable agriculture will require that more emphasis is placed on long-term solutions, even though the practical outputs are inherently simple. It must be remembered that the complexity of nature also makes it difficult to apply conventional area-wide prescriptions successfully across all systems. Equally, compensating for a lack of understanding with excessive reliance on single technologies and short-term solutions does not make sense; experience of the last 30 years has shown that the greater the apparent success in achieving pest or disease control in the short term, the greater the likelihood of a serious breakdown in the long run (Conway, 1997). Many IPM tools are available now (*eg.*, partially resistant germplasm, partially effective indigenous natural enemies, and potentially effective but unreliable microbial agents) which do not require further technological breakthroughs to be useful. The immediate challenge lies in the genuine application of ecological research to understand how these components interact and identify more effective and sustainable integrated strategies for their use.

References

Altieri, M. A. 1994. Biodiversity and pest management in agroecosystems. New York: Food Products. 185 pp.

Bartlett, B. R. 1956. Natural predators – can selective insecticides help preserve biotic control? *Agric. Chem.* 11(42):44107.

Bottrell, D. G. 1979. Integrated Pest Management. Washington, DC: Council on Environmental Quality, US Government Printing Office, 120 pp.

Conway, G. 1997. The Doubly Green Revolution: Food for All in the 21st Century (Penguin, London).

Dahlberg, K. A. 1979. Beyond the Green Revolution: The Ecology and Politics of Global Agricultural Development. New York: Plenum. 256 pp.

Dempewolf, H., Bordoni, P., Rieseberg, L.H. and Engels, J.M. 2010. Food Security: Crop Species Diversity. *Science* 328 (5975): 169-170.

Dent D. 1997. Paradigms and all that? *Integr. Pest Manage. Rev.* 2: 173–174.

Dhaliwal, G.S and Arora, R. 2001. Integrated pest management: concepts and approaches. Kalyani Publishers, New Delhi, 427 pp.FAO (2001). World Review of the State of Food and Agriculture. Food and Agriculture Organisation of the United Nations Rome, Italy,295 pp.

Gallagher, K.D., Kenmore, P.E. & Sogawa, K. 1994. Judicious use of insecticides deter planthopper outbreaks and extend the life of resistant varieties in Southeast Asian rice. In R.F. Denno & T.J. Perfect, eds. *Planthoppers; their ecology and management*. pp. 599-614. New York, USA, Chapman & Hall. 799 pp.

Higley, L. G and Pedigo, L. P., eds. 1996. Economic Thresholds for Integrated Pest Management. Lincoln, NE: University of Nebraska Press. 328 pp.

Hinrichsen, D and Robey, B. 2000. Population and the environment: the global challenge. Population Reports, Series M, No. 5, Johns Hopkins University School of Public Health, Baltimore, MD, 31pp.

Kenmore, P.E. 1980. Ecology and outbreaks of a tropical insect pest of the green revolution, the rice brown planthopper, *Nilaparvata lugens* (Stal), PhD Dissertation University of California, Berkeley,USA.

Kennedy, G.G. and Sutton, T.B. 2000. Emerging Technologies for Integrated Pest Management: Concepts, *Research, and Implementation.* APS Press, St Paul, MN.

Knipling, E. F. 1979. The Basic Principles of Insect Population Suppression and Management. Washington, DC: USDA, *Agricultural Handbook* 512. 659 pp.

Levins, R. L. and Wilson, M. 1980. Ecological theory and pest management. *Annu. Rev. Entomol.* 25:287–308.

Litsinger, J A, Canapi, B and Alviola A III, 1982. Farmer perception and control of rice pests in Solana, Cagayan Valley, a pre-green revolution area of the Philippines. *Philipp. Entomol.* 5(4):373-383.

Metcalf, R. L. and Luckmann, W. H., eds. 1994. Introduction to Insect Pest Management. 3rd edn. New York: Wiley. 650 pp.

National Research Council, 1989. Drinking water and health, selected issues in risk assessment Safe Drinking Water Committee, National Research Council, Washington, DC.

National Research Council, 1996. Ecologically based pest management: new solutions for a new century. National Academy of Sciences Press, Washington, DC.

Norton G. A., Mumford J. D (Eds). 1993. Decision tools for pest management (Wallingford, UK: CAB International, 1993). 279 pp.

Nwilene, F. E., Nwanze, K. F., and Youdeowei, A. 2008. Impact of integrated pest management on food and horticultural crops in Africa. *Entomol. Exp. Appl.* 128:355–363.

Ooi, P.A.C. 1988. Ecology and surveillance of *Nilaparvata lugens* (Stal) implications for its management in Malaysia. University of Malaya. 275 pp. (Ph.D. dissertation).

Pedigo LP 1999. Entomology and Pest Management, 3rd edn. Prentice-Hall, Englewood Cliffs, NJ, 691 pp.

Perkins, J.H. 1982. Insects, Experts, and the Insecticide Crisis: The Quest for New Pest Management Strategies. Plenum Press, New York, 304 pp.

Pimentel, D. 1997. Techniques for Reducing Pesticides: Environmental and Economic Benefits. Chichester, UK: John Wiley.

Radcliffe E. B. and Hutchison W. D. (eds.), 2002. Radcliffe's IPM World Textbook. URL: http://ipmworld.umn.edu, University of Minnesota, St. Paul, MN.

Ruberson J.R.1999. Handbook of Pest Management. Edited by J.R. Ruberson. Marcel Dekker, New York, 864 pp.

Settle, W. H., Ariawan, H., Astuti, E. T., Cahyana, W., Hakim, A. L., Hindayana, D. H., 1996. Managing tropical rice pests through conservation of generalist natural enemies and alternative prey. *Ecology* 77: 1975–1988.

Smith R.F and Michelbacher A. E. 1943. Some natural factors limiting the abundance of the alfalfa butterfly. *Hilgardia* 15:369-397.

Smith, R. F and Allen, W. W. 1954. Insect control and the balance of nature. *Sci. Am.* 190(6): 38-92.

Smith, R. F.and H. T. Reynolds. 1966. Principles, definitions and scope of integrated pest control. III. Proceedings of the FAO Symposium on Integrated Pest Control, pp. 11-17. Rome, Italy, 91 pp.

Speight MR, Hunter MD, Watt AD, 1999. Ecology of Insects: Concepts and Applications. Blackwell Science, London, 350 pp.

Turner, C. E. 1992. Biological approaches to weed management. In: Beyond Pesticides: Biological Approaches to Pest Management in California (T. Beall, Ed.) Oakland, CA: University of California, pp. 32-67.

van den Bosch, R. and Stern, V. M. 1962. The integration of chemical and biological control of arthropod pests. *Annu. Rev. Entomol.* 7:367–386.

Walter G. H. 1995. Species concepts and the nature of ecological generalizations about diversity. In Speciation and the Recognition Concept: Theory and Application (D. M. Lambert and H. G. Spencer Eds). Baltimore, MD: Johns Hopkins University Press, pp. 191-224.

Integrated Pest Management in the Tropics; pp. 717-732

New India Publishing Agency, New Delhi (India)

CHAPTER - 23

Cultural and Physical Methods for Integrated Pest Management in Tropics

Dharam P. Abrol
Division of Entomology, Sher-e-Kashmir University of Agricultural Sciences and Technology, Jammu, Jammu and Kashmir, India

23.1 Introduction

Cultural methods have been in practice for pest management in traditional systems of agriculture since time immemorial. The worldwide awareness of safe environment provided the impetus to foster non-chemical pest management strategies. Of these strategies, cultural control constitutes the most farmer-oriented approach, where the weak points in the biology and behaviour of insects are exploited and pressure is exerted on the population by manipulating the environment. In addition to the cultural practices, the other widely used approach is the use of physical control methods which hold promise and provide an attractive alternative to current practices, particularly when integrated into a general control scheme incorporating cultural methods, plant breeding, biological or behavioural methods and pesticides. In physical control methods, the physical environment of the pest is modified in such a way that the insects no longer pose a threat to the agricultural crop. This can be achieved by generating stress levels ranging from agitation to death or by using devices such as physical barriers that protect produce or plants from infestation. Many physical control methods target an ensemble of physiological and behavioural processes, whereas chemical methods have well-defined and limited modes of action. Effective physical control methods protect plants during the entire season from emergence to postharvest (Hallman, 2001).

In this chapter, cultural and physical control methods limiting/and or influencing the insect attack are described below.

23.2 Factors or Practices that may be Manipulated to Prevent or Avoid Pests

23.2.1 Sanitation

Sanitation practices, such as use of healthy seed material, disposal of fruit drops, burning and destruction of crop refuge, clean storage, etc. reduce the pest population, discourage breeding and hibernating sites and prevent carry over to the next crop/season. These methods are economical, effective and easy to adopt. Examples include: use of pest-free seed in potato against *Phthorimaea operculella* (Zeller) (Misra and Agrawal, 1991); healthy suckers of banana for *Cosmopolites sordidus* (Germar) and *Odoiporus longicollis* (Olivier) (Butani, 1987a); and in sugarcane to *Chilo auricilius* Dudgeon (Chaudhary, 1981) and *Melanaspis glomerata* (Green) (Chaudhary, 1987).

23.2.2 Tillage operations

Tillage operations help to kill the insects by exposing them to predators, sun and burying a part of them and also help to destroy the weeds. Ploughing helps to control variety of insects. For instance ploughing under the tree exposes egg clusters of *D. mangiferae* (Lakra *et al.*, 1980) which die due to exposure and or taken by other insects such as ants. Similarly, ploughing helps to control white grubs in field (Shah and Garg, 1988) in forest nursery beds (Sen-Sarma, 1987) and to kill *Heliothis* pupae before chickpea planting (Reed *et al.*, 1987). Butani (1987a, b) reported the utility of ploughing to kill grubs of *Aulacophora* spp. in cucurbits, diamond backmoth,*Plutella xylostella* in cole crops and *Dacus* spp. in fruit crops. Chhabra *et al.* (1993) reported winter ploughing against *Acherontia styx* Westwood to prevent infestation in summer greengram and blackgram.

23.2.3 Crop rotation

Crop rotation helps a greater extent to control the pests as continuous cropping encourages carry over of pest from one planting season to another. This situation favours pest build up and increased infestation and consequently severe crop damage. To break the continuity, another crop is planted in between which is not a host or a least preferred one. Crop rotations are very effective in breaking this cycle, especially for insects of limited host range. The crops in which rotation has been recommended include sugarcane and rice against *Holotrichia consangu* Thea Blanchard and *H. serrata* Fabricius (Veeresh 1977), *Leucopholis lepidophora* (Yadav and Mathur, 1987); in rice against several pests (Kalode and Krishnaiah, 1991); in cotton against *Heliothis armigera* (Hubrer) (DPR, 1990); and in potato against *P. operculella* (Misra and Agrawal, 1991).

23.2.4 Fertility management

Plant growth is dependent on the nutritional status of the soil which in turn has an indirect effect on pests. A slight change in the nutritional balance can lead to differential responses. Of the 17 essential elements, the effects of N, P and K have been studied in several cropping systems. According to studies carried out by Russell (1947), the mean nitrogen content of nearly 400 species of plants was 2.14%. This is considerably below the level needed for normal growth and development of insects. Low N supply, invariably results in increased total consumption and/or a prolonged period of feeding, digestion and development of insects (Mattson, 1980). However, increased rates of nitrogenous fertilizers result in increased tissue softness and water content, plant tissues become softer as the carbohydrates are diverted for protein synthesis rather than construction of cell walls (Tisdale and Nelson, 1975). Higher levels of N always favour insects and make plants more susceptible to severe insect infestation.

Phosphorus is generally a limiting element in most soils. The lack of P is also linked to the uptake of other nutrients (Brady, 1974). It is mostly concentrated in the younger parts of the plants, flowers and seeds. It is necessary for photosynthesis and is involved in many energy transfer reactions (Dale, 1988). P fertilization has been known to reduce the incidence of *Atherigona vane soccata* Rondani in sorghum, *Empoasca kern* Pruthi and six defoliators in cowpea, *Athalia proxima* Klug in mustard, *Dasyneura lini* Barnes in linseed and *H. armigera* in chickpea. Increased incidence was also recorded in *Mythimna separata* Walker in wheat and *H. armigera* in chickpea.

Potassium exerts a general balancing effect on nitrogen and phosphorus and this is an important component in mixed fertilizers (Brady 1974). A low potassium supply frequently favours the development of insects, whereas optimum and high K has depressant or neutral effects (Dale, 1988). In a few cases, K is known to result in temporary resistance, generally referred to as induced resistance (Painter, 1951). Increased K enables better protein synthesis, a physiological phenomenon correlated with secondarily enhanced proteolysis and reduces the amino acids and reducing sugars in the sap which were otherwise favourable for the reproduction of sucking pests (Chaboussou, 1972).

23.2.5 Planting dates and Crop duration

Date of planting takes advantage of the pest on the crop or when the susceptible stage of the crop synchronizes with the most inactive period or lowest pest population. However, manipulation of planting date is more meaningful if plantings are based on the information obtained from pest monitoring. Yadav and Lal (1988) found two larval peaks of *H. armigera*, first in 47–50 standard weeks in December and second in 11–15 standard weeks in March. It was the second peak which inflicted losses in chickpea. Early planting (early October) escaped damage and gave higher yield. This confirmed the findings of Lal *et al.*, (1980) and Dias *et al*,. (1983) in Kanpur. Early sowing of chickpea also avoids infestation of *H. armigera* in Punjab (Chhabra *et al.*, 1978) and Haryana (Chauhan 1992). Another striking example of less infestation as influenced by planting dates is known in rape and mustard against mustard aphid, *Lipaphis erysimi* Kalt. Studies by Singh *et al.*, (1992) and Bhadauria *et al.*, (1992) conclusively revealed that early planting of mustard from central to north India received less aphids and provided higher yields. Plant mortality in pea due to Agromyzid flies (*Ophiomyia phaseoli* Tryon and *Chromatomyla horticola* (Gour.) was higher in crops sown on 1 October. Crops sown in mid November escaped the damage (Kumar *et al.*, 1991).

23.2.6 Planting geometry

Planting geometry affects plant density which is related to competition for available nutrients, moisture, sunlight, *etc.* Plant density may also affect natural enemies and results in greater pest control. Logiswaran and Mohanasundaram (1985) found that mulching rice straw and closer spacing in groundnut provided a significant reduction in incidence and increased the rate of parasitism. Kalode and Krishnaiah (1991) reported higher incidence of *Nilaparvata lugens* (Stal.) at closer plantings. Provision of alleys of 30 cm width after every 2–3 m could help in reducing pest incidence in endemic areas. Yadav (1987) reported that, at closer spacing or higher plant density, chickpea had a higher incidence of *H. armigera.* However, Chauhan (1992) found a large number of larvae per unit area at a closer spacing but the reduction in pod damage or increase in yield was generally non-significant.

23.2.6.1 Distance from other crops

Infestation can be reduced if a new crop is planted away from the previous one which was infested earlier. In intensive agriculture, it is not practically possible to adopt such practices. However, the planting of a nonpreferred crop in the vicinity of a preferred one will prevent the spread of the pest to newly planted fields. In Egyptian clover, the larval population of *H. armigera* and damage to inflorescence were greater in fields adjacent to tomato, chickpea or wheat crops than in isolated fields and seed yield was 155, 149, 168 and 185 kg/ha, respectively (Dhaliwal and Sidhu, 1988). This method could be useful if immature stages and adults are feeding on different crops. The non-availability of preferred host to adults will certainly limit the future generations and may aid in a reduction in pest incidence.

23.2.6.2 Inter (Mixed) cropping

Growing two or more crops, planted simultaneously, on the same piece of land is considered to be inter or mixed cropping. In India, as many as 84 different crops are used in mixed cropping (Swindale, 1981). Mixed cropping in traditional farming systems is an insurance against the risk of crop failure, better utilization of farm resources and labour and protection from insects (Rajat De and Singh, 1981). Tragardh (1925), while comparing the outbreaks of pests in the forest ecosystem, found fewer outbreaks in Sweden because two-thirds of the forests were mixed tree species. Monocultures have been characterized for their genetic uniformity and pest susceptibility and are not sustainable for the long-term social and economic well-being of subsistent farmers (Lamb, 1978). Bhatnagar and Davies (1981) reviewed the possible factors which are likely to be important in efficient pest regulation in intercropping systems and listed them as increased parasitoid and predatory population, availability of alternative prey, decreased colonization and reproduction in pests, chemical repellency, masking, feeding inhibition of odours from non-host plants, prevention of emigration in pests and optimum synchrony in the relationship between pests and their natural enemies. Changes in the microclimate of an intercrop also influence the behaviour of insect pests. Usually temperatures and humidity in the intercrop will differ from those of the sole crop, thereby affecting pest colonization (Mehto *et al.*, 1988).

23.2.6.3 Strip cropping

This practice is followed in areas that are prone to wind erosion, and is used for moisture conservation, in sustainable management of natural enemies and in the judicious use of pesticide application to protect them. Bohlen and Barrett (1990) compared the dispersal of a released population of *Papilio japonica* in soybean monoculture, soybean stripped cut with dwarf sorghum, and sorghum stripped crop with tall sorghum in Ohio. Beetles were released in soybean strips in each treatment. Beetles remained longer in the monoculture indicating that sorghum strips inhibited their movement. Dispersal rates were similar in dwarf and tall sorghum.

23.2.7 Shifting cultivation

Shifting cultivation, also known as jhumming, padu, tire farming, slash and burn or swidden agriculture, is practised in the North East Hill region, part of Andhra Pradesh, Bihar, Madhya Pradesh and Orissa. The system involves clearing of a piece of land by cutting of forests, bushes, *etc.* up to stump level, leaving the cut material for drying and finally for burning to make land ready for dibbling of seeds before the onset of monsoon (ICAR 1983). Cultivation continues as long as the land remains productive. Once productivity becomes inadequate, the area is abandoned and a new site is selected. This practice is repeated in different regions following the *jhum* cycle prevalent in the area.

23.3 Trap Cropping

Trap cropping has been known for warding off insect pests for centuries and is still being exploited in many traditional systems. The trap crop provides protection either by preventing the pests from reaching the crop or by concentrating them in certain pockets of the field where they can be economically destroyed (Hokkanen, 1991). The trap crop must be more attractive to the pest than the main crop. The difference in attractiveness can be achieved, a) by using a more preferred cultivar of the same species as the main crop, both grown at the same time, or b) by planting the trap crop at such a time that the most attractive stage of the crop is synchronized with pest incidence and when the main crop is unattractive. The first method is often used when a very mobile pest arrives in the field at a vulnerable crop stage and the second method is used when the pest arrives in the field before the crop is vulnerable or even attractive (Hokkanen, 1991).

23.4 Destruction of Alternative Hosts

Many insects use a wide range of cultivated and uncultivated plants, especially weeds as alternative hosts for the offseason carry-over of population. Altieri *et al.* (1987) observed weeds as a major component of an agro-ecosystem which affects the biology of pests and beneficial insects by providing flowers, presence of neutral insects, modification of crop microclimate, production of chemical stimulus, alteration of colonization background, *etc.* Matteson *et al.*, (1984) reported that weeds and second growth around or within the crop can alter the proportion of harmful and beneficial insects that are present and increase or decrease crop damage. Removal of alternative hosts, such as *Achyranthus aspera* for *Caliothrips indicus* Bagnall in groundnut (Mohan *et al.,* 1984); wild rice (*Oryza nivara*) for *Orseolia oryzae* (Wood-Mason); *Leersia hexandra, Echinochioa colonum, E. crusgalli, Cynodon dactylon, Panicum repens* to *Nephotettix nigropictus* Stiel; *Chieres barbata* to *Sogetella furcifera* Horvath, *Cyperus difformis* and *E. colonum* to *N. lugens* and there are many on record for gundhy bug, rice hopper and leaf folder in rice (Kalode and Krishnaiah, 1991); solanaceous crops and weeds for *P. operculella* in potato (Misra and Agrawal, 1991); *E. colonum* for *M. separata* in rice (Kulshrestha and Nigam, 1987); water shoots and sprouts/tillers for *C. auricilius* and *Pyrilla per-pusila* Walker in sugarcane (Chaudhary, 1987), grassy weeds for *Contarinia sorghicola* Coq. in sorghum (Prem Kishore, 1987) has been recommended to minimize the population.

23.5 Destruction of Off Type and Volunteer Plants

During cultivation, a new variety becomes impure and it possesses some off types. These off type plants grow along with the standard variety and due to differences in susceptibility; they attract different types of insect pests. According to Joginder Singh *et al.* (1987), cotton crop (cv. F414) in the majority of fields contained a large number of off type plants of Jhurar, Ganganagar Ageti and arboretum varieties. Plants of different types helped differently in the population build up of various pests. Ganganagar Ageti proved more susceptible to *Amrasca bigutulla* Ishida and *P. gossypiella*, Jhurar to *B. tabac*, *H. armigera* and *Thrips tabaci* Lindemann compared to F414. These off type plants served as hot spots for the multiplication of insect pests and eventually injury to the crop.

23.6 Selection of Site

Selection of the appropriate site for the cultivation of field crops and forest trees can reduce the future infestation from insect pests. The decision demands good knowledge of crop husbandry and pest ecology and behaviour. Insects *M. separate, Spodoptera mauritia* Boisd and *S. litura* occur sporadically but in some states, particularly in areas subjected to flood epidemics, these occur after water recedes (Kulshrestha and Nigam, 1987). It would therefore, be better not to grow crops prone to attack by the aforesaid insects in these areas.

23.7 Thinning and Topping

Cultural practices such as thinning and topping greatly alter the population of insect pests if timed properly and result in less damage to the crop and subsequent crops grown in the area. In cotton, nipping of terminal buds when 16 to 18 sympods, reduces the egg load of *H. armigera* (DPR 1990).

23.8 Pruning and Defoliation

Pruning of infested parts of fruit and forest trees and defoliation in field crops for the destruction of insects help in reducing the pest population. To prevent infestation of stem borer, *Crypsonema* spp., in *Populus ciliatus*, 15-30 cm thick cuttings is recommended (Chaukiyal *et al.*, 1987). Removal of tops affected by *A. steniella* in sugarcane during July, August and September at weekly intervals resulted in a population decline (Chaudhary *et al.*, 1987; Jolly and Singh, 1990). Interestingly, infestation was higher in Banger (well irrigated soil) than in Khadar (flooded during the rainy season) soils (Chaudhary, 1987). Prasad and Singh (1986) observed that pruning malformed inflorescence in mango with 30 cm healthy portion decreased the number by the end of the fourth year.

23.9 Water Management

Availability of water in requisite amounts at the appropriate time is crucial to the lives of the plants. As a result, water affects the associated insects by way of nutritional quality and quantity, allelochemical availability, cell division and expansion, canopy and root system structure, partitioning of resources between vegetative growth and reproduction, timing of senescence, water potential, micro-environment and third trophic level interactions (Holtzer, *et al.*, 1988), Management of water can, therefore, accentuate or hinder growth and development of insect pests.

In field trials, Gopalan *et al.* (1987) found that continuous ponding at 5 cm throughout the growth period in rice recorded a lower incidence of *Brevennia rehi* Lindenger. Copius irrigation in the nurseries of fruit trees minimizes the incidence of *Odontotermes obesus* (Rambur) and *A. ipsilon* (Butani, 1987b) and standing water more than a week to kill white grubs in sugarcane (Chaudhary, 1987) reduces infestation. A substantial reduction in egg hatch of *D. maniferae* was observed in mango orchards when soil was raked four times (May, June, August and October) and irrigated compared to control (Chandra *et al.*, 1989). Draining of water to field capacity to avoid build up of *N. lugens* and caseworm, *Nymphula depunctalis* Gueanee was recommended (Thomas, 1986).

23.10 Cultural Practices and Natural Enemies

Like insect pests, predators and parasitoids are also affected by various cultural practices adopted to minimize the pest population. Practice such as mulching with rice straw and planting more closely than the normal spacing in groundnut resulted in an increase in parasitism (Logiswaran and Mohanasundaram, 1985). A higher population of *Coccinella septempunctata* L. in blackgram/sorghum intercropping and *Pollstes hebreus* F. in blackgram/pigeonpea was observed. Spider, *Acantholeptis simplex*, population was not affected by intercropping (Dhuri *et al.*, 1986). However, Pawar *et al.* (1989) did not find any difference in *Heliothis* population and its parasitism on sole sorghum and when intercropped with pigeonpea. When pearl millet was intercropped with groundnut, the population of *Coccinella* sp. and *Menochilus sexmaculatus* F. increased compared with the sole crop. Greater activity might have been due to their movement from colonizing resources to the area of activity (Kennedy *et al.*, 1990) or the system provided vital food, shelter and a resting site (Risch, 1981). Murli Baskaran and Thangavelu (1990) also observed a high mortality of leaf miner, *Aproaerema modicelia* Deventer, due to *Goniozus* sp. in the groundnut/pearl millet combination. As results have indicated, there is a great possibility to enhance parasitoid activity through inter-cropping.

23.11 Cultural Practices in IPM

Cultural practices are ideal to use in an IPM programme. Appukuttan *et al.* (1990) reported the achievements of IPM in a rice crop in Tamil Nadu. Before the implementation of IPM, the farmers were generally applying pesticides 4–5 times during the crop season. In IPM demonstration plots, application of insecticides was resorted to only if it became necessary and the pest crossed Economic Threshold Levels (ETL). IPM used non-monetary control tactics such as, a) use of resistant/tolerant varieties, b) provision of proper drainage facilities, c) clean cultivation and removal of alternative host plants, d) judicious use of nitrogenous fertilizers, e) periodic weeding and conservation of biological agents. Results revealed that spiders, coccinellids and carabid beetles played a major role in containing the population of key pests such as brown planthopper and green planthopper. In addition to proper drainage, optimum plant population and balanced use of fertilizers also helped in suppression of leaf and planthoppers build up. In addition to minimum use of pesticides, there has been no adverse effect on yield. Yield in IPM was not significantly changed but the cost of production was greatly reduced. Besides an advantageous cost–benefit ratio there were many benefits in adopting IPM.

In conclusion, cultural methods are cheapest measures and within the reach of common farmers. If timely adopted can be very effective and accurate. They have no adverse effect on natural enemies and are ecofriendly. The only disadvantage is that the farmers want quick means of pest management and have little reliance on thse methods because of slow action and uncertainty.

23.12 Physical Methods

Of the several control tactics, physical and mechanical methods are used from ancient times. However, in the last two decades due to indiscriminate use of pesticides and degradation of the environment interest has been regenerated to search for alternative and sustainable pest control methods. Physical control methods give immediate results and are generally popular among the farmers, even though they are expensive, time and energy consuming, laborious and often applied when much damage has been done. These control measures are generally ineffective and on a large scale and cannot be applied commercially (Atwal and Dhaliwal, 2002). Physical control methods hold great promise both in pre and post harvest situations of the crop (Vincent *et al.*, 2008). Vincent *et al.* (2008) reviewed the information on use of physical methods for the management of insect pests. Earlier, Banks (1976) stated that in stored commodity protection, physical methods are of increasing importance and may completely replace chemically-based current protectant methods. Direct insect control by ionising radiation has received much attention but large scale application appears doubtful. Application of many physical methods is restricted by high capital cost and lack of post-treatment.

According to their mode of action, physical control methods can be active or passive. The level of efficacy of active methods is proportional to both intensity of the energy and duration of its application to the target. Examples of active methods include thermal shock (heat, cold), electromagnetic radiations (microwaves, radiofrequencies, infrared, ionizing radiations, UV and visible light), mechanical shock and pneumatic (blowing or vacuum). Most active methods have a short persistence as the stressor effect is limited to the period of application. This could be an advantage, as in post-harvest situations or situations where the environment should be minimally disturbed, or a disadvantage, in situations where the treatment must be repeated several times to achieve prolonged control. Passive methods do not require further energy to achieve desired effect. Examples are traps, airtight or hermetic storage, barriers of various kinds and trenches (for details See Vincent *et al*., 2008).

In general, physical control methods are relatively more labor intensive and often time consuming when applied pre-harvest. The implementation of physical methods compares favorably with, and can augment, biological methods. However, most successes occur in post-harvest situations. Effective use of physical control measures, in both pre- and post-harvest control, relies on three main components: (1) good sanitation practices -which should be applied with all control methods, (2) monitoring to determine extent and stage of pest populations and (3) application of other direct or indirect physical control methods.

The effects of physical control methods with the exception of electromagnetic radiation are limited spatially. This attribute starkly contrasts with chemicals which may drift several hundred meters and may be bio-accumulated in food chains, or with biological control agents which may disperse actively or passively over long distances. Post-harvest IPM is greatly facilitated bythe fact that it takes place inside a well-defined area, delimited by the structures of industrial buildings (food processing facility, grain storage elevator, food factory, feed mill, *etc*.). As a consequence of all of these aspects, physical control methods occupy a smaller market share than pesticides worldwide. For further information on physical control methods in agricultural plant protection, we refer the reader to the reviews of Banks (1976), Oseto (2000), Vincent *et al.* (2001) and Weintraub and Berlinger (2004).

Throughout the world there is an urge to reduce pesticides; evidently the physical control methods would be widely used. However, the use of these methods both in pre and post harvest situations is not without limitations. Implementation of physical control methods in food storage and processing needs a certain amount of preparedness and investment. Besides, complete information about bionomics and behaviour of insects to be dealt must be known. Some methods, such as steaming of Colorado potato beetle in the field, simply lack technical efficacy (Lague *et al.,* 2001). Other methods (*e.g.*, trenches) are technically and environmentally acceptable but are impractical and costly relative to an efficacious chemical. Many physical methods rely on energy transfer by diffusion, convection, or radiation. In the field, this is a major obstacle because it is difficult to use the energy efficiently without excessive losses to nontarget substrates (Matthews, 1992). In physical control methods as in any components related to IPM, it is not known whether the insects become resistant to some physical control methods? Likewise, it is not clear if insects can become resistant to inert dusts (Korunic and Ormesher, 1998). If the control method allows the insect to adapt and reproduce, such as anthomyiid adults flying over fences (Boiteau and Vernon, 2001), then genetics is at play and resistance can develop. Except for a few studies related to the effects of vacuum on insect pollinators (Chiasson *et al.*, 1997) and predators (Lacasse *et al.,* 2001), little is known about the effects on nontarget organisms, especially in preharvest situations. Little is known also about the effects of physical control methods on plant diseases. Physical stresses induce general mechanisms of physiological responses that, in most cases, have nonspecific targets or receptors, in contrast to chemicals. Of particular concern is the widening technological and economic gaps between developed and developing countries (Donahaye, 2000). Because their success generally depends on context, many physical control methods should be as useful and desirable in developed and developing countries.

Hand picking is one of the most ancient method employed by man and is still being used for picking out lice from human hair. In the field, insects can be hand picked if they are large, conspicuous, available in large numbers and easily accessible to the picker (Atwal and Dhaliwal, 2002). This method is useful in case of lemon butterfly eggs and adult clusters, grubs of mustard saw fly, *Athalia lugens* and all the developmental stages of *Epilachna* sp. The use of hand nets and big nets is another method recommended for collection of adults for example pyrilla, *Pyrilla* spp when these insects are migrating from maize to sugarcane in April and May (Verma and Tewari, 1994). The field bag is a strong cloth bag, 2 metres long with its mouth measuring 1x1.5 metres and supported with bamboo sticks and two strings on the upper side. It is scrapped on the surface of the ground by two men and it is recommended for surface grasshoppers, rice grasshoppers, crickets *etc.* Six men can cover one hectare in six hours. A one man field bag can also be devised for reducing the size of the mouth. Killing house flies with fly flappers and locusts with brooms or thorny bushes is effective. On coconut palms the rhinoceros beetle can be picked out of the holes with the help of crooked hooks made of iron. Sieving and winnowing can be employed against insect pests of stored grains. A good number are removed with these operations, particularly the grubs of *Tribolium castaneum* and *Trogoderma granarium* which infest wheat. Similarly, Shaking or jarring of small trees or shrubs particularly early in the morning in the cold season when insects are benumbed can be helpful in collecting them in open tubs containing kersenized water or simply burying them in pits is effective against locusts and the defoliating beetles, *Adoretus* sp. Clipping off the top of rice seedlings containing egg masses of yellow stem borers and grubs of hispa would reduce carry over of infestation from seed bed to field. Mechanical exclusion consists of the use of devices by which insects can be physically prevented from reaching the crops and agricultural produce. The various methods include: The application of a fluffy cotton band 15 cm wide or a band of a sticky material like Ostico or a band of a folded slippery sheets like alkathene around the trunk of a mango tree to prevent the upward movement of the mango mealy bugs, *Drosicha mangiferae* (Green). Screening windows, doors and ventilators of houses to keep away house flies, mosquitoes, bugs, etc is quite helpful. In the morning and at dusk when mosquitoes gather on the screen they can be squashed with a piece of cloth. Wrapping individual fruits of pomegranate and citrus fruits and citrus with butter paper envelopes to save them from attack of the anar butterfly, *Virachola isocrates* and fruit sucking moths, *Ophideres* spp. respectively. Maize cobs can be protected from the attack of crows if the nearest leaf is wrapped around the exposed portion of the cob. Trenching fields or erecting barriers, 30 cm high, in order to save crops from the invading bands of locust hoppers or the red hairy caterpillars, is quite helpful. Placing four legs of a meat safe in vessels containing water to prevent ants from ascending is also useful. Using red light in the monsoons to keep away most of the insects, and to keep the fields well lit with white light to protect against certain insects has been quite effective. Light reflection by aluminium foil is effective against aphids. Similarly, light reflected by plastic ribbon or plastic flags hung in the ripening rice fields will protect the crop from bird attack.

Scaring birds by creating noise with explosives is quite effective; an automatic devise is available in which an explosive gas catches fire intermittently with a loud noise is produced.

References

Altieri, M.A., Trujillo, F.J., Farrell, J., 1987. Plant–insect interactions and soil fertility relations in agroforestry systems: Implications for the design of sustainable.

Atwal A S and Dhaliwal G. S. 2002. Agricultural Pests of South East Asia. Kalyani Publishers, Ludhiana, India 498p.

Banks, H. J. 1976 Physical control of insects: recent *developments. Journal of the Australian Entomological Society,* 15: 89-100.

Bhadauria, N.S., Bahadur, J., Dhamdhere, S.V., Jakhmola, S.S., 1992 . Effect of different sowing dates mustard crop on infestation by the mustard aphid, *Lipaphis erysimi* (Kalt.). *J. Insect Sci.* 5 (1): 37 – 39 .

Bhatnagar, V.S. and Davies, J.C. 1981. Pest management in intercrop subsistence farming. In: *Proceedings of the International Workshop on Intercropping, 10–13 January 1979, Hyderabad*, India, pp. 249–257.

Bohlen, P.J., Barrett, G.W., 1990 . Dispersal of the Japanese beetle (Coleoptera: Scarebaeidae) in strip-cropped soybean agroecosystems. *Environ. Entomol.* 19 (4): 955 – 960.

Boiteau, G., Vernon, R.S. 2001. Physical barriers for the control of insect pests. In *Physical Control Methods in Plant Protection,* Eds C. Vincent, B. Panneton and F. Fleurat Lessard. Berlin, Germany: Springer Verlag, pp. 224–47.

Brady, N.C., 1974. The Nature and Properties of Soils, eighth ed. Macmillan, New York. Butani, D.K., 1987 a. Key pests of important fruit crops and their management . In: Mathur, Y.K., Bhattacharya, A.K., Pandey, N.D., Upadhyay , K.D., Srivastava, J.P. (Eds.), Recent Advances in Entomology. Gopal Prakashan, Kanpur, pp. 408 – 432 .

Butani, D.K., 1987. Key pests of vegetables and their management In: Mathur, Y.K., Bhattacharya, A.K., Pandey, N.D., Upadhyay, K.D., Srivastava, J.P. (Eds.), Recent Advances in Entomology Gopal Prakashan, Kanpur, pp. 433 – 454 .

Chaboussou, F., 1972 . The role of potassium and cation equilibrium in the resistance of the plants towards diseases . *Potash Rev* . 23 (39): 10 – 15 .

Chaudhary, J.P., 1987. Key pests of sugarcane and their management . In: Matthur , Y.K. , Bhattacharya, A.K., Pandey, N.D., Upadhyay, K.D., Srivastava, J.P. (Eds.), Recent Advances in Entomology Gopal Prakashan , Kanpur , pp. 260 – 285.

Chaudhary, J.P., Yadav, S.R., Singh, R.P., 1987. Declining trend in the incidence of Gurdaspur borer, *Acigona steniella* (Hmpsn) due to its mechanical control. *Indian J. Entomol.* 49 (4): 460 – 470 .

Chaudhary, J.P. 1981. Proceedings of National Symposium on Stalk Borer, *Chilo auricillius* Dudgeon, held on 20–22 April 1981, HAU Regional Research Station, Bhiwani, Karnal (Haryana), India, 182 pp.

Chaudhary, R.N. and Sharma, V.K. 1988. Incidence of maize borer (*Chilo partellus*) in relation to fertilizer levels. *Indian J. Agric. Sci.* 58(10): 780–781.

Chauhan, R. 1992. Present status of *Helicoverpa armigera* in pulses and strategies for its management in Haryana. In: Sachan, J.N. (Ed.), *Helicoverpa Management: Current Status and Future Strategies. Proceedings of First National Workshop held at Directorate of Pulses Research, Kanpur, India*, 30–31 August 1990, pp. 49–54.

Chaukiyal, S.P., Sharma , S.R., Ombir Singh, 1987. Effect of thickness of cutting on the susceptibility of shoot borers, erectness and branchness of stem, growth and survivability of *Populus ciliates*. *Indian J. Forest.* 10 (40): 245 – 248 .

Chhabra, K.S., Kooner, B.S., Dhingra, K.K., Tripathi, H.P., 1978. Preliminary studies on the influence of sowing dates on the incidence of gram pod borer, *Heliothis armigea* (Hub.). Proceedings of All India Rabi Pulses Workshop held at Bhubaneshwar, 3–6 October, 1978.

Chhabra, K.S., Lal, S., Kooner, B.S., and Verma, M.M. 1993. Insect Pests of Pulses – Identification and Control Manual. Punjab Agricultural University, Ludhiana and Directorate of Pulses Research, Kanpur, 88 pp.

Chiasson, H., Vincent, C., de Oliveira , D., 1997 . Effect of an insect vacuum device on strawberry pollinators. *Acta Hortic.* 437: 373 – 377 .

Dale, D., 1988. Plant mediated effects of soil mineral stresses on insects. In: Heinrichs, E.A. (Ed.), Plant Stress–Insect Interactions. John Wiley & Sons, New York, pp. 35-110.

Dhaliwal, J.S., Sidhu, D.S., 1988. Effect of adjoining host crops, date of last cut and natural enemies of *Heliothis armigera* infesting seed crop of Egyptian clover (*Trifolium alexandrium*). Indian *J. Agric. Sci.* 58 (11): 832 – 836.

Dhuri, A.V., Singh, K.M., Singh, R.N., 1986 . Effect of intercropping on population dynamics of insect pests of blackgram, *Vigna mungo* (L.) Hepper. *Indian J. Entomol.* 48(3): 329 – 338.

Dias, C.A.R., Lal, S.S., Yadav, C.P., 1983. Difference in susceptibility of certain chickpea cultivars and local collection to *Heliothis armigera* (Hub.). *Indian J. Agric. Sci.* 53: 482-485.

Gopalan, M., Coumararadja, N., Balasubramanian, G., 1987. Effect of different planting dates and irrigation regimes on the incidence of rice mealybug, *Brevennia rehi Lindinger. Madras Agric.* J. 74 (4–5): 226 – 227 .

Hallman, G.J., 2001. Irradiation as a quarantine treatment. In: Molins , R.A. (Ed.) , Food Irradiation: Principles and Applications. John Wiley, New York, pp. 113 – 130 .

Hokkanen, H.M.T., 1991 . Trap cropping in pest management. *Annu. Rev. Entomol.,* 36: 119 – 138.

Holtzer, O.T., Archer, T.L., Norman, J.M., 1988. Host plant suitability in relation to water stress. In: Heinrichs , E.A. (Ed.), Plant Stress–Insect Interactions . John Wiley & Sons , New York, pp. 111 – 137.

ICAR, 1983. Shifting Cultivation in North East India. ICAR Bulletin.

Joginder Singh, Ramesh Arora, Sidhu, A.S., 1987. Impact of cotton off types on the incidence of different insect pests and their management. *Indian J. Ecol.,* 14 (2): 254–260.

Jolly, S.C., Singh, R.P., 1990. Mechanical control of Gurudaspur borer (*Accigona steniella*) (Hmpsn.) in Yamunanagar (Haryana). In: *Proceedings of the 52nd Annual Convention of the Sugar Technologists Association of India* 1989, Kanpur, India: Sugar Technologists Association of India, pp. 157–162.

Kalode, M.B., Krishnaiah, K., 1991. Integrated Pest Management in Rice. *Indian J. Plant Prot.,* 9: 117 – 132 .

Kennedy, F.J.H., Rajamanickam, K. and Raveendran, T.S. 1990. Effect of intercropping on insect pests of groundnut and their natural enemies. *J. Biol. Control* 4(1): 63–64.

Korunic, Z., 1998 . Diatomaceous earths, a group of natural insecticides . *J. Stored Prod. Res.* 34: 87 – 97.

Kulshrestha, J.P., Nigam, P.M., 1987 . Integrated management of key pests of paddy. In: Mahur, Y.K., Bhattacharya, A.K., Pandey, N.D., Upadhyay, K.D. Srivastava, J.P. (Eds.), Recent Advances in Entomology. Gopal Prakashan, Kanpur, pp. 186 – 211.

Kumar, D., Mavi, G.S., Joginder Singh, 1991. Impact of agrotechnical practices on the incidence of pea leaf miner, *Chromatomyia horticola* (Goureau) on peas . *J. Insect Sci.* 4 (2): 180–181.

Lacasse, B., Laguë, C., Roy, P.M., *et al.*, 2001. Pneumatic control of agricultural pests. In: Vincent, C., Panneton, B., Fleurat Lessard, F. (Eds.), Physical Control Methods in Plant Protection. Springer Verlag, Berlin, Germany, pp. 282 – 293.

Lague, C., Gill, J., Péloquin, G., 2001. Thermal control in plant protection. *In:* Vincent, C., Panneton, B., Fleurat Lessard, F. (Eds.), Physical Control Methods in Plant Protection. Springer Verlag, Berlin, Germany, pp. 35 – 46 .

Lakra, R.K., Kharub, W., Singh, Z., 1980 . Pest management systems for the mango mealy bug *Drosicha mangifera* Green – a polyphagous pest of fruit trees in Haryana . *Indian J. Entomol.* 42 : 153 – 165 .

Lal, S. S., Yadava, C. P. and Dias, C. A. R. 1980. Insect pests of pulse crops and their management. *Pestic. Annu.* 1980–1981: 66–67.

Lamb, K.P., 1978. Pests of winged bean and their control in Papua and New Guinea. *In:* Singh, S.R., van Emden, H.F., Taylor, J.A. (Eds.), Pests of Grain Legumes: Ecology and Control. Academic Press , London , pp. 53 – 58 .

Logiswaran, G., Mohanasundaram, M., 1985 . Effect of intercropping, spacing and mulching in the control of groundnut leaf miner, *Aproaerema modicella* Deventer (Gelechiidae: Lepidoptera). *Madras Agric. J.* 72 : 695 – 700 .

Mattson Jr, W.J., 1980 . Herbivory in relation to plantnitrogen content . *Annu. Rev. Ecol. Sys.* 11: 119 – 161 .

Mehto, D.N., Singh, K.M., Singh, R.N., 1988. Influence of intercropping on succession and population build up of insect pests in chickpea, *Cicer arientinum* Linn. *Indian J. Entomol.* 50(3): 257 – 275.

Misra, S.S., Agrawal, H.O., 1991. Potato tuber moth, *Ptithorimaea operculella* (Zeller) and its integrated management. *Pestic. Inform.* 17 (2): 3 – 7 .

Murli Baskaran, R. and Thangavelu, S. 1990. Influence of intercrops on the incidence of leaf miner (*Aproaerema modicella* Deventer) in groundnut. *J. Oilseed Res.* 7: 143–146.

Oseto, C. Y. 2000. Physical control of insects. In: Insect Pest Management, eds. J. E. Rechcigl & N. A. Rechcigl,pp. 25-100. Boca Raton, FL: Lewis Publishers.

Pawar, C.S., Bhatnagar, V.S., Jadhav, D.R., 1989 . Helicoverpa on sorghum . *Indian J. Entomol.* 51 (4) : 416 – 421.

Prasad, V.G., Singh, R.K., 1986 . Prevalence and control of malformation of mango inflorescence in Bihar . *Indian J. Entomol.* 48 (1) : 104 – 112.

Rajat De and Singh, S.P., 1981. Management practices for intercropping systems. *In: Proceedings of the International Workshop on Intercropping, 10–13 January 1979, Hyderabad, India*, pp. 17–21.

Reed, G.L., D.K. Reed, A.C. York, 1987. The effect of black plastic film mulch and furadan upon yields of muskmelon infested with larvae of the striped cucumber.

Risch, S.J., 1981. Insect herbivore abundance in tropical monoculture and polyculture on experimental test of two hypotheses. *Ecology* 62 : 1325 – 1340 .

Russell, F.C., 1947. The chemical composition and digestibility of fodder shrubs and trees . Jt. Publ. (Imp. Agric. Bur.) 10 : 185 – 231 .

Sen-Sarma, P.K., 1987 . Insect pest problems in social forestry plantation and their management. *Indian J. For.* 10 (4) : 239 – 244 .

Shah, N.K., Garg, D.K., 1988 . Exposure of root grubs during ploughing and their predation by birds. *Indian J. Entomol.* 50 (1) : 131 – 132 .

Singh, V.K., Thakur, B.S. and Bajpai, R.P. 1992. Influence of sowing dates and incidence of aphids on yield of mustard. *Curr. Res. Univ. Agric. Sci.* (Bangalore) 21(5–7): 97–98.

Swindale, L.D., 1981. Foreword. In: The Proceedings of the International Workshop on Intercropping, 10–13 January 1979, at ICRISAT, Hyderabad, India.

Thomas, M.J. 1986. A note on evaluation of integrated pest management in Kutanad rice fields of Kerala. *Indian J. Plant Prot.* 14: 27–28.

Tisdale, S.L., Nelson, W.L., 1975. Soil Fertility and Fertilizers, Third ed., Macmillan , New York.

Veeresh, G.K., 1977 . Bionomics and control of white grubs in India. *Sugar News* 9 : 44 – 56.

Verma, A. and Tewari, N.K., 1994. Recent concepts in concepts insect pest management in sugarcane. *In:* G.S. Dhaliwal and Ramesh Arora (eds.) Trends in Agricultural Insect Pest Management. Commonwealth Publishers, New Delhi, India, pp 238 – 264.

Vincent, C. and Boiteau, G., 2001. Pneumatic control of agricultural pests. In: Vincent, C., Panneton, B., Fleurat Lessard, F. (Eds.), Physical Control Methods in Plant Protection. Springer Verlag, Berlin, Germany, pp. 270 – 281.

Vincent, C., Weintraub, P.G., Hallman, G.J. and Fleurat-Lessard, F., 2008. Insect management with physical methods in pre- and post-harvest situations. *In:* Radcliff, E.B., Hutchison, W.D., Cancelando, R.E. (Eds.), Integrated Pest Management. Cambridge University Press, pp. 309-323 .

Weintraub, P.G. and Berlinger, M. 2004. Physical controlin greenhouses and field crops. *In* : Novel Approaches to Insect Pest Management, eds. A. R. Horowitz & I. Ishaaya,pp. 301-318. Heidelberg, Germany: Springer-Verlag.

Yadav, C.P. and Lal, S.S., 1988. Relationship between certain abiotic and biotic factors and the occurrence of gram pod borer, *Heliothis armigera* (Hbn) on chickpea . *Entomon* 13 (3&4): 269 – 273 .

Yadav, C.P.S. and Mathur, Y.K., 1987. White grub – a national pest and strategy of its management. In: Mathur, Y.K., Bhattacharya, A.K., Pandey, N.D., Upadhyay, K.D., Srivastava, J.P. (Eds.), Recent Advances in Entomology. Gopal Prakashan, Kanpur, pp. 1 – 20 .

Yadav, C.P. 1987. Ecological studies on gram pod borer, *Heliothis armigera* (Hubner) in chickpea in relation to its biotic and abiotic factors. Ph.D. Thesis, Kanpur University, Kanpur.

Integrated Pest Management in the Tropics; pp. 733-750
© 2016, Editor, Dharam P. Abrol
New India Publishing Agency, New Delhi (India)

CHAPTER - 24

Taxonomy as a Tool for Successful Future of Integrated Pest Management

Dharam P. Abrol, Devinder Sharma and V.V. Ramamurthy*
Division of Entomology, Faculty of Agriculture, Main Campus Chatha
Sher-E-Kashmir University of Agricultural Sciences & Technology-Jammu, 180009 (J&K)
email:dharam_abrol@rediffmail.com,honeybee_abrol@yahoo.com
*Division of Entomology, Indian Agricultural Research Institute, Pusa
New Delhi - 110012

24.1 Introduction

Taxonomy, which includes identification and classification has transformed into systematics and biosystematics in the recent times and is integrating itself into the biodiversity, which is now a global phenomenon occupying the main thrust of human life (Narendran, 2001). The magnitude and cause of this biodiversity is the central problem of taxonomy and one of the key problems of science as a whole, not to speak of pest management. The science of taxonomy started in the late eighteenth century by a Swedish biologist, Carl Linnaeus who with a small number of similar minded scientists pioneered a standard system of using two Latin names for an organism, denoting "genus" and "species" which is still in vogue. This led to referring of biological organisms by an universally agreed name and identity, and further providing means for their realistic classifications. When these identifications and classifications got strengthened further with accumulated biological knowledge, it allowed the development of modern biology on a sustainable and stable base. In addition, in the recent times taxonomy has become more demanding subject for areas in science and technology, which are dependant on access to accumulated experience and knowledge for their survival and development. Insect pest management is one such area where taxonomy provides the essential scientific basis, without having any claims of its own and integrating itself in such a manner that its role is difficult to delineate precisely (Rosen, 1986).

For scientific management of a pest, the essential requirements are that one must know what is being dealt with, and generate major concepts for dealing with the same. Concepts are generated by generalizations and generating specific theories by testable hypotheses. By providing a name and classifying a pest, taxonomy provides information on the nature of populations, life cycle, habitat and behaviour and thereby enable developing hypotheses, concepts and theories. In addition, it enables comparison and synthesis, all of which lead to refinement and fine tuning of these concepts and theories and accomplish identification of causes and problems. Insect pest management is one such branch where taxonomy has got all these general roles to play. Taxonomy is the science behind stimulating ideas, making information available, and allowing identification which are the basic requirements in any branch of science involving understanding of nature and their practical application.

24.2 Role in Insect Pest Management

It is the prerequisite for any Insect pest management system to know the proper identity of the pest. It is biosystematics that answers this question in full and its entire context and thereby forms what is called the first step in insect pest management, namely the "pest diagnostics". By knowing what is that which is causing crop losses, a pest management worker can ascribe/ foresee what will be the characters of the pest. For example, when a pest management worker in a sweet potato crop gets an answer to his question as *Cylas formicarius*, he can immediately ascribe that it shall be a destructive pest on tubers and his strategies shall be planned accordingly. By knowing the name he can derive a wealth of information that could be effectively utilized in initiating his actions and planning a long time strategy (Debach, 1960).

The second aspect is that we shall answer the questions namely "What is its native home, its distribution and biological features?" Biosystematics by providing the name of an insect will enable answering of all these questions. It will help search for the information already available on a particular insect, its bionomics, parasites, predators, vulnerable life stages/ seasons of occurrence, tested control measures, economic injury levels/ thresholds, damage vs yield equations etc. For instance, if we know the name of the pest, by knowing its classification we can assume the biology of the pest. Once such an answer is known, we can forecast the developmental stages of the pest; what is the vulnerable stage amenable for management; what are its natural enemies; whether any appropriate control measures are already available; what are the problems so far in its management; what are the factors that shall be taken into account, in managing the same, especially the problems relating to insecticide resistance *etc.* Biosystematics, if it changes its directions and pave the way for development of a comprehensive database on pests, a field level pest management specialist can derive a lot from these databases and thereby take pest management decisions, plan strategies and implement the same efficiently. Biosystematic specialists in the recent days accept this development of a comprehensive database on the biodiversity associated with an agroecosystem as their duty unlike the previous days, when they were churning out only specialized monographs which were of use only to other biosystematic specialists. The development of a crop protection compendium for the pests of crops of South East Asia by the taxonomists of CABI is one such example. It has been planned in such a manner that it will provide the basic information like diagnostics, damage, and case studies on control etc., comprehended in one place. Such databases if developed in a user friendly, interactive mode in the form of CD-ROMs will go a long way in helping the field level pest management workers.

The third and important aspect of insect pest management is that it shall know whether it could succeed in its contents and for this it shall answer the question "Did we succeed?" It is biosystematics again which helps in evaluating the pest incidence by identifying the insect and its life stages, identifying the parasites and predators introduced, help evaluate the pheromones/ other species specific strategies employed etc., and thereby answer the question "Did we succeed?" This role is very important in case of any insect pest management strategy involving release of parasites/ predators/ other biological agents. In case of the recently developed species specific management measures like utilization of pheromones, insect transgenic plants and their use in insect pest management, this postmortem step is very essential. Biosystematics has got to play a very major role in these, as development of biotypes is going to be a serious handicap. It has been the experience in India and abroad that the host plant resistance/ Bt transgenic based pest management measures had led to unresolved complex situations baffling the pest management workers. This is mostly due to the development of biotypes/ races/ other intraspecific populations in many pest species. Only biosystematics can resolve these complex situations and pave the way for successful reorientation of pest management strategies. The burning problems the world over, in the management of pests namely, cotton boll worms like *Helicoverpa armigera*, leaf feeders like *Spodoptera litura*, whiteflies like *Bemisia tabaci*, leafhoppers like *Nephotettix* spp., planthoppers like *Nilaparvata lugens*, diamond back moth *Plutella xylostella etc.*, have been presumed to be due to development of biotypes/ complex intraspecific populations. It is imminent that biosystematics will have a greater role to play in the management of these and other problem pests (Norlund and Lewis, 1985).

In addition to the above general role, taxonomy has specific roles to play in studying biological diversity, phylogeny, evolutionary perspectives, faunal synthesis, ecology, including habitats and habits, adaptations, bionomics, associations among organisms- especially natural enemies, plant insect associations, vectors of pathogens and parasites, ecosystems and communities, physiology, biochemistry, toxicology, ethology, morphology, embryology, molecular biology and biotechnology, all of which now have an increased impact on the insect pest management strategies, either directly or indirectly. In view of globalization and other developments, it will become indispensable on the part of insect pest management to lean more on biodiversity and thereby get more demanding on biosystematics. The recent WTO guidelines relating to the International plant protection convention of the FAO stipulate many conditions and many of these relate to the pests, and other faunal biodiversity associated with a crop. The successful implementation of these guidelines will envisage greater emphasis on the biosystematics.

24.3 General Role of Taxonomy

Historically taxonomy played an important part in the development of biological thought, because information on diversity provided by the means for biologists to discover major concepts, notably evolution. Studies by taxonomists of large samples of specimens likewise helped in the development of concepts related to populations of organisms, which have been fruitful especially in genetics. Taxonomy continues to contribute to these fields that it helped to found. Currently taxonomy provides a framework for understanding organic diversity and for organising the variety of organisms to assist co-operative work and stimulate ideas in numerous disciplines. It contributes the information for studies of the broadest scope in faunistics and biogeography, providing answers and suggesting new questions. Taxonomy also has important connections with ecology and other fields. Practically, of course taxonomy makes it possible to identify organisms in nature and supports the growth of knowledge and understanding in every basic and applied field. In brief, taxonomy permits us to organise available information, and directly provides some of the information, for understanding nature, which is the basic purpose of scientific enquiry. The key to all of the diverse facts about the insect world is identification and classification.

24.4 Specific Role of Taxonomy

Taxonomy has ample specific roles to play in biological control, by way of providing information on morphology, ecology, physiology, behaviour *etc.*, of the target organisms and their natural enemies. It has also provided information on communities and the associations between the host organism and their natural enemies. Comparative taxonomy supported by morphology, physiology, biochemistry, molecular biology have all contributed to biology of living organisms and these contributions are of immense value in handling specific situations in biological control. Useful clues on morphology, physiology, behaviour, biochemistry provided by taxonomists have revolutionized the application aspects of many biological strategies and biological control is one among them. There have been plenty of specific examples where substantial disadvantages have resulted due to inadequate taxonomy and associated information, and biological control strategy failed or was misdirected owing to lack of good taxonomic support.

24.5 Role of Taxonomy in Documentation

Documentation in taxonomy is most important both for research on diversity as well as variation and irreplaceable reference materials of type specimens. Therefore the importance of taxonomic collections and the continuing research on them has been emphasized. Collections also are important stores of voucher material for verifying the identity of species studied in ecological or other types of work. Unless voucher specimens have been preserved, detailed data often cannot be assigned to the correct species, especially when subsequent taxonomic or ecological work demonstrates the existence of several species within a complex originally though to be a single species. Taxonomy, if it changes its directions and pave the way for development of a comprehensive database on pests, a field level biological control can derive a lot from these databases and thereby take decisions, plan strategies and implement the same efficiently. Taxonomic specialists in the recent days accept this development of a comprehensive database on the biodiversity associated with an agroecosystem as their duty unlike the previous days, when they were churning out only specialized monographs which were of use only to other taxonomists.

24.6 Changing Role of Taxonomy

Taxonomy is really useful in understanding biology used for biological control. There is a need of sound and logical basis for establishing and ranking the organisms into taxonomic unit. Evidently, taxonomy has got a major role to play in biological control or any other practical aspects of pest management. Many times the changing role of taxonomy and its integration into biodiversity is ignored which involves much more than naming, describing and classifying species- the details of ecology, cytology, chorology, biometry, evolution, genetics, palaeontology, parasitology, physiology, anatomy, climatology, geography, biochemistry, molecular biology, biotechnology, information technology are all now getting intertwined with taxonomy. Evidently, the role of taxonomy has become more complex for biological control studies as a consequence of above changes. Hence the role of taxonomy can be fully understood only if these changing roles are well defined and understood. It is well known that taxonomy and biological control are interdependent as many times the parasitological and ecological characters are integrated with taxonomy, and many times these form the sole taxonomic characters, especially in the complex parasitoid groups of insects (Miller and Wenzel, 1995).

24.7 Taxonomy and Pest Identification

The most important feature in the concept of pest management is the emphasis on knowing the entire ecosystem. Since pest management deals with the populations and communities, the accurate identification of pests and their parasite-predator complex is critical. Implementation of IPM and management of invasive pests requires correct identification of species. CABI has been promoting free taxonomic and pest identity service through a global plant clinic in collaboration with number of organizations in developing countries. CABI's UK office has maintained authoritative identifications and reference collection of almost half a million fungi and bacteria. Besides, CABI also runs training programmes and user friendly identification keys and involved in development of novel pest identification methods based on molecular techniques.

24.8 Taxonomy and Biological Control

Biological control, through the action of indigenous natural enemies, is the basis of most integrated pest management (IPM) systems (Debach, 1960, Beard 1999, Cresswell, 2011). Biological control can be successful only if the exact country of origin and associated natural enemies such as parasites, predators and parasitoid is known. Since the biological control is absolutely dependent upon the accurate identification of the pests, the natural enemies and the hyperparasitiods. A mistake in the identity of the host may completely nullify the hard work of foreign field agents years of work and the useless expenditure of large amounts of money. The misidentification of insects or natural enemies mostly results due to either lack of proper keys, or inappropriate morphological characters. The other reasons include absence of voucher specimens, lack of skilled taxonomists and molecular techniques which could be useful to differentiate between cryptic/sibling species.

In correct identifications may result in management problems of the pests. For example, if a pest is of oriental origin is mistakenly identified as a closely related European species, the search for the natural enemy in Europe and the collections, rearing and colonization for biological control might well prove futile. Improper identification of California red scale is one such example, which infact is a complex of species distributed in tropics and subtropics of Africa and Asia (McKenzie, 1937) and became a serious pest in new world in absence of natural enemies (Compere, 1961). Many parasitoids available under natural conditions for *Aonidiella* were not effective against *A. auranti.* This was due to the fact that parasitoids were effective against *A. citri* and not *A. auranti.* Similarly, some parasitoids introduced into California from oriental region failed to control *A. aurantii* as they were specific to *A. citri.* This became clear when the genus *Aonidiella* could be separated on the basis of microscopic differences (McKenzie, 1937) and led to the exploration of host specific parasitoids from the genus *Aphytis*. Evidently, correct identification of pest species is of paramount importance in the selection of parasitoids for biocontrol. Pembertan (1941) gave an outstanding example of the importance of an identified taxonomic collection in biological control. The weevil, *Syagrius fulvitarsus* Pascoe, became a very destructive pest of *Sadleira ferns* in a forest reserve in Hawaii. It was not known that this species also existed outside Hawaii except in greenhouse in Australia and Ireland. From this knowledge the origin of this species could not be traced. In 1921 when Pembertan examined the insect collection of Sydney, he found a single specimen of *S. fulvitarsus* with the date of collection of 1857 and its locality in Australia. This led to the search of the beetle in the mentioned locality and a braconid parasitoid attacking larvae was found. The parasitoid was shipped to Hawaii and the pest was effectively controlled.

Another example of wrong identification refers to mealy bug (*Planococcus kenyae*) problem on Arabica coffee in the Kenya causing considerable losses. This pest first appeared in Kenya in 1930 causing severe losses to coffee. It was incorrectly identified as *P. citri* Risso in place of *P. lilacinus* Cockerell. But later, both the identifications proved wrong and the problem was solved by LePelley (LePelley 1935, 1943) who stated that it was a undescribed species having inconspicuous but consistent morphological differences. This mealy bug also occurred in Uganda and Tanzania where biological control through naturally occurring parasitoids was very successful. When these parasitoids were imported to Kenya from these areas, the biological control program proved very effective. It was estimated that by 1959 the project had saved Kenya at least £10 million for an outlay of less than £30,000 (Melville , 1959).

Genus *Aphytis* and *Marietta* belonging to family Aphelinidae, are so closely related that it is difficult to distinguish them on the basis of morphology. However, biologically the differences between the genera are so great that *Aphytis* species are primary parasitoids of armored scale insects while *Marietta* species are hyperparasitoids. A minor mistake in proper differentiation could lead to serious consequences as hyperparasitiods (*Marietta*) are deleterious to biological control. Similarly, family Encyrtidae contains both a large number of primary parasitiods and secondary hyperparasitiods of phytophagous insects. Recognition of these parasitiods requires proper knowledge of taxonomy and any failure thereof is likely to result in the establishment of undesirable species. Earlier examples of misidentification of *Aphytis holoxanthus* n. sp. as *Aphytis lingnanensis* Comp. retarded its use and distribution in various countries for the biological control of *Chrysomphalus aonidum* (L.). Same was true for *Aphytis cohenin.* sp., a parasite on *Aonidiella aurantii* (Mask.) in Israel, which was thought to be *Aphytis lingnanensis* (Debach, 1960). Similarly, the predatory mite *Neoseiulus cucumeris* (Oudemans) (Acari: Phytoseiidae) which is an important biological control agent of thrips. Yet, despite its economic importance, this species is poorly defined taxonomically and cannot be reliably separated from other species on the basis of morphology alone *Neoseiulus cucumeris* has been reported from Australia, although considerable confusion exists as to whether the Australian material is actually *N. cucumeris* or *Neoseiulus bellinus* (Womersley) (Beard 1999).

Zeddies *et al.* (2001) reported that correct identification of a new mealy bug species *Phenacoccus manihoti*, causing 80 per cent losses in cassava crop in republic of Congo resulted in a highly effective biological control using a hymenopteran parasitoid *A. lopezi* with a cost benefit ratio between 1:1200. Similarly, in Namibia maize production was seriously affected by false cutworm damage to seedlings and seed treatments were ineffective till the pest was properly identified and correct pesticide formulation evaluated to prevent the beetle damage to sprouting maize (Penrith, 1977). In a similar study, Hancock *et al.* (2003) reported that Namibia's export of agricultural produce to USA was threatened by the risk of occurrence of quarantine pest Mediterranean fruit fly in the area, but appropriate measures at early stages for early recognition of the pest appearance, helped solve this problem. Viscarret *et al.* (2003) in Argentina reported that proactive taxonomy allows identification of biotypes of whiteflies (*Siphonius phillyreae* Haliday; *Bemisia tabaci*) for effective implementation of biological control with natural enemies such as *Encarsia hispidia; E. protransvena* Viggiani and *E. transvena* Timberlake. In another example, oil palm introduced in Malaysia failed to produce fruit in the absence of pollinators, but when the weevil, *Elaeidobius kamerunicus* was introduced from Cameroon fruit yields increased quickly (Dave, 2011). Similarly, exact taxonomic identifications have helped save million of dollors throughout the world by successful implementation of biological control programmes (Drew and Hancock, 1994, Julien and Griffiths, 1998., Lindquist, *et al.*, 1992, Waterhouse and Sands, 2001, Calder and Sands, 1985, Tricio *et al.*, 2002 , Mervyn, 2011, Robert, 2011, Wagener, *et al.*, 2006, Cresswell, 2011, Naggs, 2002, Kirton and Brown, 2003, Cooke, 1992. Miyamoto *et al,*, 1983. Fiaboe *et al.*, 2006 Frana, 2011a, Frana, 2011b).

The above examples clearly indicate that for successful biological control, the correct identification of both pest and their natural enemies is essential as most of the natural enemies are highly specific (Gauld 1986). Furthermore, the correct identification is particularly more important when either the pest or natural enemies are closely associated and morphologically indistinguishable. The incorrect identication could lead to costly failure of different biological control initiative (Rosen, 1986). Morphological characters are not always reliable to distinguish species in mixed populations where the species may differ in their behavior and physiology (Fernando and Walter, 1997).

The need for accurate identification is of great importance in biological control. The species which are similar morphologically may not be similar biologically and even the minor structural differences can mean a lot between pest and non pest species.

Like other scientific studies, taxonomic findings are never static and are subject to change particularly when new characters are discovered. For instance, in case of minute parasitoids such as *Trichogramma,* lack of specific morphological characters has made the taxonomy difficult and several conflicting assessments have been made (Pinto, 2006a). The identification of this genus is based on male antenna and genetalia but the existence of closely related species could be easily determined by reproductive characters, allozymes and minor morphological traits (Basso *et al.*, 1999). Allozyme technique though is sensitive enough to detect allozyme variation in small insects or first instars but allow the detection of only some (if any) of the most active enzymes (Gotoh *et al.*, 1993), thereby impending the application of electrophoresis analysis in important pest species (spider mites) or parasitoids (*Trichogramma* spp.) used in pest management programmes. Kazmar (1991) used cellulose acetate gel electrophoresis technique for the multilayered electrophoresis and obtained quality results with various insect pests and biocontrol agents (Pinto *et al.*, 1991, Oudman, 1992). The molecular studies in taxonomy help to differentiate the species (Rosen, 1986; Stouthamer *et al.*, 1999; Silva *et al.*, 1999).

24.9 Role of Biological Characters in Systematics

Pinto (2006b) in a comprehensive review on the importance of systematics in biological control emphasized that successful biocontrol programmes depend upon the accurate identification of insects and their natural enemies, cryptic species or intra specific entities. He stressed on the selection of proper traits for accurate identification. Since most of the taxonomic studies depend upon keys based on morphological characters which are easy to to observe than biological traits which are more informative to define relationship between species and group of species (Shah, 1988). Shah (1988) stated that biological characters give proper identification of cryptic species of natural enemies of insects as they take into account ecological, behavioural and reproductive data. According to him several species of entomophagous insects were identified based on their ecological data and microhabitat preferences which differe fron one species to another. For instance, all the trichgramma species are known to parasitize the eggs of wide variety of insects, but even within a particular habitat there may be significant differences in their preference (Pinto and Stouthamer, 1994).

Mode of reproduction is another important parameter used for taxonomic identification. In parasitic hymenoptera, thelytoky and arrhenotoky is prevalent. For instance, in case of genera *Encarsia, Aphytis* and *Trichogramma* various morphologically indistinguishable forms have been distinguished based on their differing mode of reproduction. The biological characters should be assessed in as many populations as possible for their variations.

24.10 Challenges in Insect Taxonomy

Insects will soon reach one million known species worldwide. The number of described species on the planet is about 1.9 million, with ca. 17,000 new species described annually, mostly from the tropics. Although society has a growing need for credible taxonomic information in order to allow us to conserve, manage, understand, and enjoy the natural world, support for taxonomy and collections is failing to keep pace (Godfray, 2002a, b; Wheeler *et al.*, 2004) and passing through a world crisis (Agnarsson and Kuntner, 2007). Taxonomy is a science in crisis, lacking manpower and funding, a politically acknowledged problem known as the Taxonomic Impediments.

24.10.1 Lack of national repository centre and museum

In insect systematic, proper repository for entomological collections is essential. Most of the institutions because of financial constraints and shortage of technical persons face difficulty in maintaining the valuable collections. The lack of proper support from policy makers and administrators is another impediment for the growth of the taxonomy. Collaborative approach on the part of National and International organizations is essential for advancement of taxonomy to truly understand the biodiversity.

24.10.2 Taxonomic collections and maintenance

The lack of proper housing and well-curated insect collections is one of the major impediments. Most of the valuable specimens are damaged for lack of trained handling staff. Where these specimens have been lost, for example by museum pests or acts of war, this information is lost forever. Descriptions of species and records by earlier workers are often inadequate or unreliable without the specimens available for examination. Specimens from many collections are used frequently for current research in the most of the countries. Collections can be broken down into national, local, university and private collections. University museums vary greatly in size and coverage and can be considered to maintain some of most important collections. Very few taxonomists are currently employed in museums cross the world. They are dependent on grant funding, which was always erratic due to the apathy of Government.

24.10.3 Declining expertise

Most of the universities do not have positions of taxonomists or curators to maintain the museums. Where ever available there number is very less. For example only 17 insect taxonomists are available in Malaysia, 32 taxonomists in Chile (Simonetti, 1997) and 140 taxonomists in Brazil (Rafael *et al.*, 2009) and hardly 100 taxonomists in India (Narendran, 2001) are available to study insects. Most of the post graduate students are more inclined to study subjects relating to more advanced or frontier areas of science that are applied or job oriented than opting for taxonomic work which is time consuming,labourious with less dividends. Consequently, the taxonomists of entomophagous species are rare and becoming more rare in times to come. University teachers often find students who wish to become taxonomists and their general advice is to take a different job, as there is no money and precious few jobs in taxonomy.

24.10.4 Lack of funding for taxonomic research

Lack of proper funding is a major cause for decline in taxonomic studies. Most of the funding is diverted to fronrier areas of science such as genome projects, thereby neglecting it as the science of nineteenth century. It is beyond doubt that thousands of species which are at the verge of extinction may disappear before they are identified (Wheeler, 2004; Wheeler *et al.*, 2004). Most of the taxonomic projects do not attract proper funding because of lack of clearly achievable goals that are realistic and relevant (Godfray, 2002a). The most of the research Councils fund only "hypothesis-driven" research; taxonomic publications are judged to have low impact; and because funding is scarce and new money tends to go to fields that are perceived to be innovating.

24.10.5 Lack of training in taxonomy

The discipline of taxonomy lacks proper training (Nelson and Platnick, 1981). Most of the young recruits do not find taxonomy as a lucrative subject of their choice as it involves long years of working to attain perfection and excellence in the area and have limited job opportunities. It takes years to attain perfect level of excellence to describe complex characters. Evidently, the transfer of taxonomic expertise and skills to a new generation is decreasing.

24.10.6 Lack of taxonomic literature

Furthermore, there is lack of literature in taxonomy and most of the modern libraries do not subscribe to the taxonomic journals. There is no proper documentation of taxonomic literatures and it is very difficult to know the holdings of a particular library. Relevant data basis are absent in most of the institutions and if available are incomplete.

24.10.7 Difficulty in publishing taxonomic work

Currently, there are very few quality journals dealing with taxonomic research and exclusively publishing taxonomic papers. Furthermore, impact factor of most of the journals publishing taxonomic work is relatively low and peer reviewers are also scanty and scattered in different areas. Unfortunately, when the biodiversity is under threat the taxonomy of important biological resources is under siege and still considered a science of 18th century.

24.10.8 Lack of taxonomy as subject in universities curricula

There is a lack of well-organized, systematic education and research in basic Entomology. There is no satisfactory training in modern systematic biology available even at university level, where the subject should be taught in conjunction with insect ecology, evolutionary biology, genetics, molecular biology and biological illustrations. With the advent of more fashioned, demanding and employment oriented study areas, taxonomy as a basic study is losing its luster in university curricula. Regular grants and faculty position as taxonomists is the other barriers/hurdles which enable taxonomy as less lucrative studies. In order to generate interest and develop the science of taxonomy, it needs to be incorporated as core subject at graduate and post graduate levels. It could also be blended with much demanding integrated pest management and ecological studies to make them more popular among students and researchers.

24.10.9 Lack of identification service

There is no specialized centre in the country fully equipped to provide identification services. Even the available centres do not have expertise for all the group of insects and services provided by them are inadequate. Earlier the free identification services available with CAB International Institute of Entomology have now imposed heavy charges for each identification have virtually worsen the situation to keep ward off many workers to bear such expenses. It is, therefore necessary to establish at least one or two identification centre in each region to cater the needs on regional basis. Adequate funding should be provided for establishment of such identification centre fully equipped with adequate manpower, a library, collections and necessary equipments. These centers should also coordinate with the specialists available in the country and can serve as national repositories for the betterment of society.

24.10.10 International cooperation

There is a need for international cooperation among the different institutions and organizations involved in taxonomic studies to develop net working for caring of collections and their sharing to study insect diversity. National consortium of insect systematic collections or organizations could help sharing of specimens as well as expertise to train new taxonomists. Such integrated efforts would help to identify and document insect biodiversity on the brink of extinction.

Table 24.1 : Selected instances in which substantial disadvantages resulted from inadequate systematic information

Problem	Result	References
Inadequetly known species or biotypes of *Trichogramma* used for biocontrol often did not attack the target host.	Costly introduction and releases of the wrong species gave no control.	Knutson, 1981
Species of *Aphytis* parasitoids were not correctly recognized ; potential control agents were sought in wrong places.	Delayed releases for biocontrol in California.	Rosen, 1986
Cultures of San Jose Scale parasitoid *Prospaltella perniciosi* effective for biocontrol, were invaded by ineffective (but vigorous in culture) *Prospaltella fasciata*, which was not recognized.	Four million individuals of the wrong species were released in Germany, 1956-58.	Delucchi *et al.*, 1976; Rosen and Debach, 1973
Parasitoids of Gypsy moth, *Lymantria dispar*, were introduced in to eastern North America without proper taxonomic analysis, and no voucher specimens were kept.	Inability to understand what happerned; past and future research rendered ineffective.	Sabrosky, 1955
Natural enemies of pests and wild pollinators or crops are inadvertently destroyed by insecticides, because knowledge of biology and systematic of these insects is inadequate.	Estimated annual increase of over $300 million in the cost of pest control in the US.	Pimentel *et al.*, 1980
Control of malarial mosquitoes was attempted by draining swamps in Trinidad, but the vector species actually bred in air plants, where control should have been applied.	Millions of dollars wasted on swamp drainage ; control of vector delayed.	Cushing, 1957; Rosen, 1978
Detailed mathematical and energtics analysis for arctic communities were based on inadequate taxonomy and ecological information for invertebrates, so that the results are likely to be in error.	A very costly project gave insufficiently reliable results.	Ryan, 1977; Danks, 1981
Legislations and regulations that require evaluation of potentially adverse impacts presupposes a knowledge of the biota, which is illusory.	Legislation is unworkable.	Watts, 1979

24.11 Conclusions

To conclude, the role of biosystematics in insect pest management can be explained in a simple way as follows. There are two kinds of sciences, namely speculative sciences -whose objective is discovery of truth, new facts or new laws and practical sciences whose object is production of facts, causing something new to happen. Economic entomology falls under the latter category and it provides the building blocks for insect control, from which the technology of insect pest management has emanated. Such practical sciences are nothing but arts and these cannot exist and survive without a speculative and pure science backing them. Thus, for the art of insect pest management, the science of taxonomy provides the essential backing. Its general, specific provides these backing, changing roles in addition to the recently acquired role in documentation or bioinformatics.

The primary need in the integrated pest management is to manage the human resources to cater the needs of trained professionals, well-supported museums and specialized and regular publications. To avoid the constraints of parataxonomists for in-house identification services, pest managers and agricultural scientists have to be equipped with basic knowledge of available taxonomic tools and taxonomic resources. There is a dire need for updating the collection of insect pests and biocontrol agents in reference libraries and museums, upgrading the reference libraries, high speed internet facility for prompt access to web based taxonomic portals in the institutes dealing with taxonomic research and teaching. For accurate and prompt identification of insect pest and biocontrol agents, Agricultural Universities/Institutes must strengthen and should focus on capacity building of identification services, automated species identification tools and DNA barcode reference study.

Some of the grave issues such as biotypes development, biodiversity crisis, WTO guidelines on phytosanitary requirements for exports and imports, IPR regime, molecular biology and biotechnology have pave the way for more demand of taxonomy and it must be addressed properly as future approach in pest management. It is of course the high time for biosystematic specialists to mould themselves in the ever changing and demanding scenario to keep contributing to insect pest management and above all to society in an efficient manner.

References

Agnarsson, I and Kuntner MZ. 2007. Taxonomy in a Changing World: Seeking Solutions for a Science in Crisis. *Systematic Biology* 56(3):531–539.

Basso, P B and Grille G. 1999. Taxonomic study of two Trichogramma species from Uruguay (Hymenoptera.: Trichogrammatidae). *Boletin de sanidad vegetal, Plagas*, 25: 373-382.

Beard JJ. 1999. Taxonomy and biological control: *Neoseiulus cucumeris* (Acari: Phytoseiidae), a case study. *Australian Journal of Entomology.* 38 (2) :51–59.

Calder, AA and Sands, DPA. 1985. A new Brazilian *Cyrtobagous* Hustache. (Coleoptera: Curculionoidae) introduced into Australia to control *Salvinia Journal of Australian Entomological Society*. 24: 57–64.

Compere H. 1961. The red scale and its natural enemies. *Hilgardia*, 31: 173-278.

Cooke, M.J. 1992. Turnabout is fair play – Cape bee invades African bee territory. *American Bee Journal* 132: 519 – 521.

Cresswell I. 2011. Biocontrol of a red scale boosts citrus industry in North America. Case study 28.bionet@bionet intl.org | www.bionet intl.org. page 31.

Cushing, EC. 1957. History of Entomology in World War II. Ashington DC; Smithsonian Institute. 117p.

Danks, HV. 1981. Arctic Arthropods . A review of systematics and ecology with particular reference to the Noth American Fauna. Ottawa. *Entomological Society of Canada*. 608p.

Dave M. 2011. The Oil Palm pollination mystery and the $370 million/year yield increase. Case study 14.bionet@bionet intl.org | www.bionet intl.org. page 16.

Debach, P. 1960. The importance of taxonomy to biological control as illustrated by the cryptic history of *Aphytis holoxanthus* n. sp. (Hymen-optera: Aphelinidae), a parasite of*Chrysomphalus aonidum*, and *Aphytis coheni* n. sp., a parasite of *Aonidiella aurantii. Annals of the Entomological Society of America*. 53 (6) : 701-705.

Delucchi, V., Rosen D., Schlinger, EI. 1976. Relationship of systematics to biological control. In: Theory and Parctice in Biological Control (ed.) Huffaker, CB., Messsenger, PS. Pp. 81-91.

Drew, R.A.I. and Hancock, D.L. 1994. The Bactrocera dorsalis complex of fruit flies in Asia. Bulletin of Entomological Research: Supplement Series. Supplement No. 2. CAB International, Wallingford, UK.

Fernando, L. C. P. and G. H. Walter. 1997. Species status of two host-associated populations of Aphytis lingnanensis (Hymenoptera: Aphelinidae) in citrus. *Bulletin of Entomological Research* 87: 137-144.

Fiaboe, K.K.M.; Fonseca, R.L.; Moraes, G.J. de; Ogol, C.K.P.O. and Knapp, M. 2006. Identification of priorty areas in south America for exploration of natural enemies for classical biological control of Tetranychus evansi (Acari: Tetranychidae) in Africa. *Biological Control* 38: 373-379.

Frana J.E.. 2011a. Targeting one larvae in nine saves money and protects health and the environment in Argentina. Case study 43. bionet@bionet intl.org | www.bionet intl.org. page 46.

Frana. J.E. 2011b. Spotting the assassin in the garden. Case study 44. bionet@bionet intl.org | www.bionet intl.org. page 47.

Gauld, I. D. 1986. Taxonomy, its limitations and its role in understanding parasitoid biology. In : Insect Parasitoids (ed : Waage, J. and D. Greathead), Academic Press, New York.

Godfray, H.C.J. 2002a. How might more systematics be funded? *Antenna: Bulletin of the Royal Entomological Society* 26: 11-17.

Godfray, HCJ. (2002b). Challenges for Taxonomy. *Nature*, 417: 17-19.

Gotoh, T, Bruin J, Sabelis MW and Menken SBJ. 1993. Host race formation in *Tetranychus urticae* : genetic differentiation , host plant preference and mate choice in a tomato and cucumber strain. *Entomologia Experimentalis et Applicata* 68 : 171-178.

Hancock, D. H., Kirk Spriggs, A. H. and Marais, E. 2003. New records of Namibian Tephritidae (Diptera: Schizophora), with notes on the classification of subfamily Tephritinae. Cimbebasia 18.

Julien, M.H. and M.W. Griffiths (eds.). 1998. 4th edition. Biological control of weeds. A world catalogue of agents and their target weeds. CABI Publishing, Walllingford, UK.

Kazmar D.J. 1991. Isoelectric focusing procedures for analysis of allozyme variation in minute arthropods. *Annals of the Entomological Society of America* 84:332-339

Kirton, L.G. & Brown, V.K. 2003. The taxonomic status of pest species of *Coptotermes* in Southeast Asia: Resolving the paradox in the pest status of the termites, *Coptotermes gestroi, C. havilandi* and *C. travians* (Isoptera: Rhinotermitidae). *Sociobiology* 42(1): 43 63.

Knutson, LV. 1981. Symbiosis of biosystematics and biological control. In : Beltsville Symposia in Agricultural Research. 5. Biological control in crop production (ed.) Papavizas, GC. 61-78. Totowa, NJ. Allanheld, Osmun. 461p.

LePelley RH. 1935. The Common Coffee Mealy-bug of Kenya (Hem. Coeeidae). *Stylops,*. 8:185-188.

LePelley, R. H. 1943. The biological control of a mealybug on coffee and other crops in Kenya. *Empirical Journal of Experimental Agriculture* 11(42): 78-88.

Lindquist, D.A., Abusowa, M. and Hall, M.J.R. 1992. The New World screwworm fly in Libya: a review of its introduction and eradication. *Medical and Veterinary Entomology* 6: 2-8.

McKenzie, H.L. 1937. Morphological differences distinguishing Californian red scale, yellow scale and related species. *University of California Publications in Entomology* 6: 323-326.

Melville A. R. .1959. The place of biological control in the modern science of entomology. *Kenya Coffee* 24: 81 85.

Melville, A. R. 1938. Kenya coffee mealybug research. *East African Agricultural Journal.* 3: 411-422.

Mervyn W. M.. 2011. Use of taxonomy leads to criminal convictions. ARC –Plant Protection Research Institute, P/B X134, Queenswood, Pretoria, 0121, South Africa. Case study 24.bionet@bionet intl.org I www.bionet intl.org. page 26

Miller, J. S. and J. W. Wenzel. 1995. Ecological characters and phylogeny. *Annual Review of Entomology* 40:389-415.

Miyamoto, S., N.A. Torres, B.H. Habibuddin. 1983. Emerging problems of pests and diseases The black bug in Southeast Asia. International Rice Research Conference, IRRI, Los Banos, Laguna, Philippines, 33pp.

Naggs, F. 2002. Molluscan pests in Sri Lanka: voracious exotics having a major and rapidly increasing impact on agriculture. Zoology Department, The Natural History Museum, London. Land snail diversity in Sri Lanka: Darwin Initiative project information leaflet.

Narendran TC, 2001. Taxonomic Entomology: Research and Education in India. *Current Science* 8 (5): 445-447.

Nelson, G. and Platnick, N. 1981. Systematics and Biogeography: Cladistics and Vicariance. Columbia University Press, New York.

Neuenschwander, P. 2003. Biological control of cassava and mango mealybug in Africa. Pp.45 59 in: Biological Control in IPM Systems in Africa. Neuenschwander, P., Borgemeister, C. & Langewald, J. (eds). CABI Publishing, Wallingford, UK.

Nordlund, D.A. and Lewis, W.J. 1985. Response of females of the braconid parasitoid *Microplitis demolitor* to frass of larvae of the noctuids *Heliothis zea* and *Trichoplusia ni* and to 13 methylhentriacontane. *Entomologia Experimentalis et Applicata* 38: 109-112.

Norlund, D. A. 1994. Habitat location by *Trichogramma*. In, Wajnberg, E. and S. A. Hassan (eds.), Biological control with egg parasitoids. CAB International, Wallingford.

Oudman, L. 1992. Identification of economically Liriomyza species (Diptera: Agromyzidae)and their parasitoids using enzyme electrophoresis. *In : Proceedings of Netherland Entomological Society* 3:315-319.

Pembertan, CE. 1941. Contributions of the entomologists to Hawaii's welfare. *Hawaii Plant Records* 45 : 107-119.

Penrith, M.L. 1977. The Zophosini (Coleoptera: Tenebrionidae) of western southern Africa. *Cimbebasia Memoir* 3: 1 -291.

Pimentel, D., Andow, D., Gallahan, D., Schreiner, I., Thompson, TE., *et al.*, 1980. Pesticides : environmental and social costs. In : Pest control: cultural and environmental aspects (ed.) Pimentel, D., Perkins, JH. pp99-158.

Pinto, J. D., Stouthamer, R., Platner, G.R., Oatman, E.R. 1991. Variation in reproductive compatibility in Trichogramma and its taxonomic significance (Hymenoptera: Trichogrammatidae). *Annals of the Entomological Society of America*, 84: 37-46.

Pinto, J.D. 2006a. A review of the New World genera of Trichogrammatidae (Hymenoptera) *Journal of Hymenopteran Research*. 15 : 38-163

Pinto, J.D. 2006b. Systematics and Biological Characteristics In : Radcliffe, E. B., Hutchison, W. D and Cancelado, R.E. [eds.], Radcliffe's IPM World Textbook, URL: http:// ipmworld.umn.edu, University of Minnesota, St. Paul, MN.

Rafael, J.A., Aguiar, A.P., Amorim, D de S. 2009. Knowledge of insect diversity in Brazil: challenges and advances. *Neotropical Entomology*. 38 (5); 565-70.

Robert, H. R. 2011. Biodiversity and waterways win in a weevil's battle against *Azolla.* Case study 26.bionet@bionet intl.org | www.bionet intl.org. page 29.

Rosen, D. 1986. The role of taxonomy in effective biological control programs. *Agriculture Ecosystems and Environment* 15: 121-129.

Rosen, D. 1978. The importance of crop species and specific identifications as related to biological control. In: Beltsville symposia in Agricultural Research. 2, Biosystematics in Agriculture (ed.) Romberger, JA. pp23-35. Monteplair NJ, Allanheld, Osmun, 340p.

Rosen, D., Debach, P. 1973. Systematics, morphology and biological control. *Entomophaga.* 18 : 215-222.

Ryan, J.K. 1977. Synthesis of energy flows and population dynamics of truelove low land invertebrates. In True Love Lowland. Devon Island, Canada : A high Arctic ecosystem (ed.) Bliss, L.C. pp325-46. Edmonton, Alberta University, Alberta Press. 714p.

Sabrosky, C.W. 1955. The interrelation of biological control and taxonomy. *Journal of Economic Entomology*. 48 : 710-14.

Shaw, S. R. 1988. Euphorine phylogeny: the evolution of diversity in host-utilization by parasitoid wasps (Hymenoptera: Braconidae). *Ecological Entomology* 13:323-335.

Silva, I. S., Honda, J., Kan, Van F. M., Hu,J., Neto, L., Pintureau, B. and Stouthamer, R. 1999. Molecular differentiation of five Trichogramma species occurring in Portugal. *Biological Control*, 16: 177-184.

Simonetti, J.A. 1997. Biodiversity and a taxonomy of Chilean taxonomists. Biodiversity and Conservation. 6:633-637.

Stouthamer, R., Hu, J., Van Kan, F.J.P.M., Platner, G.R., and J.D. Pinto 1999. The utility of internally transcribed spacer 2 DNA sequences of the nuclear ribosomal gene for distinguishing sibling species of Trichogramma. *BioControl.* 43 : 421-440.

Tricio, A., Morawicki, P.M., Fernández Díaz, C. I., Krsticevic, F. and Araki, S. 2002. Monitoreo de dípteros vectores hematófagos en el área de influencia de la represa Yacyretá. Periodo, 2000 2001. In: Oscar Salomón (ed.) Actualizaciones en Artropodología Sanitaria Argentina. Serie Enfermedades Transmisibles. RAVE. Fundación Mundo Sano, Buenos Aires, 301 pp.

Viscarret, M. M., Torres Jerez, I., Agostini de Manero, E., López, S.N., Botto, E.E.Y J.K. Brown. 2003. Characterization of Non B Biotype Populations of the *Bemisia tabaci* (Genn.) (Hemiptera/Homoptera:Aleyrodidae) Species Complex from Argentina and Bolivia and First Report of the B Type in Argentina. *Annals of Entomological Society of America*: *Volumen* 96(1): 65-72.

Wagener, B., A. Reineke, B. Löhr, C.P.W. Zebitz. 2006. Phylogenetic study of Diadegma species (Hymenoptera: Ichneumonidae) inferred from analysis of mitochondrial and nuclear DNA sequences. *Biological Control* 37:131-140.

Waterhouse, D.F. and Sands, D.P.A. 2001. Classical biological control of arthropods in Australia. CSIRO Entomology, Australian Centre for Agricultural Research, Canberra. p. 559.

Watt, J.C. 1979. Biosystematics; the neglected science. New Zealand Science Review. 36 : 68-72.

Wheeler, Q.D., Raven, P.H., Wilson, E.O. 2004. Taxonomy: Impediment or Expedient? *Science*, 303: 285.

Wheeler, Q.D. 2004. Taxonomic triage and the poverty of phylogeny. Phylosophical Transactions of the Royal Society of London, 359: 571-583.

Zeddies, J., Schaab, R.P., Neuenschwander and Herren, H.R. 2001. Economics of biological control of cassava mealybug in Africa. *Agricultural Economics* 24: 209-219.

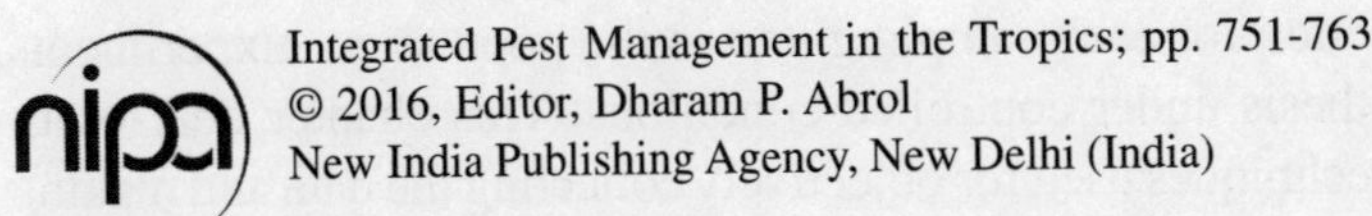

Integrated Pest Management in the Tropics; pp. 751-763

New India Publishing Agency, New Delhi (India)

CHAPTER - 25

Statistical Techniques for Insect-Pest Management

Manish Sharma*, Banti Kumar and S.E.H. Rizvi
Division of Statistics and Computer Science, FBSc, SKUAST-Jammu
*Corresponding author *email: manshstat@gmail.com*

Like other sciences, the insect–pest management research is also based on scientific methods requires formulation of hypothesis from observed facts through logical arguments which in turn, verified through objective methods. For instance, in integrated pest management the entomologists are primarily concerned with the estimation of insect-pest population, its distribution pattern over time and space, its relationship and fluctuation in population with the effect of weather parameters and ultimately its control measure through the judicious use of chemicals, biological control agents and resistant varieties etc. Two major practical aspects of scientific investigation are as

- collection of data and
- interpretation of the collected data.

The data may be generated through a sample survey or designed experiment. Simple, descriptive statistics are essential for interpreting any collected data. Regardless of how data is gathered whether as continuous measurements (e.g. leaf area consumed), in the form of numerical counts (e.g., number of beetles per plant), as ordinal ratings (e.g., on a scale from 1 to 10), or in binomial form (e.g., presence/ absence), there is always some degree of uncertainty about its accuracy. Therefore, it is need to use of the proper statistical techniques in order to give the most reasonable estimate of the data. Also, statistical theory of surveys rely on random sampling which assigns known probability of selection for each sampling unit in the population. Experiments serve to test hypothesis under controlled conditions. This chapter deals with various statistical techniques used for objectively collecting the data and making valid inference out of the same. The common terminologies in statistics used are: Average is the comparisons between the similar treatments will be made easier with averages. The different types of averages are Arithmetic mean, Mode, Median, Geometric Mean and Harmonic Mean. The most popular measures for representing the entire entomological data are the Arithmetic mean (A.M.), Median and Mode. In case of qualitative data, mode can be used. If the data are of count, skewed type or consisting outliers, median or mode can be used. The most popular measure is arithmetic mean. The average performance of the two pesticides can be same but they can be varying w.r.t consistency. So averages alone are not sufficient to compare the performance of the two treatments. In this case, Standard deviation is the most reliable measure of dispersion to compare two pesticides on the basis of variability. The square of the standard deviation is called as Variance (used by R.A. Fisher in 1913).Coefficient of variation (CV) or Coefficient variability is a relative (unit free) measure of dispersion. It is used when we have to compare the variability of two or more series. A series which have more variability as compared to the other is less consistent. It is always presented in percentage, CV = $\frac{S.D}{Mean}$ x 100 $= \frac{\sigma}{\bar{x}}$ x 100. Higher CV having less consistency, less reliability and less uniformity whereas Lower CV having higher consistency, higher reliability and higher uniformity. Standard error of mean is used because the mean is a random variable; therefore it would be different every time during the sampling process. The sampling error of the mean is just the estimated standard deviation of the sample mean or standard error of mean. It is used to report inferences about the population mean. It is denoted by S.E. S.E $(\bar{x}) = \sqrt{\frac{\sigma^2}{n}} = \frac{\sigma}{\sqrt{n}}$. It is also used to find the precision or reliability of an estimate, Precession $= \frac{1}{StandardError}$. Point Estimation provides a single numerical estimate of the parameter whereas Interval Estimation provides an interval which will cover the parameter value with specified probability. For example, in order to estimation of biological population (Insect-Pest) which are generally mobile in nature so point estimation cannot be a rely estimation. In this case, instead of providing a precise numerical estimate of the same it might be good if we can provide limits between which the total population will lie with specified probability. The confidence interval includes the upper and lower boundaries of our faith in the reliability of the samples. A 99 per cent confidence interval means that the probability of data falling outside a given range of values is only 1 in 100. It can be set at any level of certainty, but in general, most pest management decisions are based on 95 to 99 per cent confidence intervals. A lower confidence interval is associated with increased risk of uncertainty in the development or outcome of a pest outbreak.

Correlation and Regression

Correlation may also be interpreted as "when two variables are so related that a change in one accompanied by a change in the other in such a way that an increase in one is accompanied by an increase or decrease in the other, or decrease in one is accompanied by an increase or decrease in the other, and greater the magnitude of change in one, the greater the amount of change in the other, then the variables are correlated". For example, an investigator is interested in measuring the correlation between the population of insect-pest with the temperature or/relative humidity maximum and minimum. If one of the variables among these increases (decreases) and other variable also increases (decreases) then the two variables are positively correlated. Whereas in opposite direction result to negative correlation. Correlation is said to be linear when the amount of change in one variable leads to a constant ratio of change in the other variable over the entire range of values. The range of simple correlation co-efficient is from -1 to +1. Partial correlation co-efficient is a measure of the degrees of linear relationship between any two variables out of a set of variables, when the influence of the remaining variables is eliminated from both of them e.g., to measure the correlation between the population of insect-pest with the temperature maximum by eliminating the effect temperature minimum, relative humidity maximum, relative humidity minimum, rainfall etc. from both the variables. The range of partial correlation co-efficient is from -1 to +1.Whereas, multiple correlation coefficients is a numerical expression of the degree of relationship existing between one variable (dependent) and the combined effect of other variables (independent) in the whole group e.g., to measure the correlation between the population of insect-pest with the temperature maximum, temperature minimum, relative humidity maximum, relative humidity minimum and rainfall. The range of multiple correlations co-efficient is from 0 to 1. Therefore, the multiple correlation is a positive correlation.

To regress means to 'return' or 'move backward' and for statistical purpose 'to return to the mean value'. In short, regression analysis is an important statistical technique with the help of which we are in a position to estimate or predict the unknown values of one variable from the known values of another variable. Thus, it indicates the average relationship between two or more variables. One of these variables is called the 'dependent' or the 'explained' variable and the other variable is called 'independent' or the 'explaining' variable.

Thus, the variable which is used to predict the variable of interest is independent variable and the variable which we are trying to estimate is dependent variable. It is also called as cause and effect relationship. If the regression analysis considers only two variables at a time then it is described as simple regression. On the other hand, regression analysis for more than two variables at a time is described as multiple regressions. As an example one may interested to study the effect on the production of quality mulberry leaf and subsequent production of quality silk is hampered due to the disease severity by incidence of various insect pests during rainy season. The regression model will support to study and predict the relationship between the pest population build up due to the weather factors. The prediction equation can be used as $Y = a + b_1x_1 + b_2X_2 +.... b_5X_5 + e$ Where, Y = Predicted variable; a = Intercept; b_1 to b_6 = Regression coefficients; X_1 = Maximum temperature; X_2 = Minimum temperature; X_3 = Maximum relative humidity; X_4 = Minimum relative humidity and X_5 = Rainfall.

Testing of Hypothesis

In many experimental works, estimation of the numerical value of a parameter is difficult to estimate or not of direct interest to us. Thus, we may be interested in making decisions about populations on the basis of sample selected or collected. For example, we may wish to decide on the basis of the sample data whether a new pesticide is really effective in controlling incidence of pest as compared to the old one, a particular crop management method will result in better yield of a crop, the cross fertilization of two specific crop varieties will results in a better hybrid etc. For this, we need statistical test to support such decisions. Generally, the results of the experiments are usually ambiguous. Before moving to the tests it needs to understand the basic terminologies use in the test: Statistical hypothesis is the statistical statement regarding the form of a parameter of the distribution. Null hypothesis is hypothesis of no difference or no effect among the treatments, it is denoted by Ho. Alternative statements to null hypothesis that be believe is true, is called alternative hypothesis denoted by H_1. During the course of investigation, one may commit two types of errors as:

Decision	Actually in the population	
	H_0 True	H_1 True
Reject H_0	Wrong decision (Type I error)	Correct decision
Accept H_0	Correct decision	Wrong decision (Type II error)

The size of type I error denoted by Greek letter α (alpha) is the probability of rejecting null hypothesis when actually it is true. i.e., $\alpha = P\ [\text{Reject } H_0/H_0]$, is also known as level of significance. The level of significance is taken as 5%

(0.05) in case of entomological field experiments and 1% (0.01) for laboratory experiments. The researchers have the freedom to choose his own criterion in fixing level of significance due to availability of material, money to pursue an experiment, time, seriousness of type I error and nature of the problem. The size of type II error is equal to the probability of accepting null hypothesis when actually H_1 is true and is denoted by Greek letter β (beta) i.e., β = P [Reject H_0/ H_1]. The numbers of observations that are free to vary after certain restriction have been placed in on the data known as degree of freedom. It is abbreviated as d.f. Test of significance is the procedure through which sample results are used to throw light on the truth or falsity of a null hypothesis regarding the form of population parameter under consideration. The value of test statistics which separates the rejection region and the acceptance region is called critical value or significant value. This value depends on the level of significance as well as on the form of alternative hypothesis. Now let we discuss the tests that are commonly used for testing the significance. These are as:

Student's t-test: Student's t is defined as the deviation of sample mean from its population mean expressed in terms of standard error. The credit of t-test goes to W.S.Gosset who published it in 1908 in research paper entitled, "The probable error of the mean". Professor R.A. Fisher (1926) defined correctly. Since, in case of standard normal variate (SNV), $Z = \frac{\bar{x} - \mu}{\sigma/\sqrt{n}}$. Where, σ is not known it can be substituted by the sample standard deviation(s). It has been generally observed that, in case of large sample, $\sigma \cong s$ E.g.: to study the significant effect of insulin hormone on diapauses behavior of males and females in terms of percentage. But, in case of small samples it makes a substantial difference. Thus the ratio $\frac{\bar{x} - \mu}{s/\sqrt{n}}$ is not being a SNV. This ratio is called student's t with n-1 degree of freedom. Thus, $t = \frac{\bar{x} - \mu}{s/\sqrt{n}}$ is a statistic. The t-test is used in testing the significance of :

(i) the sample mean confirms to a specified value of population mean or not.

(ii) the equality of two populations mean based on sample means e.g. to test the significant difference bin beetle density for population buildup in water stressed and non-water stressed conditions.

(iii) correlation coefficient and regression coefficient e.g.to test the significance of correlation and regression coefficients between mean number of aphids and maximum temperature.

(iv) partial correlation coefficient.

In case of large sample and test relates to proportion we used Z-test.

F-test: G. W. Snedecor founded (1934) F- Statistics, it is the ratio of two sample variances and prepared. It is named so in honor of R.A. Fisher who first developed and described. It follows F-distribution with $n_1 - 1$ $(=v_1)$ and $v_2 - 1$ $(=v_2)$ degree of freedom. The F-distribution is used in testing the significance of :

(i) equality of two population variance e.g. For testing H_o: $\sigma_1^2 = \sigma_2^2$.

(ii) used in analysis of variance.

(iii) the equality of several regression coefficient.

(iv) the equality of several population means e.g. to study the significant effect on natural infestation of parthenium at grazing area, crop fields, home yards, road sides etc.

Chi-square (χ^2) test: The chi-square distribution was first discovered by Helmert in 1876 and later independently by Karl Pearson in 1900. The χ^2 is square of the Z-test. If x ~ N (μ, σ^2).Then Z ~ N (0, 1)

$$\left(=\frac{x-\mu}{\sigma}\right) \text{and } Z^2 = \left(\frac{x-\mu}{\sigma}\right)^2 \sim\chi^2_{(1)} \Rightarrow \quad \sum_{i=1}^{n} Z^2 = \sum_{i=1}^{n}\left(\frac{X_i-\mu}{\sigma}\right)^2 \sim\chi^2_{(n)}.$$

The chi-square distribution is used to test:

(i) a hypothetical value σ_0^2 of the population variance is true or not.

(ii) a test of goodness of fit.

(iii) the independence of attributes.

(iv) validity of a hypothetical ratio e.g., to test the difference between percent flight of muscles degeneration and cessation of oviposition in diapausing females and non-diapausing females of *Z.bicolorata*.

(v) the homogeneity of several population variances e.g., For testing H_o: $\sigma_1^2 = \sigma_2^2 ... = \sigma_n^2$

(vi) equality of several population correlation coefficients. For testing H_o: $\beta_1 = \beta_2 \cdots = \beta_n$

While applying the above test in testing hypotheses you need to have formulated the steps: stating the null hypothesis, choosing a statistical test for testing the null hypothesis, specifying the level of significance, finding the sampling distribution of the test static under the null hypothesis and computing the value of test statistic using the data obtained from the sample(s) and making a decision based on the value of the test statistic and the predefined region of rejection which leads to a conclusion of either to reject the null hypothesis, or not to reject the null hypothesis.

Experimental Design for Entomological Experiments

Design of experiment, an important branch of statistics discipline, plays vital role in research experiments in order to have relevant information from it with accuracy and reliability. To achieve this goal the experimenter has to plan his experiment more carefully in order to reduce the experimental errors because the field trials are costly and may take many months and years to complete. The structuring of the dependent and independent variables, the choice of their levels in the experiment, the type of experimental material to be used, the method of the manipulation of the variables on the experimental material, the method of recording and tabulation of data, the mode of analysis of the material, the method drawing sound and valid inference etc. are all intermediary details that go with the design and analysis of an experiment. The objectives of the experiments must be carefully defined before conducting, for the crop and the environment in which the pest is to be tackled. Generally, the designs are the extensions of the t-test. The most common terminologies used in the experimental designs are as:

Coefficient of variation: In order to find out the reliability of the experiment, the coefficient of variation (CV) is used. It is computed as

$$\text{CV}=\sqrt{\frac{\text{Error MS}}{\text{Overall mean}}}\text{X}\,100\,.$$

If the CV is 20% or less, it is an indication of better precision of the experiment. When the CV is more than 20%, the experiment may be repeated and efforts made to reduce the experimental error.

Critical difference: The Critical Difference (CD) is a form of t test. Its formula is given by

CD = t. SE (d) Where, SE (d) = $\sqrt{EMS\left(\frac{1}{r_i}+\frac{1}{r_j}\right)}$. In case of equal replications

$$\text{SE (d)} =\sqrt{\frac{2EMS}{r}}$$

In the formula, t is the critical (Table) value of t for a specified level of significance and error degrees of freedom; and r_i and r_j are the number of replications for i^{th} and j^{th} treatments, respectively. Two treatments are declared significantly different at a specified level of significance if their difference exceeds the calculated CD value; otherwise they are not significantly different.

There are three basic principles of experimental designs which are Replication, Randomization and Local Control. The repeated application of the treatments under investigation is known as replication. The major functions are to provide an estimate of experimental error and to reduce the standard error of the treatment mean(s). A satisfactory precision can be obtained by taking the number of replications in such a way that the error degree of freedom should not be less than 12. The allocation of the treatments in the experimental units in such a way that all the treatments have equal chances of being allocated to different experimental units is known as randomization. The major functions: are to eliminate the human biases and to makes the experimental error independent.The principle of local control is to achieve a greater homogeneity by grouping the experimental units into a block such that there is minimum variation within a block and maximum variation between the block. The major functions are to reduce the experimental error and to makes design more efficient. Depending upon the different manners in which the experimental material is grouped, various statistical designs are used. The Commonly used experimental designs in entomology trials are as follows:

Completely randomized design (CRD)

If the experimental material is homogenous and does not exhibit any variability then the treatments are applied to the experimental units. Such designs are known as zero-way elimination of heterogeneity design or completely randomized design CRD is applied when the experimental material is completely homogeneous and dividing them into blocks is of no use e.g. pot experiment, laboratory experiment and green house experiment. In case of nursery experiments some plants may not germinate during course of experiment resulting in unequal number of replications. So, CRD is used as it is the only basic design which can accommodate unequal replications. Consider an experiment with four treatments A, B, C and D, each replicated five times. Thus the total number of experimental plot n = $\sum_1^t ri$ = 20 needed

A	B	A	D	B
C	D	C	A	D
B	A	D	B	C
D	C	B	C	A

The experimental units will be randomized over the whole experimental material. While working with this design one has to keep in mind the points, this design is suitable for small numbers of treatments because as the number of treatments increases the experimental error increases as the heterogeneity may increase, this design is very less sensitive if there is a fertility gradient present in the soil and the soil need to be tested before the experiment commences. CRD is based on two principles only viz. replication and randomization.

Randomized block design (RCBD)

If the whole experimental material is not uniform but it can be divided into homogenous groups (called blocks) in such a way that there is maximum heterogeneity between blocks. We use randomized block design (RCBD). Each block consists of as many experimental units as the number of treatments. Here the randomization of the whole treatments within the block. While working with this design one has to keep in mind the points, if we increase the number of treatments, the block size increases and there by the assumptions of homogeneity within a block is violated. Therefore the intra block variance or variances per plot become large. To overcome this problem, we use incomplete block designs (IBD). Some common IBD are BIBD, PBIBD, Lattice designs, cyclic designs, alpha designs etc., if the observations are missing, one has to first estimate the missing observations and then analyze the data or one has to go with the RCBD design with missing plot techniques. This design is based on all the three principles of design of experiment. Consider an experiment involved six methods for controlling the severity of the insect-pest on crop with three replications.

B_1	B_2	B_3
T_1	T_2	T_6
T_4	T_6	T_2
T_2	T_3	T_4
T_5	T_4	T_1
T_6	T_5	T_5
T_3	T_1	T_3

Factorial Experiments

In many situations in agricultural experimentation for critical studies we wish to investigate the effect of several factors like fertilizers, pesticides, varieties etc. Then instead of conducting separate experiments for each factor, a single factorial experiment is conducted with treatments as all possible combinations of levels of various factors, since it is easy to handle single factorial experiment rather than several experiments for each factor separately. In addition to operational convenience the factorial experiment provide useful information on various interactions (while separate experiments do not provide) and at the same time provide higher precision for various factors and their interaction since effective no. of replications increase in manifold. Consider a two factor experiment on insect-pest management involving two level of (a_1, a_2) of controlling insect-pest methods (Factor A) and three levels of (b_1 , b_2, b_3) insecticides (Factor B) laid out in field with three replications. The total treatment considerations will be six. They will be a_1 b_1, a_2 b_2, a_1 b_3, a_2 b_1, a_2 b_2 and a_2 b_3.

R_1	R_2	R_3
a_1 b_2	a_2 b_3	a_1 b_1
a_1 b_3	a_1 b_3	a_2 b_1
a_2 b_1	a_2 b_2	a_2 b_3
a_2 b_3	a_1 b_1	a_1 b_3
a_1 b_1	a_2 b_1	a_2 b_2
a_2 b_2	a_1 b_2	a_1 b_2

Factorial experiments are symmetrical factorial experiments in which all the factors have same no. of levels. These are denoted by p^n where n is the no. of factors each at p levels. If all the factors do not have equal no. of levels they are asymmetrical factorial experiments. These are denoted by p x q x r etc. where p, q and r are no. of levels of 3 factors respectively. These find wide applications in agricultural, industrial and animal science experimentations for testing several factors at unequal no. of levels.

Sampling Techniques

The object of sampling is to secure a sample which will represent the population and under study important characteristics of the population as closely as possible. For example, in estimating the whole population of the pest in the specified area, one may select a few pests which may appear to be average of the pest under study. However, it is clear that such methods of selection are likely to be biased by the investigator's judgment and the results will thus be biased and unreliable. When sampling is performed so that every unit in the population has some chance of being selected in the sample and the probability of selection of every unit is known, the method of sampling is called probability sampling. The important point to note is that the precision and reliability of the estimates obtained from a sample can be evaluated only for a probability sample. The object of designing a sample survey is to minimize the error in the final estimates. Any survey involving data collection and analysis of the data is subject to a variety of errors. The errors may be classified into two groups viz., (i) non-sampling errors (ii) sampling errors. The non-sampling errors are due to location of the units, measurement of the characteristics, recording mistakes, biases on enumerators etc. They are presents in both complete enumeration and sample surveys. The magnitude is likely to be larger in complete enumeration. Sampling errors arise from the fact that only a fraction of the population is enumerated.

Simple Random Sampling

In this design, there is only one type of sampling unit and, hence, the sample size (n) i.e., total number of sampling units is to be selected from plot consisting of N units. The selection of the n sampling unit is done in such a way that each of the N units in the plot is given the same chance or equal probability of being selected. This sampling technique ensures that bias is not introduced regarding who is included in the survey. It is referred as sample random sampling/ un–restricted random sampling. If a unit is selected and noted and then returned to the population before the next draw is made and this procedure repeated n times, it gives rise to a simple random sample of n units. This procedure is known as simple random sampling with replacement. If the selected unit is not returned to the population before the draw of next sample the procedure is known as simple random sampling without replacement. The advantage of simple random sampling is that it is simple and easy to apply for small population. However, a large population it is very cumbersome because every person or item in the population has to be listed before the draw of the sample. This sampling technique is so sensitive in case of small sample, huge variability in the universe and the units are spread over large area.

Stratified Random Sampling Design

As we know the precision of a simple random sampling design can be improved by increasing the sample size. However, it is not the only way. A very popular method to increase the precision is stratification of the main population into number of sub-population or groups called Strata each of which is homogenous with respect to one or more characteristics and then to select sample from each sub-population. The stratified sampling is more suitable to use in case of geographical area and for administrative convenience. For example, in order to survey for pest attack the strata can be formed through geographical regions according to area, stand density, site quality etc. or through the staff already available in each area may have to supervise the survey in the area under jurisdiction. This sampling technique is so sensitive if complete knowledge about the diverse characteristics of the population is not known.

Systematic Sampling

Systematic Sampling is used when the complete and upto date list of the sampling units is available and the units of the population to be sampled are ordered in a specific manner. The procedure for selecting of the sample is chosen by taking a random starting point and then picking every k^{th} element in succession from the sampling frame. The sampling interval, k, is determined by dividing the population size N by the sample size n and rounding to the nearest integer. When the ordering of the elements is related to the characteristic of interest, systematic sampling increases the representativeness of the sample. If the ordering of the elements produces a cyclical pattern, systematic sampling may decrease the representativeness of the sample. For example, there are 10,000 numbered infected plants in the area and a sample of 100 is desired. In this case the sampling interval, k, is 100. A random number between 1 and 100 is selected. If, for example, this number is 33, the sample consists of plants number 33, 133, 233, 333, 433, 533, and so on.

In survey for estimation of effect of pest on the plants, where infected plants are not numbered but length of a field is known suppose it is 500 m. If we sample a group of plants after a definite distance, say 2 m then we can take a sample of n=250.

Sequential Sampling

Since it is fairly easy to sample for some insects, many pest management systems utilize different sample techniques that are fairly time consuming and labour intensive. Whenever large numbers of samples are needed for an adequate level of confidence, it may be possible to use a sequential sampling system that saves time and effort by concentrating mostly on populations that are closest to the economic threshold. Sequential sampling system is relatively new in pest management, but they are based on well-established rules for determining required confidence intervals for sample data. There are three different possible outcomes in case of sequential samplings:

- If the total cumulative total of the pest exceeds an upper threshold value, than conclude that population is large enough to warrant control actions. Stop the sampling and prepare to enact control measures.
- If the cumulative total of the pest is lower than the threshold value, than conclude that the population is small and warrants no control actions. Stop sampling and leave the population untreated.
- If the cumulative total of the pest is between the upper and lower threshold values, than no conclusion is possible yet. Sampling should continue until cumulative total reach the upper or lower threshold.

References

Gomez, K.A. and Gomes, A.A. (1984). Statistical Procedures for Agricultural Research. The International Rice Research Institute, Philippines.

Panse, V.G. and Sukhatma, P.V. (1985). Statistical Methods for Agricultural Workers. ICAR, New Delhi.

Snedecor, G.W. and Cochran, W.G. (1989). Statistical Methods. Iowa State University Press, Ames Iowa.

Integrated Pest Management in the Tropics; pp. 765-774

New India Publishing Agency, New Delhi (India)

CHAPTER - 26

Future of Integrated Pest Management in Tropics - A Critical Analysis

M. Raghuraman[1] and Dharam P. Abrol[2]
[1]Department of Entomology & Agricultural Zoology, Institute of Agricultural Sciences, Banaras Hindu University, Varanasi, Uttar Pradesh - 221 005
e-mail: raghu_iari@yahoo.com
[2]Department of Entomology, Faculty of Agriculture, Sher-e-Kashmir University of Agricultural Sciences and Technology of Jammu, Chatha, Jammu-180 009

26.1 Introduction

The green revolution brought dramatic changes in productivity of rice and wheat and the production increased tremendously using high yielding wheat and rice varieties coupled with fertilizers and pesticide use to meet urgent demands of food. The world population is growing rapidly at an alarming rate. It has crossed 7 billion mark and is expected to become between 8.3 and 10.9 billion by 2050 (United States Census Bureau (USCB). India is second largest populous country of the world with 17.4 % of world population. Such an enormous overpopulation will result in extreme pressure on environment, global food supplies and energy resources. The less arable land and fewer renewable and non-renewable resources, the food production has to come from increases in production rather than expansion of cultivable areas. The rate of production of staple food crops has declined from 2.9% to 1.9% following green revolution (Rosegrant *et al.*, 2013, 2014). Increase in food productivity has to be in conjunction with environmental protection and ecological sustainability. To meet this challenge integrated pest management is such a tool that has reduced pesticide pressure without compromising agricultural productivity (Perkins, 1982, Walter, 2003, Kogan, 1998, Pedigo, 2002).

The use of pesticides for crop productivity has serious consequences for stability of food production and human health. Various insect pests have developed resistance and natural enemies have started disappearing from agricultural ecosystems. The pesticide resistance was first reported in Sanjose Jose scale in 1908 and since then more than 500 (Atwal and Dhaliwal, 1997) to 1000 species of insects and mites (Miller 2004) have developed resistance.

Successful implementation of IPM has helped to reduce the pesticide pressure and encouraged the use of biocontrol agents (Binns and Nyrop,1992, Stern *et al.*, 1959, Metcalf and Luckman, 1994). There is a demand throughout the world for safe foods produced by environmentally friendly way (Barfield and Swisher, 1994). In several countries of Asia, Africa and Latin America, the IPM has been successfully implemented and adopted by the farmers for single pest species on a crop. Developing IPM for a multitude of pests and diseases has been difficult.

26.2 Pesticide Use in India

Herbicides and pesticides are less commonly used in India as compared to other countries of the world. Of the total pesticides used, the proportion of insecticide used is much higher (76%) in India as against 44% globally (Mathur, 1999). Cotton and paddy are the two major crops where more than 45% pesticides are used.

Table 26.1 : Consumption and demand of pesticides in India in metric tones (2005-2010) (http:/ppqs.gov.in)

	2005-06	2006-07	2007-08	2008-09	2009-10
Consumption	39773	41515	43630	43860	41822
Demand	44324	43718	42827	45180	42264

However, with the popularization of Integrated Pest Management pesticide consumption has considerably decline in recent years. Moreover, use of *Bt* cotton is also responsible for decline in pesticides (Table 26.1).

26.3 IPM Research in India

Due to continuous and unilateral use of synthetic insecticides, many insect pests have started developing resistance, due to which a large number of insecticides are not able to contain the pest menace effectively. This has further led to the destruction of the natural enemies, emergence of new pests and minor pests becoming major pests. Use of alternative control methods such as biopesticides and genetically engineered crops may provide promising results. Different technologies which use the plant derived products and pathogens are available but they fail to provide promising results in the field due to their short shelf life and less economic feasibility. Biological control programmes with egg parasitoids such as *Trichogramma* and predators like *Chrysoperla canea* have been found effective on some selected crops. Genetically modified crops are better alternatives exemplified by *Bt* cotton which is grown successfully (Chaudhary and Gaur, 2010).

In participatory programmes with farmers Integrated Pest Management has shown considerable success in India. National Centre for Integrated Pest Management (NCIPM), National Information System for Pest Management (NISPM), Crop Pest Surveillance and Advisory Project (Maharashtra) (CROPSAP), National Food Security Mission (NFSM) and Pest dynamics in relation to climate change under National Initiative on Climate Resilient Agriculture (NICRA) are some of the programmes initiated in India for sustainability of crop production. These programmes have established 36 real time pest surveillance centres across 12 States among researchers of five crops *viz.*, rice, pigeon pea, groundnut, tomato and mango for investigation into climate change effects on pests and disseminate IPM (NCIPM, 2012).

26.4 Education and Training

The popularization of IPM among farmers is done by The Directorate of Plant Protection and Quarantine by conducting pest surveillance, training resource personal for production of biocontrol agents and promote biopesticides through farmer field schools (FFSs). The different course are being offered for marketing personnel, non governmental organisations, students at graduate and post graduate levels, farmers and entrepreneurs. They also conduct trainings for long duration on major crops. Popularization of IPM is done by department of agriculture & cooperation through 31 centres located in 28 states and one union territory. These centres are responsible for monitoring pest/disease incidence, conservation of biocontrol agents and officers human resource development through agricultural extension officers and farmers field schools at grass root levels (Dubey and Sharma, 2004)) (Table 26.2).

Table 26.2 : Farmers and functionaries trained by an FAO project through Directorate of Plant Protection & Quarantine, India

	No. trained/produced		**No. organised/trained**	
	Functionaries	**Facilitators**	**Farmer Field Schools**	**Farmers**
FAO-EU	5	163	358	13836
Through States	21	524	1098	21992
Grant Total	26	687	1456	35828

Source : http://ppqs.gov.in
Success stories in different Crops

26.5 IPM in Cotton

Cotton requires more than 12-13 sprays to control the insect pests. The pesticide consumption was 55% in cotton alone when the cropped area is just 5%. When the farmers were involved in IPM approach, the profit was doubled in the IPM crop as compared to non-IPM crop using neem seed kernel extract *Helicoverpa,* Nuclear Polyhedrosis Virus, field sanitation and deep pouching and use of *Trichogramma chilonis,* the profit was doubled in the IPM crop as compared to non-IPM crop. This system has become self sustainable as the farmers have themselves become decision makers and on their own have started adopting many of the IPM practices (Singh *et al.,* 2003). Biopesticides like neem, *Callistemon, Eucalyptus* and *Bacillus thuringiensis* have been reported as an important component of IPM in cotton(Gupta and Raghuraman, 2004; Raghuraman *et al.,* 2008).

Most of the research and developmental activities of cotton IPM in India have arisen in response to pest outbreaks/ resistance with no holistic approach. Currently practiced pest management packages in India are specific to insect (s), disease (s), and weed (s) arising out of unidisciplinary research results involving integration of different control tactis against single pest and cotton crop is no exception. Development of insecticide resistance in *Helicoverpa armigera* made high dosages and number of sprays ineffective resulting in unprofitable cotton production. This has been the driving force for using other pest control options ranging from conservation of natural enemies, cultural practices, use of biorationals such as botanicals, *Bacillus thuringiensis* and HaNPV besides products of biotechnology *viz.*, *Bt.*transgenic cotton (Vennila *et al.* 2000).

26.6 IPM in Rice

Due to the advent of fertilizer responsive varieties, the farmers began to apply large amounts of nitrogenous fertilizers to realize better yields. But this had adverse effect on rice yield due to pest outbreaks in several areas of the country (Rao and Muralidharan, 1977; Chelliah *et al*., 1989). A severe damate due to gall midge outbreak was reported to a tune of Rs 6 crores (Devi *et al*., 1998). Different factors have been attributed for the outbreak of rice pests including monocropping, susceptible varieties, continuous rice crop throughout the year and application of heavy doses of nitrogen, due to which there was an increase use of pesticides (Chelliah *et al.*, 1989).

Directorate of Rice Research (DRR), Hyderabad, India initiated pest scouting, encouraging the natural enemies, altering cultural practices which are unfavourable for insect pests, use of resistant varieties *etc.* which resulted in adoption of suitable IPM practices in rice, thereby increasing rice yield from

3488 to 4983 kg/ha in Andhra Pradesh during 1981-86 (Krishnaiah and Reddy, 1989).

The Indian Council of Agricultural Research (ICAR) initiated 6 Operational Research Projects (ORPs) on integrated pest management for rice in 1975 under the oversight of the Directorate of Rice Research (DRR), Hyderabad, Kerala Agricultural University and Department of Agriculture, West Bengal. These programmes envisaged pest scouting, encouraging the natural enemies, altering cultural practices which are unfavourable for insect pests, use of resistant varieties *etc.* Due to adoption of suitable IPM practices in rice, there was an increase in the rice yield from 3488 to 4983 kg/ha in Andhra Pradesh during 1981-86 (Krishnaiah and Reddy, 1989). Similarly, in Kerala, adoption of IPM technology, resulted in better yields to the farmers even with an average of two sprays of the insecticides, where they used to spray 4-5 times previously (Sankaran, 1987). Similar, examples of successful adoption of IPM in rice are available in states like Haryana, Tamil Nadu, Andhra Pradesh, Uttar Pradesh, Kerala and Madhya Pradesh (Razak, 1986, Sardana and Kanwar, 2012) (Table 26.3).

Table 26. 3 : Yield and economics in IPM and non-IPM rice fields at Singarawan and Pandani in Jharkhand, India (Sardana and Kanwar, 2012)

Parameters	Singarawan		Pandani	
	IPM	Non-IPM	IPM	Non-IPM
Total cost of cultivation (Rs/ha)	7280	15700	27220	15600
Grain yield (quintals/ha)	22.1	13.2	35.8	18.6
Straw yield (quintals/ha)	50.0	22.5	65.0	35.0
Total returns (Rs/ha)	33868	18756	51664	27088
Net returns (Rs/ha)	6588	3056	24384	11388
Benefit : Cost ration	1.24	1.19	1.89	1.73

26.7 IPM in Vegetables

The vegetables available in market are found to be contaminated with pesticides. IPM practices has not been able to fully eliminate the use of pesticides. Natural enemies do exist for some key pests, but this delicate balance is disrupted largely due to misuse of insecticides and monocropping of vegetables in specialized production centres (Srinivasan, 1994). This has resulted in emergence of several new pests on different groups of vegetable crops in India. Among these, the diamondback moth is one of the most serious pest on cabbage and cauliflower in India. The South Asian Research Network (SARNET) has linked different countries like Bangladesh, India, Nepal, Pakistan, Sri Lanka and Bhutan since 1992 for extension and adoption of IPM technology by farmers (Shanmugasundaram, 1994).

Successful IPM for diamondback moth has been developed by Indian Institute of Horticultural Research (IIHR) utilizing mustard as trap crop and neem seed kernel extract which is widely adopted by farmers (Srinivasan, 1994). In Punjab, where diamondback moth has developed resistance to all the major groups of conventional insecticides, IPM was considered to be the only solution to manage this pest (Arora *et al.,* 2000). In tomato also successful IPM technology for *Helicoverpa armigera* has been developed. Biological control as a major component of the IPM in different vegetable crops is attaining utmost significance (Dhandapani *et al.,* 2003). It was found that by adopting IPM technology in vegetables, yield could be increased to a tune of more than 50% coupled with drastic reduction in use of pesticides. IPM technology developed for cauliflower has been disseminated in various districts of Haryana (Sardana and Kanwar, 2012)

In all these crops discussed above, a little area only has received the IPM technology, due to various limiting factors like awareness among the farmers, constraints in adoption *etc.* Adoption of these technologies in larger areas for wider acceptance should be practicable, economically feasible and sustainable.

26.8 Problems in Biopesticides Promotion

Biopesticides comprise only 2% of the total agrochemical consumption in India. This is largely due to the fact that most of the biopesticides and or technologies are produced by private sector and agrochemicals also produced by private sector indicating tough competition for bioagents. However, there is good scope for biopesticides if quality control is followed (Dubey and Sharma, 2004). Intellectual property rights and ownership issues need to be considered to overcome technology transfer barriers and delivery to the farmers (Erbisch and Maredia, 2003). The private sector plays a significant role in the trade and export of commodities such as vegetables for the niche market. Networking with NGOs and private sector can provide vital input into technology development and transfer. Strong links with private industry and NGOs can help for successful implementation of IPM programmes.

26.9 Economic Feasibility of IPM Programmes

IPM programmes has given encouraging results under controlled conditions, however, in field most of the techniques have failed to give desired results. Furthermore, to make the techniques acceptable by the farmers, demonstrated in form of first line and front line demonstrations is necessary to convince the farmers and to assure them that the techniques are economically feasible and environmentally sound. Evidently, farmer's participatory research programmes and the involvement of the research and extension personnel is most essential (Dubey and Birthal, 2004).

26.10 Areawide Adoption

The information on the successful adoption of IPM programme in India is incomplete. Most of the information is based on arbitrary estimates. One of the major problem in successful implementation of IPM programmes is the lack of availability of pesticides to meet the demans of farmers and their proper information. More attention needs to be given to areawide IPM systems in future global IPM programs (Knipling, 1979, 1980). Some IPM strategies are more likely to succeed if implemented over large areas, rather than on an individual farmer field basis. If implemented in large areas, they should be implemented with by the network programmes involving different organizations ultimately targeting the management of major key pests of a geographic area. Government extension services or private agencies should conduct area-wide monitoring and provide the appropriate information on pest outbreaks to farmers on a regular basis. Forecasting of pest outbreaks especially migratory pests is of immense value in implementation of successful IPM programmes. The extension workers and farmers must be fully trained to monitor and identify the pest for its efficient management. Major problem with the implementation of IPM programme is that it is difficult to convince the farmers as the biocontrol control options such as use of parasitoids, predators and microbial agents is slow in action as compared to quick results from pesticides. Success of IPM can be acheived through participation of farmers.

26.11 Emerging Issues and Challenges

Some of the central institutes are involved in promotion of IPM, but for the IPM to be more effective and successful there is need that all the state governments be fully involved with allocation of funding to encourage the farmers to adopt this technology. Another difficulty is the registration of biopesticides by the small entrepreneurs who are interested in promoting them. Even the banned pesticides are still vogue in several parts of the country and their elimination will take long time from the markets. Promoting new and safer insecticides will help the farmers to promote IPM.

26.12 Future Thrust

With sufficient food grain production due to favorable monsoon and available technology, the policy makers envisages the production of sufficient food without pesticides. There are some evidence that decline in the pesticide usage in the country is due to awareness created among farmers. The consumers are also very aware now a days and prefer food which is free of pesticides (Dubey and Birthal, 2004). Hence it becomes important that the future programs should be directed on producing pesticide free food which is of great concern in the coming years. The entomologists have to design and implement new agricultural production systems in cooperation with farmers' community where tools of IPM like cultural and biological controls are enhanced to have a self sustainable system. The agro-ecosystem, then, will act as an area of refuge for predators and pests to exist in ecological balance. The goal of ecologically based IPM systems, is not to eliminate pests but to maintain them below economic levels. The integration of more organic or sustainable tactics into agriculture production systems will be a major part of the solutions toward sustainable agricultural production in the future. Integrated Pest Management promotion has not gained momentum as expected among the farmers due to certain constraints in technology, socio-economic, institutional and infrastructure related factors (Dubey and Birthal, 2004). If they are alleviated, then it would be possible to achieve the success in IPM programs at a greater level.

IPM adoption has not achieved the desired success because of (1) lack of national policy in most of the countries (2) lack of institutions to help and coordinate the IPM Programmes (3) lack of multi-disciplinary approach to improper identification of the problem (4) lack of research on integration of all tactics, rather than focusing on one or two tactics leading to resistance among pests due to selection pressure (5) lack of human resources, financial resources and collaboration among different governmental and other IPM promoting agencies (6) lack of farmers participation. Farmers need to be seen as equal partners in development of IPM technologies through their knowledge on ITK, socio-economic factors and ecological conditions (Maredia *et al.*, 2003). Farmers and scientists need to work together to characterize the agroecosystem flowed by farmer led testing, monitoring and evaluation of different interventions. Farmer field school is another innovative method where farmers learn by doing and ultimately become the IPM experts in their field (Gallagher, 1992).

IPM in the future shall have to create a balance between basic and applied research. Most of the IPM research programs in India are time based and short term programs. Area-wide long-term ecological programmes will give a better understanding of the biological and ecological interactions within the larger areas. There is a need to understand the basic and applied aspects of pest management which will lead to development of totally novel approaches for pest management. The approach to IPM technology development must be farmers participatory, multidisciplinary and multi-pest oriented and must integrate all available tactics, including cultural, biological control (Neuenschwander *et al.*, 2003) and genetic approaches including both biotechnology and conventional breeding. The approach must be ecologically based. Novel tactics that favour biodiversity should be explored and may include transgenic insects, biorational approaches, botanicals and management of resistance to pesticides and to resistant varieties and Genetically Modified (GM) crops.

References

Arora, R., Battu, G.S. and Bath, D.S. (2000).Management of insect pests of cauliflower with biopesticides. *Indian Journal of Ecology*, 27: 156-162.

Atwal, A.S., and Dhaliwal, G.S. (1997). Agricultural pests of South Asia and their management. Kalyani Publishers, Ludhiana. 486p.

Barfield C.S., and Swisher, M.E., (1994). Integrated pest management: ready for export? Historical context and internationalization of IPM. *Food Review International* 10: 215–267.

Binns MR & Nyrop JP. (1992). Sampling insect populations for the purpose of IPM decision making. *Annual Review of Entomology* 37: 427–453.

Chelliah, S.S., Bentur, J.S., and Prakasa Rao, P.S. (1989). Approaches in rice pest management achievements and opportunities. *Oryza,* 26: 12-26.

Choudhary, B. & Gaur, K. (2010). *Bt* Cotton in India: A Country Profile. ISAAA Series of Biotech Crop Profiles. ISAAA: Ithaca, NY

Devi., D. Nair, K.P.V., George Thomas, and Rama Devi. (1998). Status of rice gall midge, *Orseolia oryzae* (Wood-Mason) in Kuttanad (abst.) National Symposium on Rice Research for 21st Century, Challenges, Priorities and Strategies, organised by the Association of Rice Research Workers, Cuttack, India. Feb 5-7, pp.96-97.

Dhandapani, N., Umeshchandra R., Shelkar and Murugan, M. (2003). Bio-intensive pest management (BIPM) in major vegetable crops: An Indian perspective. *Food, Agriculture & Environment* : 333-339.

Dubey, O.P and Sharma O.P. (2004). www.ncap.res.in/upload_files/workshop/wsp11.pdf

Erbisch, F. E., and Maredia, K. M. (2003). Intellectual Property Rights in Agricultural Biotechnology.Wallingford, UK: CABI Publishing.

Gallagher, K.D. (1992) IPM development in the Indonesian National IPM programme. In: Aziz, A , Kader, S.A and Barlow, H.S (eds) Pest management and the environment in 2000. CAB International Wallingford, UK pp 82-89. .

Gupta, G.P., and Raghuraman, M. (2004). Utilization of biopesticides in sustainable pest management in cotton. In: Biopesticides (ed.), Tata Energy Resource Institute (TERI), New Delhi. 288pp.

Knipling, E. F. (1979). The Basic Principles of Insect Population Suppression and Management, Agricultural Handbook No. 512. Washington, DC: US Department of Agriculture.

Knipling, E. F. (1980). Areawide pest suppression and other innovative concepts to cope with our more important insect pest problems. In Minutes of the 54th Annual Meeting of the National Plant Board, pp. 68–97. Sacramento, CA: National Plant Board.

Kogan M. (1998). Integrated pest management: historical perspectives and contemporary developments. *Annual Review of Entomology* 43: 243–270.

Krishnaiah, K., and P.C. Reddy. (1989). Operational research project on integrated control of rice pests in Medchal Area, Ranga Reddy district, A.P.: Achievements and constraints. Annual Rice Workshop, HAU, Hissar, April 19-22, 1989, pp.9.

Maredia K.M., Dakouo D., Mota-Sanchez D. and Raman K.V. (2003). Making IPM Successful Globally: Research, Policy, Management and Networking Recommendations In: *Integrated Pest Management in the Global Arena* (eds K.M. Maredia, D. Dakouo and D. Mota-Sanchez) published by CAB International, Wallingford, UK pp 501-506

Mathur S.C. (1999). Future of Indian pesticides industry in next millennium. *Pesticide Information*. 24: 9–23.

Metcalf, R.L and Luckman, W.H. (1994). Introduction to insect pest management (3rd edition). John Wiley and Sons, Inc., New York.

Michelbacher, A.E. and Smith, R.F. (1943). Some natural factors limiting the abundance of the alfalfa butterfly. *Hilgardia* 15: 369–397.

Miller GT (2004). Sustaining the Earth, 6th edition. Thompson Learning, Inc. Pacific Grove, California. Chapter 9, Pages 211-216.

NCIPM, (2012). www.ncipm.org.in/NCIPMPDFs/Annual Report 2012/Annual Report 2011-12.pdf

Neuenschwander, C., Borgemeister, C. and Langewald, J. (2003). Biological Control in IPM Systems in Africa., CABI Publishing Wallingford, UK.

Pedigo LP. (2002). Entomology and Pest Management, 4th edn. Pearson Education, Singapore.

Perkins J.H. (1982). Insects, Experts and the Insecticide Crisis: the Quest for New Pest Management Strategies. Plenum, New York, USA.

Raghuraman, M., Birah, A., and Gupta, G.P. (2008). Management of *Helicoverpa armigera* in chickpea with botanical formulations. *Indian Journal of Entomology*, 70: 118-122.

Rao, V.G. and K. Muralidharan. (1977). Outbreak of rice hispa in Nellore district, Andhra Pradesh, India. *International Rice Research News* 2: 18-19.

Razak, R.L. (1986). Integrated pest management in rice crop. *Plant Protection Bulletin* 38: 1-4.

Rosegrant, M. W., Tokgoz, S and Bhandary, P (2013). 'The New Normal' A tighter global agriculturally supply and demand relation and its implications for food security. *American Journal of Agricultural Economics*. 95(2):303-309.

Rosegrant Mark W., Koo Jawoo, Cenacchi Nicola, Ringler Claudia, Robertson Richard, Fisher Myles, Cox Cindy, Garrett Karen, Perez Nicostrato D., and Sabbagh Pascale (2014). Food Security in a World of Natural Resource Scarcity The Role of Agricultural Technologies. International Food Policy Research Institute, Washington, DC pp 171.

Sankaran, T. (1987). Biological control in integrated pest control – Progress and perspectives in India. *Proceedings of the national symposium on Integrated Pest Control Progress and Perspectives*, October 15-17. (Eds. N. Mohan Das and George Koshy): pp 545.

Sardana, H.R., and Kanwar, V. (2012). Annual Report 2011-12, National Centre for Integrated Pest Management, LBS Building, Pusa Campus, New Delhi - 110 012 India.

Shanmugasundaram, S. (1994). IPM of diamondback moth in South and South-east Asia. *IPM Working Development,* 3: 8.

Singh, A., Trivedi, T.P., Sardana, H.R., Sharma, O.P., and. Sabir, N. (eds.) (2003). Recent advances in integrated pest management. National Centre for Integrated Pest Management, New Delhi.

Srinivasan, K. (1994). Recent trends in insect pest management in vegetable crops. *In:* Dhaliwal and Ramesh Arora (eds.). *Trends in Agricultural Insect Pest Management*, Commonwealth Publishers, New Delhi, India, pp.103-105.

Stern V.M., and Smith, R.F., van den Bosch, R. and Hagen, K.S. (1959). The integrated control concept. *Hilgardia* 29: 81–101.

US Census Bureau – World POP Clock Projection" July 2012–July 2013 data.

Vennila, S., Ramasundaram, P., Raj, S., and Kairon, M.S. (2000). Cotton IPM and its Current Status. *Technical Bulletin No. 8.* Central Institute for Cotton Research, Nagpur.

Walter GH. (2003). Insect Pest Management and Insect Ecology Research. Cambridge University Press, Cambridge, UK.